Springer Complexity

Springer Complexity is an interdisciplinary program publishing the best research and academic-level teaching on both fundamental and applied aspects of complex systems—cutting across all traditional disciplines of the natural and life sciences, engineering, economics, medicine, neuroscience, social and computer science.

Complex Systems are systems that comprise many interacting parts with the ability to generate a new quality of macroscopic collective behavior the manifestations of which are the spontaneous formation of distinctive temporal, spatial or functional structures. Models of such systems can be successfully mapped onto quite diverse "real-life" situations like the climate, the coherent emission of light from lasers, chemical reaction-diffusion systems, biological cellular networks, the dynamics of stock markets and of the internet, earthquake statistics and prediction, freeway traffic, the human brain, or the formation of opinions in social systems, to name just some of the popular applications.

Although their scope and methodologies overlap somewhat, one can distinguish the following main concepts and tools: self-organization, nonlinear dynamics, synergetics, turbulence, dynamical systems, catastrophes, instabilities, stochastic processes, chaos, graphs and networks, cellular automata, adaptive systems, genetic algorithms and computational intelligence.

The two major book publication platforms of the Springer Complexity program are the monograph series "Understanding Complex Systems" focusing on the various applications of complexity, and the "Springer Series in Synergetics", which is devoted to the quantitative theoretical and methodological foundations. In addition to the books in these two core series, the program also incorporates individual titles ranging from textbooks to major reference works.

Editorial and Programme Advisory Board

Understanding Complex Systems

Founding Editor: J.A. Scott Kelso

Future scientific and technological developments in many fields will necessarily depend upon coming to grips with complex systems. Such systems are complex in both their composition—typically many different kinds of components interacting simultaneously and nonlinearly with each other and their environments on multiple levels—and in the rich diversity of behavior of which they are capable.

The Springer Series in Understanding Complex Systems series (UCS) promotes new strategies and paradigms for understanding and realizing applications of complex systems research in a wide variety of fields and endeavors. UCS is explicitly transdisciplinary. It has three main goals: First, to elaborate the concepts, methods and tools of complex systems at all levels of description and in all scientific fields, especially newly emerging areas within the life, social, behavioral, economic, neuro and cognitive sciences (and derivatives thereof); second, to encourage novel applications of these ideas in various fields of engineering and computation such as robotics, nano-technology and informatics; third, to provide a single forum within which commonalities and differences in the workings of complex systems may be discerned, hence leading to deeper insight and understanding.

UCS will publish monographs, lecture notes and selected edited contributions aimed at communicating new findings to a large multidisciplinary audience.

For further volumes:
www.springer.com/series/5394

Rush D. Robinett III • David G. Wilson

Nonlinear Power Flow Control Design

Utilizing Exergy, Entropy, Static and Dynamic Stability, and Lyapunov Analysis

Springer

Dr. Rush D. Robinett III
Sandia National Laboratories
PO Box 5800
Albuquerque, NM 87185
USA

Dr. David G. Wilson
Sandia National Laboratories
PO Box 5800
Albuquerque, NM 87185
USA
dwilso@sandia.gov

ISSN 1860-0832
e-ISSN 1860-0840
ISBN 978-0-85729-822-5
e-ISBN 978-0-85729-823-2
DOI 10.1007/978-0-85729-823-2
Springer London Dordrecht Heidelberg New York

British Library Cataloguing in Publication Data
A catalogue record for this book is available from the British Library

Library of Congress Control Number: 2011934507

Sandia National Laboratories is a multiprogram laboratory operated by Sandia Corporation, a Lockheed Martin Company, for the U.S. Department of Energy's National Nuclear Security Administration under contract DE-AC04-94AL85000.

Cover design: VTeX UAB, Lithuania

Printed on acid-free paper

Springer is part of Springer Science+Business Media (www.springer.com)

R & D Exploration:

"3 man fire team only."
"Prepare to deploy."
KANE 607
"Sounds reckless."
"3 against 50."
"I can tell these men have no combat experience."
"The old ones."
"The Veterans."
"They would have called for support, just in case."
CAPTAIN
Recommendation:
"Load them up with heavy artillery . . . pound that place with mortars, rockets, and cannons."
"Everything we've got."
"From a safe distance."
"It ain't fancy, but it will sure take care of your hostile military force."
CAPTAIN
"Alright do it."
"Withdraw the men."
COLONEL

SOLDIER
Excerpts from, ©1998 Warner Bros. and Morgan Creek Productions, Inc

Sandia
Z Division
1st Battalion
RECON
Three Man Fire Team

The authors dedicate this book to the memory of J. Arlin Cooper

Foreword

Common problems and challenges that the United States and other countries must deal with involve integration of their green renewable resources on an existing aging electric power grid infrastructure. Faced with quickly approaching deadlines, many countries are trying to retrofit and patch in renewables in the best way possible. Given this, many of the proposed "future" smart grids are overlaying/marrying information networks with existing electric power grid infrastructure. What is needed is a paradigm shift in current practices in power engineering. At the heart of the electric power grid is the coordination and control of generation to meet customer loads.

Globally, there is awareness of the challenges associated with realization of the next generation of green grid. In the U.S., much work is being planned to achieve high penetration (more than 20 percent) of renewable resources by 2030. Some key issues are how to best take advantage of the high-wind corridors and transmission (currently nonexistent) of the power in the Midwest, where to install solar, how best to use storage, and what incentives can be offered. In some local cases, penetration can be greater than 40 percent, a serious technical challenge to overcome.

Traditionally, this has consisted of large portions of generation that is called on in an "open-loop" fashion to be dispatched to service random load needs. The grid of the future will require much improved automation with an efficiently integrated "closed-loop" configuration.

New breakthroughs and advances in our tools and methodologies based on sound scientific and engineering practices will need to be developed to meet these challenges and problems. A new approach will be required to formally address the green grid of the future with distributed variable generation, buying and selling of power (bidirectional flow), and decentralization of the electric power grid.

Many researchers are attempting to address this problem. While I was at the U.S. Department of Energy's Office of Energy Efficiency and Renewable Energy, a key program I led to address issues associated with increasing the role of renewable energy on our grid was the Solar Energy and Grid Integration Systems (SEGIS) program. The concepts developed under the SEGIS program are key to achieving high penetration of photovoltaic (PV) systems into the utility grid. Advanced, integrated inverter/controllers will be the enabling technology to maximize the benefits of residential and commercial solar energy systems, both to the system owners and to the utility distribution network as a whole. The value of the energy provided by these solar systems will increase through advanced communication interfaces and controls, while the reliability of electrical service, both for solar and nonsolar customers, will also increase.

In my current role as the Research and Development Program Manager in the Office of Electricity Delivery and Energy Reliability, I have further developed programs directed at integration of renewable energy into our electric grid with concentration on the perspective of the electric grid. In both my SEGIS program and Office of Electricity renewable to grid activities, Sandia National Laboratory has provided a leadership role in developing and demonstrating the all important control system architectures necessary to increase the role of renewable energy penetration on our electric grid with respect to stability and performance.

The goal of this book is to present a step toward addressing the integration of renewable resources into the electric power grid by applying new nonlinear power flow control techniques to the analysis of renewable generators connected to the electric power grid.

Washington, DC, USA
March 2011

Dan T. Ton

Program Manager
Office of Electricity Delivery & Energy Reliability
Smart Grid Research & Development

Preface

This book presents an innovative control system design methodology that is based on the latest research and development results at Sandia National Laboratories within the renewable energy electric power grid integration program. The inspiration for these research and development results are from three problems. The first problem is to find a unifying metric to compare the value of different energy sources to be integrated into the electric power grid such as coal-burning power plants, wind turbines, solar photovoltaics, etc., instead of the typical metric of costs/profits. The unifying metric of choice turns out to be exergy which is effectively negative entropy since price, in economic terms, has too many unaccounted for externalities. The second problem was to develop a new nonlinear control tool that applies power flow control, thermodynamics, and complex adaptive systems theory to the energy grid in a consistent way. The third problem is mathematically formulating how a person effectively navigates a time, spatially varying environment from a robotic engineer's point of view such that collective robotic theories can be used to create optimal individuals and high-performance teams of people. This problem is far from solved, however many of the basic concepts in this book are a result of how a person regulates power flows. These insights will be used in the future to account for the effects of individuals and groups of people that will be controlling and selling power into a distributed, decentralized electric power grid. In addition, these problems have several key concepts in common: exergy flow, limit cycles, and balance between competing power flows.

The main goal of the research effort was to develop a methodology that provides a unique set of criteria to design nonlinear controllers for nonlinear systems. This methodology addresses both stability and performance as well as seamlessly integrating information theory concepts instead of following the typical linear controller zero-sum design trade-off process between stability and performance. It works by combining concepts from thermodynamic exergy and entropy, Hamiltonian mechanics, Lyapunov's direct method and Lyapunov optimal analysis, electric AC power concepts, power flow analysis, and Fisher Information. The thermodynamic concepts are employed to allow the control design to be viewed as a power flow approach. This power flow approach balances power generation and power dissipation subject to the power storage (i.e., kinetic and potential energies) for a special class of dynamical systems called Hamiltonian natural systems and adiabatic irreversible work processes. This approach provides both necessary and sufficient conditions for stability while simultaneously allowing for performance specifications.

This book has been subdivided into three parts; theory, applications, and advanced concepts. In the first part (Chaps. 1–5), the necessary theoretical developments are presented that include: nonequilibrium thermodynamics, Hamiltonian mechanics, stability principles, and advanced control design concepts. In part two (Chaps. 6–13), the methodology is demonstrated through multiple case studies ranging from control design issues, collective plume tracing, nonlinear aeroelasticity and wind turbine control, fundamental power engineering, renewable energy microgrid design, robotic manipulator control, to satellite reorientation control. In part three, Chap. 14, advanced concepts are introduced that employ the fundamental theory, from part one as a foundation, and demonstrates how to extend it sustainability of self-organizing systems.

Research scientists, practicing engineers, applied mathematicians, physicists, and engineering students with a background in and basic understanding of thermodynamics, dynamics, and controls will be able to develop and apply this methodology to their particular problems. Considerable emphasis is placed on the necessary design steps for which the concepts are introduced and explained with numerous examples and a variety of case studies. The organization of the book makes it possible to be used as a first-level graduate course on nonlinear control design or as a reference or supplemental textbook for a special topic in support of a broader control theory course. In addition, the book has been developed in such a fashion that the interested reader, through self-study, could broaden their understanding of the analysis, design, and synthesis of nonlinear control systems.

Several distinctive features that are developed in this book are:

- Our approach provides both necessary and sufficient conditions for stability of a class of nonlinear systems while simultaneously allowing for performance specifications.
- Our approach provides seamless connections between information theory and nonlinear control by demonstrating the equivalence of physical and information exergies by applying nonlinear equilibrium thermodynamics, Hamiltonian mechanics, quantum mechanics, and Fisher Information.
- The material has been subdivided into three parts: theory, applications, and advanced concepts. This allows the reader to progressively move through the material, such as in a classroom environment or selectively investigate chapters most related to their own interests, working at their own pace. Many of the case studies are provided with explicit design steps to illustrate the ideas and principles behind the nonlinear power flow control methodology.
- Several of the case studies have been selected surrounding advanced and future control system designs and issues associated with the current integration of renewable energies (wind and solar) and interlaced with conventional energy sources and operations.

Acknowledgements The authors would like to thank the many sponsors and stake-holders who have contributed to the sustained effort of this work.

Sandia National Laboratories provides a unique atmosphere for the advancement of control research and the development of needed control tool sets through continued commitment and support. In particular, the Laboratory Directed Research and Development (LDRD) projects *Design Tools for Complex Dynamic Security Systems—Adaptive Complexity* and *Innovative Control of a Flexible, Adaptive Energy Grid* have very much facilitated the outcome of this research. Sandia is currently supporting a Grand Challenge LDRD project associated with our advanced microgrid control architectures. In addition, work completed for the following external projects: the Office of Force Transformation, Office of the Secretary of Defense, *Implementing the Power of the Collective* project and DARPA Mechanics Based Analogies for Swarms, *Collective Systems: Physical and Information Exergies* also played a key role.

The authors wish to acknowledge those who have been helpful in the preparation of this book and our many collaborators and coauthors over it's duration.

Special thanks to Dr. Al Reed for his support in the early work in thermodynamics and exergy analysis most notable in Chaps. 2 and 14 (simple spacecraft example).

Many thanks to Dr. Don Hardesty for his discussions and support in energy and exergy concepts.

Our appreciation to the management and project PIs at Sandia both in the early work at the robotics center by Dr. John Feddema, Dr. Ken Groom, Dr. Ray Byrne and currently by the energy security and renewables focus areas: Margorie Tatro, Juan Torres, and Jose Zayas.

Special thanks to Professor Chaouki Abdallah at the University of New Mexico, EECE department for his encouragement, energetic discussions, and support in the technical community arena.

Special thanks to Professor Gordon Parker and his team; Matthew Heath and Lecturer Charles Margraves at the Mechanical Engineering–Engineering Mechanics Department, Michigan Tech for their support of the technical review and providing valuable feedback.

Finally, a special thanks to Marna Wilson for providing a much needed editorial and detailed manuscript review.

Albuquerque, NM, USA
November 2010

Rush D. Robinett III
David G. Wilson

Contents

List of Figures

List of Tables

Chapter 1
Introduction

1.1 Background

Today's engineering systems sustain desirable performance by using well-designed control systems based on fundamental principles and mathematics. Many engineering breakthroughs and improvements in sensing and computation have helped to advance the field. Currently control systems play critical roles in many areas. These include automation, manufacturing, electronics, communications, transportation, computers, and networks, as well as many commercial and military systems [1]. Traditionally, almost all modern control design is based on forcing the nonlinear systems to perform and behave like linear systems, thus limiting its maximum potential. A novel nonlinear control design methodology is introduced in this book. It overcomes that limitation by identifying the stability boundaries for a class of nonlinear systems, Hamiltonian natural systems, and adiabatic irreversible work processes.

In general, this new approach has the capability of providing a completely different method of designing controllers based on static stability and dynamic stability by locating stability boundaries: rigid body modes and limit cycles. This methodology is directly applicable to the design and control of complex, multicomponent and often adaptive systems of arbitrary purpose, design and underlying physics, such as critical infrastructure (for example: electric power grids, SCADA systems, telecommunication and satellite systems, oil and gas pipelines) and military systems (individual vehicles and formations of vehicles). The secure and reliable operation for all these systems is vital to a healthy economy for both civil and military infrastructure. In addition, the methodology is inclusive of exergy sustainability for most open systems which may include mass, energy, entropy, and exergy flow analysis. For example, this approach can be used to design efficient satellite spacecraft systems, distributed/decentralized energy system utilization, and vehicle survivability subject to maximizing finite power limitations during the course of the mission.

The concepts developed and presented in this book resulted from the convergence of three research and development goals that were pursued within the renewable energy electric power grid integration program at Sandia National Laborato-

R.D. Robinett III, D.G. Wilson, *Nonlinear Power Flow Control Design*, Understanding Complex Systems,
DOI 10.1007/978-0-85729-823-2_1,

ries. The first goal was to create a unifying metric to compare the value of different energy sources integrated into the electric power grid such as a coal-burning power plant, wind turbines, solar photovoltaics, etc., instead of the typical metric of costs/profit. The second goal was to develop a new nonlinear control tool that applies power flow control, thermodynamics, and complex adaptive systems theory to the energy grid in a consistent way. The third goal was to apply collective robotics theories to the creation of high-performance teams of people and optimal individuals in order to account for the effects of individuals and groups of people that will be controlling and selling power into a distributed, decentralized electric power grid. These three goals have several key concepts in common: exergy flow, limit cycles, and balance between competing power flows.

The development of a unifying metric began with entropy since entropy is a measure of disorder across all energy systems. Unfortunately as Greven, Keller, and Warnecke [2] state in `Entropy`,

> *They will realize that entropy appears to be an important link between their respective fields of research. But very soon they realized that they have major difficulties in understanding each other ... The large number of different meanings attributed to this word within the community of mathematicians and scientists from fields close to mathematics.*

The search turned to free energy as described by Schrödinger [3] and ultimately focused on exergy as the metric of choice. Exergy is also known as availability, the maximum useful work possible during a process that brings the system into equilibrium with a heat reservoir, and is notionally negative entropy.

The development of a new nonlinear control tool began with irreversible entropy production as a Lyapunov function following the lead of Kondepudi and Prigogine [4]. This function provides an excellent Lyapunov function, but it is difficult to tie it to Hamiltonian mechanics which provides the nonlinear dynamical systems tools of choice because Hamiltonians are typically derived for conservative systems. Fortunately, we realized that the Hamiltonian is stored exergy and the time derivative of the Hamiltonian is directly related to the irreversible entropy production rate which is known as the exergy destruction rate. We had found a missing link between Hamiltonian mechanics, irreversible and nonequilibrium thermodynamics, and complex adaptive systems (which are defined in Exploring Complexity [5]). In addition, we had developed a nonlinear control tool that expands on Merovitch's concepts [6] while providing necessary and sufficient conditions for stability of a class of nonlinear systems, Hamiltonian natural systems and adiabatic irreversible work processes. This control tool finds the generalized stability boundaries, rigid body modes, and limit cycles. A rigid body mode is a result of a singular stiffness operator. The limit cycle is a result of balanced generative and dissipative power or exergy flows into and within a system.

The creation of high-performance teams of people is based on collective robotics theories [7] which are described in nonmathematical terms in a companion book, Collective Groups [8]. The basis of these concepts is the model of an individual that is the balance of exergy flows that leads to limit cycles: a lifestyle.

In Part I, the theoretical building blocks are developed and described to provide a solid foundation for later applications in Parts II and III. Several references are provided for those not familiar with thermodynamics, classical mechanics, nonlinear control stability principles, and advanced mathematics. However, these chapters will provide a condensed review of familiar concepts for the more experienced reader. The authors have not attempted to be exhaustive in these developments and descriptions, but to instead provide a thread that runs through and ties together many tools to design nonlinear controllers in innovative ways.

The remainder of this chapter starts to identify the relationships of each of the respective engineering topics: thermodynamics, Hamiltonian mechanics, static stability, dynamic stability, limit cycles, and information theory, with the new nonlinear control design methodology.

1.2 Thermodynamics

As described earlier, the development of a unifying metric began with entropy, and this concept meshed well with the developments of Schrödinger [3]:

> *Schrödinger (1945) suggested that all organisms need to import "negative entropy" from their environment and export high entropy (for example, heat) into their environment in order to survive. This idea was developed into a general thermodynamic concept by Prigogine and his co-workers who coined the notion of "dissipative structures" (Nicolis and Prigogine, 1977 [9]; Prigogine and Stengers, 1984 [10]), structures of increasing complexity developed by open systems on the basis of energy exchanges with the environment. In the self-organization of dissipative structures, the environment serves both as a source of low-entropic energy and as a sink for the high-entropic energy which is necessarily produced [11].*

Prigogine and his coworkers developed a way to analyze the stability of dissipative structures based on choosing irreversible entropy as the Lyapunov function. This function provides an excellent Lyapunov function, but it is difficult to tie it to Hamiltonian mechanics which provides the nonlinear dynamical systems tools of choice because Hamiltonians are typically derived for conservative systems. Fortunately, a relationship between thermodynamics and Hamiltonian mechanics was developed during our research. We realized that the Hamiltonian is stored exergy and the time derivative of the Hamiltonian is directly related to the irreversible entropy production rate which is known as the exergy destruction rate. Robinett and Wilson had found a missing link between Hamiltonian mechanics, irreversible and nonequilibrium thermodynamics, and complex adaptive systems [12, 13].

In Chap. 2, the basic concepts of thermodynamics required to support the development of necessary and sufficient conditions for stability of a class of nonlinear systems, Hamiltonian natural systems and adiabatic irreversible work processes, are reviewed. The First Law of Thermodynamics is a statement of the existence of a property called Energy which is a state function that is independent of the path and, in the context of mechanics, a conservative (potential) function such as gravity. The Second Law of Thermodynamics guarantees the existence and uniqueness of stable equilibrium states of a large class of nonlinear systems. Exergy is the availability, the maximum amount of work that can be extracted reversibly from an energy flow.

Exergy is the unifying metric of choice to develop the equations of motion and the power flow models. Local equilibrium, a way to utilize the mathematics of equilibrium thermodynamics, is applied to nonequilibrium irreversible system models to produce nonlinear rate and power equations. The First and Second Laws of Thermodynamics are reviewed in detail and applied to these models to generate energy and entropy balance equations. In turn, these balance equations are written in rate equation form in order to produce the exergy rate equation which provides the fundamental connection to Hamiltonian mechanics and stability analysis.

1.3 Hamiltonian Mechanics

As described earlier, it was time to develop an unique nonlinear control tool that takes advantage of the unifying metric, exergy, which means picking the Hamiltonian as the Lyapunov function instead of irreversible entropy. This process requires formally relating irreversible thermodynamics to Hamiltonian mechanics in order to find the stability boundaries for a class of nonlinear systems, Hamiltonian natural systems and adiabatic irreversible work processes.

In this book, the concepts of energy, work, and power will be reviewed. It will set the stage to relate thermodynamics to Hamiltonian mechanics. In Chap. 2, the energy balance equation from the First Law of Thermodynamics combined with the entropy balance equation from the Second Law of Thermodynamics are written in rate equation form to create the exergy rate equation. In Chap. 3, exergy rate is equated to power flow, the first time derivative of the Hamiltonian for natural systems. To equate these two quantities, one must define the types of thermodynamic systems that will be analyzed and modify the exergy rate equation accordingly. The thermodynamic systems of interest are adiabatic irreversible work processes where work is done on the system, no entropy (or heat) is exchanged with the environment, and irreversible entropy is produced through dissipation. The recognition that the Hamiltonian is exergy provides the basic relationship necessary to apply the Second Law of Thermodynamics to the power flow to sort terms into three classes: storage, generation, and dissipation. This sorting process defines the dynamic stability boundary, a limit cycle, while the Hamiltonian surface defines the static stability boundary, a rigid body mode. The combination of the static stability and dynamic stability of a Hamiltonian system defines the necessary and sufficient conditions for a class of nonlinear systems which will be developed in Chap. 4.

1.4 Static Stability and Dynamic Stability

Now that the Hamiltonian is the Lyapunov function of choice, it is time to formally relate the stability concepts of thermodynamics to nonlinear control design given four basic building blocks. First, the Hamiltonian is not just a Lyapunov function. It is a constraint surface that the system trajectory must traverse as power flows into, power dissipates within, and power is stored within the system as time progresses. Second, the Hamiltonian can be shaped to provide desired equilibrium points, and the power flow can be controlled to match the desired system trajectories. Third, Lyapunov analysis provides sufficient conditions for stability, so static stability and dynamic stability provide the basic format required to develop necessary and sufficient conditions for stability of a class of nonlinear systems, Hamiltonian natural systems and adiabatic irreversible work processes. Fourth, the nonlinear stability analyses of Chetayev [14] and Merovitch [6] will be extended to develop the necessary and sufficient conditions for stability of a class of nonlinear systems based on static stability and dynamics stability including the generalized stability boundaries, rigid body modes, and limit cycles. Bottom-line: this book will explain that the Hamiltonian and its time derivative, power flow or work-rate described by Junkins [15], determine necessary and sufficient conditions for stability of a class of nonlinear systems.

This book reviews the concepts of static stability, dynamic stability, eigenanalysis, and Lyapunov analysis required for the development of necessary and sufficient conditions for stability of a class of nonlinear systems. The combination of static stability of conservative systems and dynamic stability of adiabatic irreversible work processes in the form of exergy-rate equations will be utilized to develop Hamiltonian Surface Shaping and Power Flow Control (HSSPFC).

HSSPFC is a two-step control law design and analysis process. The first step treats the Hamiltonian system as if it were a conservative system with no externally applied nonconservative forces (static stability). This step shapes the Hamiltonian surface with acceleration feedback and/or proportional feedback to create an isolated minimum (stable) energy state. Static stability is utilized to find the first stability boundary, a rigid body mode (singular stiffness matrix), which defines the point of static neutral stability.

The second step analyzes and designs controllers for a Hamiltonian system with externally applied nonconservative forces by applying dynamic stability concepts to modify the power flow with dissipation and generation feedback. In the thermodynamics chapter, the energy balance equation from the First Law of Thermodynamics combined with the entropy balance equation from the Second Law of Thermodynamics are written in rate equation form to create the exergy-rate equation. In the mechanics chapter, exergy rate is equated to power flow, the first time derivative of the Hamiltonian for natural systems (the work/rate equation). The recognition that the Hamiltonian is stored exergy provides the basic relationship necessary to apply the Second Law of Thermodynamics to the power flow to sort terms into three classes: storage, generation, and dissipation. This sorting process identifies the second stability boundary, a limit cycle, which defines the point of dynamic neutral stability.

It will be explained that dynamically stable systems must be statically stable, but the converse is not true. Therefore, static stability is a necessary condition for stability, while dynamic stability is a sufficient condition for stability which produces a maximum entropy state as well as a minimum energy state.

1.5 Limit Cycles

Researchers have been investigating limit cycle behavior for many different engineering fields. Specific applications that relate to the category of time periodic systems include helicopter blades in forward flight, wind turbine blades, and airplane wing flutter, all of which Limit Cycle Oscillations (LCO) may become present. The prediction and control of LCO in a system continues to be a challenging area of research. For example, Gopinath, Beran, and Jameson [16] explore various methods in the computation of time-periodic solutions for autonomous systems. The goal was to determine the range of applicability of models of varying fidelity to the numerical prediction of LCOs including related evaluations. A simple aeroelastic model of an airfoil with nonlinear structural coupling was used to show the efficacy of the procedure. Several researchers are investigating cyclic methods to compute limit-cycle oscillations for potentially large, nonlinear, systems of equations. One such method by Hall, Thomas, and Clark [17] introduces a harmonic balance technique for modeling unsteady nonlinear flows in turbomachinery. For the example presented in [17], a transonic front stage rotor of a high-pressure compressor was found to flutter in torsion, but reached a stable limit cycle, demonstrating that strongly nonlinear flows can be modeled accurately with a small number of harmonics. Additional wing flutter LCO identification and control investigations by others are further discussed in the following references [18–22]. In more general mathematical descriptions of limit cycles, Sabatini [23] discusses a uniqueness theorem for limit cycles of a class of plane differential systems. The main result is applicable to second-order systems with dissipative terms which depend on both position and velocity. In addition, Carletti and Villar [24] consider the Líenard equation for which a sufficient condition to ensure the existence and uniqueness of limit cycles is given.

This book will explain how limit cycles are the stability boundaries for linear and nonlinear control systems. The Poincaré–Bendixson Theorem is a good place to begin showing this property of limit cycles. Boyce and DiPrima [25] provide some insight into the proof of this theorem in the form of Green's Theorem applied to a line integral over a closed curve that equals zero or

$$\dot{x} = F(x, y), \qquad \dot{y} = G(x, y),$$

$$\oint_C \left[F(x, y)\, dy - G(x, y)\, dx\right] = \int\!\!\int_R \left[F_x(x, y) + G_y(x, y)\right] dA = 0.$$

This line integral will be modified for Hamiltonian systems to determine the limit cycles resulting from power flow control. In particular, the work per cycle [26–29] defined by the line integral of the power flow

$$W_{\text{cyclic}} = \oint_\tau F \cdot \dot{x}\, dt = 0$$

is the modified form of choice since the time derivative of the Hamiltonian is the generalized power flow for natural systems [30].

A specific application of these concepts is the analysis of classical flutter. Classical flutter is a linear limit cycle that is a result of the coalescence of a bending mode and a torsional mode to produce a self-excited oscillation. Abramson [31] describes the existence of a flutter mode as

> *We see now that negative work is done on the wing by part of the torsional motion, by the flexural motion, and by the elastic restoring forces; positive work is done on the wing by part of the torsional motion. The motion will maintain itself (the condition for flutter) when the net positive work just balances the dissipation of energy due to all the damping forces. The magnitude of the positive work done by the additional lift due to the twist is directly dependent upon the phase relationship between the coupled torsional and flexural motions* ...

This book develops the existence of limit cycles based on power flows that leads to a balance between positive work and energy dissipation due to damping. This approach is generalized to nonlinear systems with nonlinear limit cycles where power flows are balanced over a cycle instead of point-by-point cancellation.

A linear limit cycle is a strange concept to most people since limit cycles are typically associated with nonlinear systems. A limit cycle is defined by Wikipedia as a closed trajectory in phase space having the property that at least one other trajectory spirals into it either as time approaches infinity or as time approaches minus-infinity. By the end of Sect. 3.6, it will be explained how a center [32] of a second-order system can be interpreted as a linear limit cycle; for example, a goal of power engineering.

At the end of Chap. 3, the concepts of cyclic equilibrium thermodynamics will be extended to cyclic nonequilibrium thermodynamics to limit cycles of Hamiltonian systems. These extensions and connections will enlarge the concept of cyclic work of equilibrium thermodynamic systems to nonequilibrium thermodynamic systems. This will enable the power sorting process required to find the dynamic stability boundary, a limit cycle.

1.6 Information Theory

Researchers are developing collective systems to generate solutions to many types of problems. Collective systems are characterized by a team or group of agents (physical and/or cyber) that sample and share information to generate feasible solutions to complex problems. The collective behavior often produces a value greater than the sum of the individual parts. This amplification or synergy can be harnessed by solving an inverse problem via an information flow/communications grid: given a desired macroscopic/collective behavior, find the required microscopic/individual behavior of each agent and the required communications grid. To make multiagent collective systems efficient, new challenges need to be addressed in coordinated distributed control strategies. These strategies need to take advantage of the large number of agents, robustness to single-point failures, and optimal use of the information that is communicated amongst the individuals in the collective system.

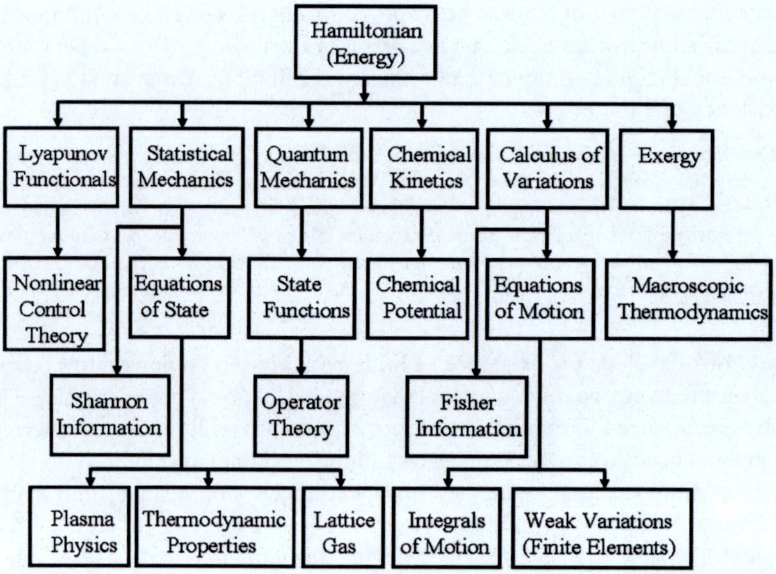

Fig. 1.1 Flowchart describing mechanics based approaches for collective systems. Robinett III, R.D. and Wilson, D.G. [37], reprinted by permission of the publisher (Interscience Publishers)

A contribution of this book is to present the fundamental nature of the Hamiltonian function (see Fig. 1.1) in the design of collective systems and the connections between and the values of physical and information exergies (see Chap. 7 and references [7, 12, 33, 34]) inherent to collective systems. In particular, physical and information exergies are shown to be equivalent based on thermodynamics, Hamiltonian mechanics, and Fisher Information.

In the Sandia National Laboratories' portion of the DARPA Distributed Robotics program [35, 36], the results of collective plume tracing can be interpreted with respect to information theory. The goal of the project was to build the smallest, dumbest robot that individually could not do anything other than random motion, but when put in a team of like robots could solve a complicated problem: track a chemical plume to its source while demonstrating emergent behavior and providing a visual presentation of the theoretical underpinning. Consequently, a fundamental information flow trade-off for the design of each robot of the collective to carry out the individual behavior that generates the collective behavior is between processing, memory, and communications. To make the design the smallest, dumbest robot, the minimum values for processing, memory, and communications simultaneously are an 8-bit processor, no memory, and 3 words communicated. It was shown with Shannon Entropy (Information) that the performance of the robot collective is limited by the sensors. Further depth and development that summarizes the critical systems functions of communication, computation, and memory are given in [38].

From a communications point of view, Fisher Information is a measure of how well the receiver can estimate the message from the sender where as Shannon Information/Entropy is a measure of the sender's transmission efficiency over a communications channel [39–42]. As a result, Fisher Information is a measure of order while Shannon Information is a measure of disorder, which is analogous to exergy being a measure of order while entropy is a measure of disorder. There exists a direct relationship between Boltzmann entropy and Shannon entropy [41]. In this book, a direct relationship between Fisher Information and exergy is developed with Hamiltonian Surface Shaping and Power Flow Control and exploited to create the Fisher Information Equivalency (FIE). FIE provides a direct connection between stability of, performance of, and information flow within the collective since physical exergy is directly related to the information exergy (virtual potential) flow and the Fisher Information flow. This relationship is developed by showing how Fisher Information is the mean kinetic energy of the Hamiltonian within quantum mechanics and utilizing the "classical limit" (as in classical mechanics).

For the development and implementation of the Smart Grid, the integration of high percentages of renewable energy power sources is fundamentally a trade-off between information flow and additional energy storage to replace fossil fuel and enable time correlation between power supply and load in the electric grid. The FIE provides the theoretical foundations required to perform these trade-offs between information flow and additional energy storage. These FIE trade-offs are formalized in Chap. 14.

1.7 Chapter Summary

Chapter 1 has provided a brief overview of the contents of this book and the motivation for its creation: the convergence of three research and development goals within the renewable energy electric power grid integration program at Sandia National Laboratories. The first goal was to create a unifying metric to compare the value of different energy sources integrated into the electric power grid such as a coal-burning power plant, wind turbines, solar photovoltaics, etc., instead of the typical metric of costs/profit. The second goal was to develop a new nonlinear control tool that applies power flow control, thermodynamics, and complex adaptive systems theory to the energy grid in a consistent way. The third goal was to apply collective robotics theories to the creation of high-performance teams of people and optimal individuals in order to account for the effects of individuals and groups of people that will be controlling and selling power into a distributed, decentralized electric power grid. It turns out that all three of these goals have several key concepts in common: exergy flow, limit cycles, and balance between competing power flows.

This chapter began identifying the relationships of each of the respective engineering topics: thermodynamics, Hamiltonian mechanics, static dynamics and dynamic stability, limit cycles, and information theory, with the new nonlinear control design methodology. These relationships will provide the required tools to develop necessary and sufficient conditions for stability of a class of nonlinear systems, Hamiltonian natural systems and adiabatic irreversible work processes, as well as their application to self-organizing systems.

This chapter has provided the basic layout for the rest of the book. Concepts from thermodynamics, mechanics, stability and control, and advanced control design are presented in Part I to provide the reader with the fundamental building blocks of the Hamiltonian Surface Shaping and Power Flow Control (HSSPFC) design and analysis process. HSSPFC is applied to eight case studies in Part II that range from integrating renewable energy generators into the grid and microgrids to satellite reorientation maneuvers to control of robotic manipulators. The last part, Part III, develops design and analysis tools for self-organizing systems.

Chapter 2
Thermodynamics

2.1 Introduction

The First Law of Thermodynamics is a statement of the existence of a property called Energy which is a state function that is independent of the path and, in the context of mechanics, a conservative (potential) function such as gravity. The Second Law of Thermodynamics guarantees the existence and uniqueness of stable equilibrium states of a large class of nonlinear systems. Exergy is the availability, the maximum amount of work that can be extracted reversibly from an energy flow with respect to a dead state, as well as the unifying metric of choice to develop the equations of motion and the power flow models. Local equilibrium, a way to utilize the mathematics of equilibrium thermodynamics, is applied to nonequilibrium irreversible system models to produce nonlinear rate and power equations. The First and Second Laws of Thermodynamics are reviewed in detail and applied to these models to generate energy and entropy balance equations. In turn, these balance equations are written in rate equation form in order to produce the exergy rate equation which provides the fundamental connection to Hamiltonian mechanics and stability analysis. The development of this chapter will follow and combine the writings of Gyftopoulos [43], Prigogine [4], and Scott [44, 45].

2.2 First Law (Energy)

The First Law of Thermodynamics is a statement of the existence of a property called Energy [43]. Energy is a state function, $\mathcal{E}$, that is path independent which implies a conservative function in the context of mechanics. An example of path-independent work is adiabatic reversible work where no heat is exchanged with the environment: raising and lowering a mass in a friction-less way referred to as a weight process. In this process, no entropy is transferred or created in this reversible system; then

$$\mathcal{E}_2 - \mathcal{E}_1 = -\mathcal{W}_{12}, \tag{2.1}$$

R.D. Robinett III, D.G. Wilson, *Nonlinear Power Flow Control Design*,
Understanding Complex Systems,
DOI 10.1007/978-0-85729-823-2_2,

where

$\mathcal{E}_i$ = energy at state i of the system,
$\mathcal{W}_{12}$ = work done by the system between states 1 and 2.

A work process transfers energy but transfers no entropy. Irreversible entropy can be produced within a system during a work process. A work-only process is an adiabatic process.

A corollary of the First Law that is often referred to as the First Law is

$$d\mathcal{E} = \delta\mathcal{Q} - \delta\mathcal{W}, \tag{2.2}$$

where

$d\mathcal{E}$—change of state, path-independent,
$\delta\mathcal{Q}$—flow of heat, path-dependent,
$\delta\mathcal{W}$—work done by the system, path-dependent,

which is the energy balance equation for work and heat interactions and identified as the conservation of energy. Note that energy is conserved by the First Law, so energy is not a useful economic metric since economics is the allocation of scarce resources. Exergy will be shown to be an ideal economic metric since it is consumed and most likely scarce.

2.3 Second Law (Stability/Entropy/Available Work)

The Second Law of Thermodynamics is a statement of the existence and uniqueness of stable equilibrium states and that reversible processes connect these states together with any arbitrary state. A nonequilibrium state is one that changes spontaneously as a function of time, that is, a state that evolves as time goes on without any effects on or interactions with any systems in the environment [43]. An example is the time evolution of a undamped or damped system with initial conditions away from an equilibrium state. An equilibrium state is defined as a state that does not change with time while the system is isolated from all other systems in the environment: a state that does not change spontaneously [43]. Specifically, an isolated system cannot exchange energy, entropy, etc. with any other system. An unstable equilibrium state is an equilibrium state that can be caused to proceed spontaneously to a sequence of entirely different states by means of a minute and short-lived interaction that has an infinitesimal temporary effect on the state of the environment [43]. An unstable state is effectively the same as an unstable state described by Lyapunov stability and will be expanded upon when discussing static stability and dynamic stability in the following chapters. A metastable equilibrium state is an equilibrium state that can be altered without leaving net effects in the environment of the system and without changing the values of amounts of constituents and parameters to an incompatible set of values, but this can be done only by means of interactions that have a finite temporary effect on the state of the environment [43]. The metastable equilibrium state is described by Lyapunov stability as a locally asymptotically stable

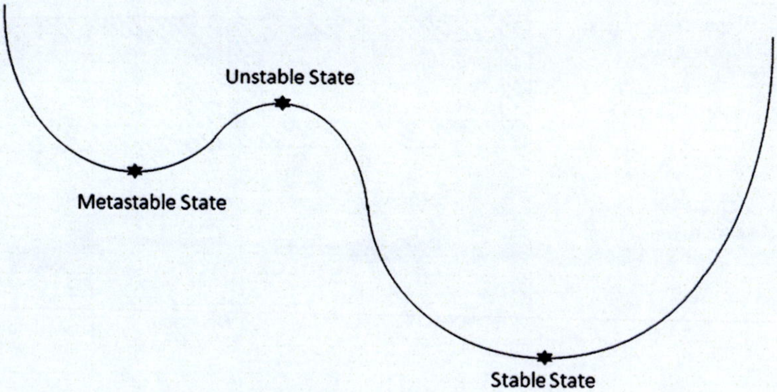

Fig. 2.1 Equilibrium states (potential energy)

state because damping is intrinsic to the system due to entropy production. A stable equilibrium state is an equilibrium state that can be altered to a different state only by interactions that either leave net effects in the environment of the system or change the values of amounts of constituents and parameters to an incompatible set of values [43]. The stable equilibrium state is described by Lyapunov stability as a globally asymptotically stable state because damping is intrinsic which leads to the minimum energy state of the system due to entropy production. These equilibrium states are presented in Fig. 2.1. We will introduce additional definitions of equilibrium states that do not include entropy production in later chapters which are referred to as Lagrange and Lyapunov stable when defining static and dynamic stabilities.

Second Law States Among all the states of a system with a given value of the energy and given values of the amounts of constituents and the parameters, there exists one and only one stable equilibrium state; and, starting from any state of a system, it is possible to reach a stable equilibrium state with arbitrarily specified values of amounts of constituents and parameters by means of a reversible weight process [43]. Notice that the Second Law states that a stable equilibrium state exists for each value of the energy while mechanics defines only a stable state at the minimum energy state or ground state. The stable equilibrium states of thermodynamics have three requirements:

1. Each stable equilibrium state is the state of lowest energy among all the states with the same values of entropy, volume, and parameters of the system.
2. Each stable equilibrium state is the state of highest entropy among all the states with the same values of energy, volume, and parameters of the system.
3. Temperature is positive and increasing with energy which implies the energy versus entropy curve for stable equilibrium states is convex.

All three requirements imply that the state of the system is static which leads to why classical thermodynamics is referred to as Equilibrium Thermodynamics or Thermostatics. Although the energy and entropy balance equations apply to any

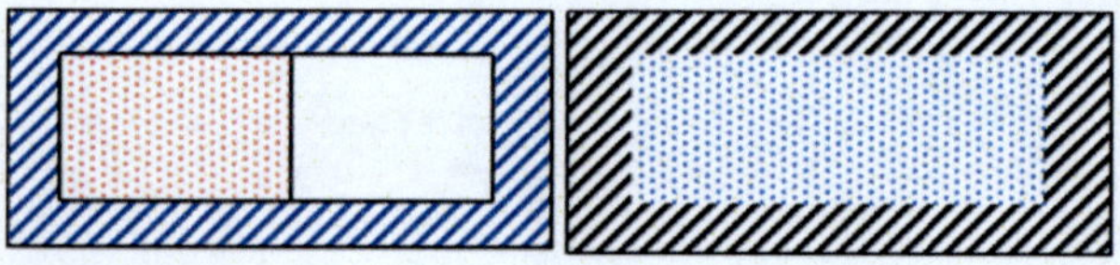

Fig. 2.2 Expansion of gas: adiabatic irreversible expansion

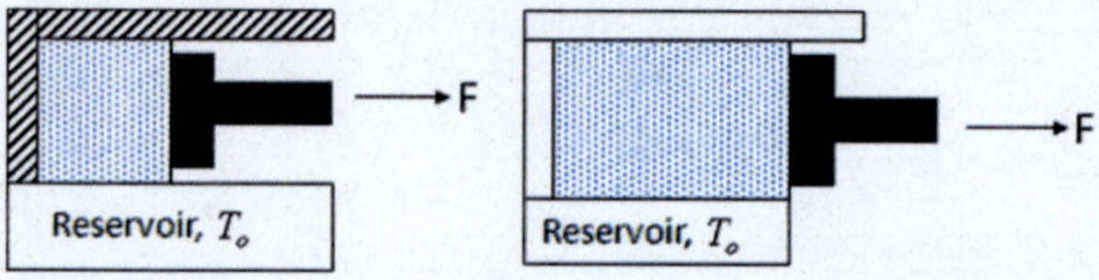

Fig. 2.3 Expansion of gas: reversible expansion

states of the system, when applying Equilibrium Thermodynamics, the system is effectively moving between equilibrium states since many of the quantities such as temperature and pressure are only defined for equilibrium states.

The Second Law helps differentiate between reversible and irreversible processes and provides a limit on the maximum amount of work that can be done with an energy flow. A process is reversible if the system and its environment can be restored to their initial states, except for differences of small order of magnitude than the maximum changes that occur during the process; otherwise the process is irreversible [43]. The classic example used to discuss the differences between irreversible and reversible processes is the expansion of a gas. Figure 2.2 shows the irreversible expansion of a gas by pulling a divider from a container that is half full to create a container with the gas fully expanded into the entire container.

This is a process that is operating between two equilibrium states. The system changes due to a constraint change: pulling the divider. To return the gas to its original condition will require additional work: irreversible. Figure 2.3 shows the reversible expansion of a gas by letting the gas do work against a friction-less piston at a constant temperature. This work can be reversed to return the gas to its original state: reversible.

To mathematically quantify these examples, the available work is defined as Ω, where

$$\Omega_2 - \Omega_1 = -(\mathcal{W}_{12})_{\text{rev}}, \quad (\mathcal{W}_{12})_{\text{rev}} \geq \mathcal{W}_{12}, \tag{2.3}$$

and the maximum available work occurs when the process is reversible. Otherwise, part of the work is lost through entropy production

$$d\mathcal{S} \equiv c_R(d\mathcal{E} - d\Omega), \tag{2.4}$$

and entropy is given the symbol $\mathcal{S}$, while c_R is a scaling factor. An interesting case of entropy production occurs for adiabatic processes such as shown in Fig. 2.2. For a reversible adiabatic process,

$$(d\mathcal{S})_{\text{rev}} = 0. \tag{2.5}$$

For an irreversible adiabatic process,

$$(d\mathcal{S})_{\text{irrev}} \geq 0. \tag{2.6}$$

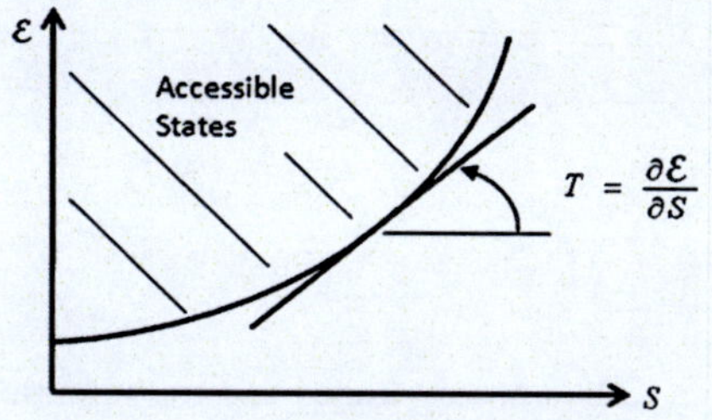

Fig. 2.4 Energy versus entropy curve

Returning to the three requirements, a corollary to the Second Law emerges from this discussion: At a stable equilibrium state, the entropy will be at its maximum value for fixed values of energy, number of particles, and constraints [43]. Also, the energy will be at a minimum value, the temperature is positive and increasing, plus the energy versus entropy curve for stable equilibrium states is convex. This is presented in Fig. 2.4.

A reversible process can be defined as interacting systems quasi-statically passing only through stable equilibrium states. The heat flowing between these interacting systems during a reversible process is

$$
\begin{aligned}
d\mathcal{Q} = \delta\mathcal{Q}_{\mathrm{rev}} &= T\,d\mathcal{S} \\
&= d\mathcal{E} + \delta\mathcal{W}_{\mathrm{rev}} \\
&= d\mathcal{E} + d\mathcal{W} \\
&= d\mathcal{E} - d\Omega \\
&= \frac{1}{C_R}\,d\mathcal{S},
\end{aligned}
\tag{2.7}
$$

where T is temperature. The entropy production during a reversible process is

$$
d\mathcal{S} = \frac{d\mathcal{Q}}{T}; \qquad \mathcal{S} = 0 = \oint \frac{d\mathcal{Q}}{T}. \tag{2.8}
$$

The entropy production during an irreversible process is [4]

$$
d\mathcal{S} \geq \frac{d\mathcal{Q}}{T}, \tag{2.9}
$$

where

$$
d\mathcal{S} = d_e\mathcal{S} + d_i\mathcal{S} \tag{2.10}
$$

with

$d_e\mathcal{S}$—entropy change due to exchange of energy and matter (entropy flux),
$d_i\mathcal{S}$—entropy change due to irreversible processes.

This is the entropy balance equation. The combination of the energy and entropy balance equations is the basic set of tools used to solve most problems in thermodynamics. These will be used to derive the exergy rate equation.

Fig. 2.5 Entropy changes

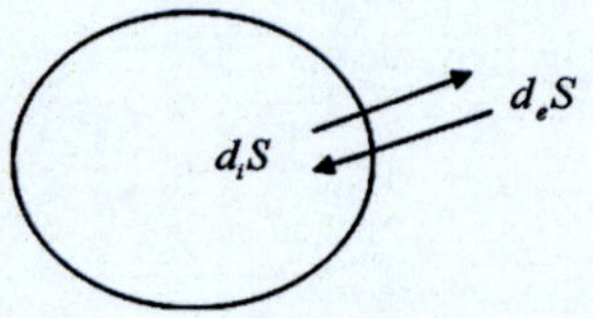

The entropy fluxes and irreversible entropy production are presented in Fig. 2.5. The irreversible entropy production can be written as

$$d_i\mathcal{S} = \sum_k \mathcal{F}_k \, d\mathcal{X}_k \geq 0, \tag{2.11}$$

where

$\mathcal{F}_k$—kth thermodynamics force,
$\mathcal{X}_k$—kth thermodynamics flow,

which appears to be a scaled power flow or work rate (discussed in the next chapter). The scaling factor is $1/T_0$.

2.4 Equilibrium Thermodynamics (Reversible/Irreversible Processes)

Returning to reversible processes one more time, in classical thermodynamics (equilibrium thermodynamics [4]), it is assumed that every irreversible transformation that occurs in nature can also be achieved through a reversible process (guaranteed by the Second Law) where

$$\mathcal{S}_2 = \mathcal{S}_1 + \int_1^2 \frac{d\mathcal{Q}}{T}, \tag{2.12}$$

and Fig. 2.6 presents a sketch of the processes of Figs. 2.2 and 2.3; note that entropy is a state function. This is the real strength of equilibrium thermodynamics. By analyzing systems in this way, one can calculate the changes in thermodynamic states via a straightforward procedure even if the processes may be quite complex such as an internal combustion engine. The amount of work that can be done by a cyclic heat engine, extracting work from a flow of heat, is limited by the Carnot efficiency of a reversible cycle (refer to [4]).

A reversible cyclic process can be analyzed with the First and Second Laws applied as

$dU = T\,dS - p\,d\bar{V}$	Energy balance equation
$dU = d\mathcal{E}$	Internal energy
$T\,d\mathcal{S} = d\mathcal{Q}$	Entropy balance equation
$p\,d\bar{V} = d\mathcal{W}$	p—pressure, $\bar{V}$—volume

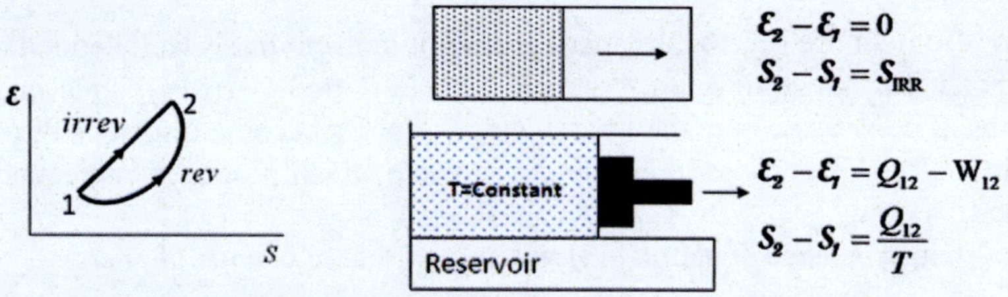

Fig. 2.6 Equilibrium thermodynamics

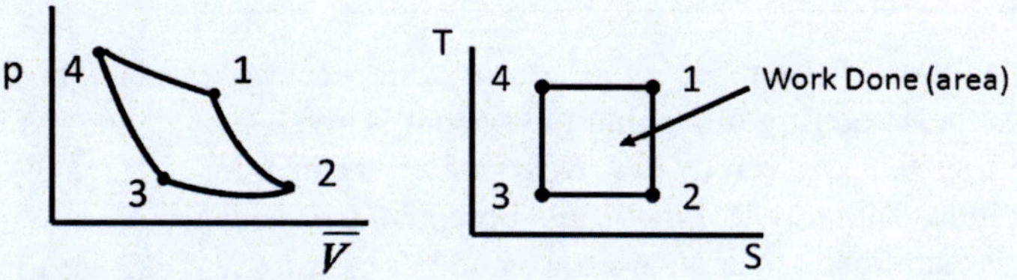

Fig. 2.7 Reversible cyclic work

Then

$$\boxed{\oint dU = 0 = \oint T\,dS - \oint p\,d\bar{V}.} \tag{2.13}$$

This implies that

$$\boxed{\oint T\,dS = \oint p\,d\bar{V}} \tag{2.14}$$

where a uniform temperature exists throughout the system during a process, and the change in energy over a cycle is zero since energy is a state function. The work done over a cycle shown in Fig. 2.7 can be calculated by the pV-work or the area in the TS plane since they balance one another. We will expand on this cyclic balance in later chapters to demonstrate the concepts of limit cycles and power flow balances that arise from nonequilibrium irreversible thermodynamic systems.

2.5 Local Equilibrium (Nonequilibrium Thermodynamics; Energy, Entropy, and Exergy Rate Equations)

The analysis of nonequilibrium irreversible systems is a complicated problem which requires large numbers of parameters and variables to accomplish. Quantities such as temperature and pressure are only defined in the context of equilibrium thermodynamics. The concept of "Local Equilibrium" can be used to extend the application of equilibrium thermodynamic quantities to nonequilibrium irreversible thermodynamic systems by providing well-defined quantities locally within each elemental volume. To be more specific, temperature is not uniform but is well defined locally. In general, for nonequilibrium systems, we can define thermodynamic quantities in terms of densities where thermodynamic variables become functions of position and time. For this purpose, the time dependence which produces rate equations is retained.

The energy balance equation [45] in terms of rate of change of energy is

$$\boxed{\dot{\mathcal{E}} = \sum_j \dot{\mathcal{Q}}_j + \sum_k \dot{\mathcal{W}}_k + \sum_l \dot{m}_l (h_l + ke_l + pe_l + \cdots).} \tag{2.15}$$

The term on the left represents the rate at which energy is changing within the system. The heat entering or leaving the system is given by $\dot{\mathcal{Q}}_j$, and the work done by or on the system is given by $\dot{\mathcal{W}}_k$. Next, material can enter or leave the system by $\dot{m}_l$ that includes enthalpy, h, kinetic and potential energies, ke, pe, etc. In addition, each term is "summed" over an arbitrary number of entry and exit locations j, k, l.

The entropy balance equation (2.10) in terms of rate of change of entropy is

$$\boxed{\dot{\mathcal{S}} = \sum_j \frac{\dot{\mathcal{Q}}_j}{T_j} + \sum_k \dot{m}_k s_k + \dot{\mathcal{S}}_i = \dot{\mathcal{S}}_e + \dot{\mathcal{S}}_i, \quad \dot{\mathcal{S}}_i = \sum_l \mathcal{F}_l \dot{\mathcal{X}}_l \geq 0} \tag{2.16}$$

where the left-hand term is the rate entropy changes within the system, and the right-hand terms represent, in order, the rate heat conducts entropy to and from the system and the rate material carries it in or out. These two terms can be combined into one term $\dot{\mathcal{S}}_e$, the entropy exchanged (either positive or negative) with the environment. $\dot{\mathcal{S}}_i$ is the irreversible entropy production rate within the system which is the sum of all the entropy changes due to the irreversible flows $\dot{\mathcal{X}}_l$ with respect to each corresponding thermodynamic force $\mathcal{F}_l$ (also see (2.11)).

Next, for systems with a constant environmental temperature, a thermodynamic quantity, called the availability function which has the same form as the Helmholtz free energy function, is defined as [4]

$$\Xi = \mathcal{E} - T_0 \mathcal{S}, \tag{2.17}$$

where T_0 is the reference environmental temperature. The availability function is the system exergy, described as the maximum theoretically available energy that can do work with respect to a given state via a reversible process. Exergy is also known as negative entropy [44, 45]. By taking the time derivative of the exergy in (2.17) and substituting into the expressions for energy (2.15) and entropy rates (2.16) results in the exergy rate equation

$$\boxed{\dot{\Xi} = \dot{\mathcal{E}} - T_0 \dot{\mathcal{S}} = \sum_j \left(1 - \frac{T_0}{T_j}\right) \dot{\mathcal{Q}}_j + \sum_k (\dot{\mathcal{W}}_k - p_0 \dot{\mathcal{V}}) + \sum_l \dot{m}_l \zeta_l^{\text{flow}} - T_0 \dot{\mathcal{S}}_i} \tag{2.18}$$

where $\dot{\Xi}$ is the rate at which exergy stored within the system is changing. The terms on the right, in order, define the rate exergy is carried in/out by: (i) heat, (ii) work (less any work the system does on the environment at constant environmental pressure p_0 if the system volume $\bar{V}$ changes), and (iii) by the material (or quantity known as flow exergy). The final term, $T_0\dot{S}_i$, is the rate exergy is destroyed (dissipated) within the system. This is the thermodynamic equation that will be used throughout the rest of the book to support the development of necessary and sufficient conditions for stability for a class of nonlinear systems, Hamiltonian natural systems. During the development of the connections between thermodynamics and Hamiltonian mechanics section, the class of thermodynamic systems that are analyzed will be adiabatic irreversible work processes: systems that have work done on them and have internal irreversible entropy production in the form of dissipation mechanisms that transfer no heat or material.

2.6 Chapter Summary

This chapter reviewed the basic concepts of thermodynamics required to support the development of necessary and sufficient conditions for stability of a class of nonlinear systems, Hamiltonian natural systems and adiabatic irreversible work processes. The First Law of Thermodynamics was reviewed and related to conservative mechanical systems. The Second Law of Thermodynamics was reviewed and related to the equilibrium of nonlinear systems. Exergy was defined as the availability, the maximum amount of work that can be extracted reversibly from an energy flow, and presented as the unifying metric of choice to develop the equations of motion and the power flow models. Local equilibrium was explained as a way to utilize the mathematics of equilibrium thermodynamics in order to analyze nonequilibrium irreversible system models to produce nonlinear rate and power equations.

To be more specific, the First and Second Laws of Thermodynamics were applied to system models to generate energy and entropy balance equations. In turn, these balance equations were written in rate equation form to produce the exergy rate equation which provides the fundamental connection to Hamiltonian mechanics and stability analysis.

Chapter 3
Mechanics

3.1 Introduction

In the previous chapter, the energy balance equation from the First Law of Thermodynamics combined with the entropy balance equation from the Second Law of Thermodynamics are written in rate equation form to create the exergy rate equation which is equivalent to power flow, the first time derivative of the Hamiltonian for natural systems. The next step is to define the types of thermodynamic systems that will be analyzed from this point forward and modify the exergy rate equation accordingly. The thermodynamic systems of interest are adiabatic irreversible work processes where work is done on the system, no entropy (or heat) is exchanged with the environment, and irreversible entropy is produced through dissipation. The recognition that the Hamiltonian is exergy provides the basic relationship necessary to apply the Second Law of Thermodynamics to the power flow to sort terms into three classes: storage, generation, and dissipation. This sorting process defines the dynamic stability boundary, a limit cycle, while the Hamiltonian surface defines the static stability boundary, a rigid body mode. The combination of the static stability and dynamic stability of a Hamiltonian natural system defines the necessary and sufficient conditions for a class of nonlinear systems, adiabatic irreversible work processes. These stability concepts will be developed in the next chapter. The development of this chapter will follow and combine the writings of Meirovitch [6], Junkins [15], and articles by Robinett and Wilson [13, 46].

3.2 Work, Energy, and Power

The incremental work done by a force on a mass, m, through an incremental distance $d\mathbf{r}$ is

$$dW = \mathbf{F} \cdot d\mathbf{r}, \tag{3.1}$$

R.D. Robinett III, D.G. Wilson, *Nonlinear Power Flow Control Design*,
Understanding Complex Systems,
DOI 10.1007/978-0-85729-823-2_3,

where

W—work,
$\mathbf{F}$—force vector,
$\mathbf{r}$—absolute position vector of m from the origin of an inertial frame.

The power generated by a force is

$$P = \frac{dW}{dt} = \frac{\mathbf{F} \cdot d\mathbf{r}}{dt} = \mathbf{F} \cdot \dot{\mathbf{r}}, \tag{3.2}$$

where

P—power,
$\dot{\mathbf{r}}$—velocity vector.

The kinetic energy can be derived from Newton's second law, $\mathbf{F} = m\ddot{\mathbf{r}}$,

$$dW = \mathbf{F} \cdot d\mathbf{r} = m\ddot{\mathbf{r}} \cdot d\mathbf{r} = m\dot{\mathbf{r}} \cdot d\dot{\mathbf{r}} = d\left(\frac{1}{2}m\dot{\mathbf{r}} \cdot \dot{\mathbf{r}}\right) = d\mathcal{T}, \tag{3.3}$$

where

$$\mathcal{T} = \frac{1}{2}m\dot{\mathbf{r}} \cdot \dot{\mathbf{r}} = \text{kinetic energy.}$$

The potential energy is a state function that is independent of time:

$$\begin{aligned} &\mathbf{F}(\mathbf{r}) \cdot d\mathbf{r} = -d\mathcal{V}(\mathbf{r}), \\ &\mathcal{V}(\mathbf{r}_2) - \mathcal{V}(\mathbf{r}_1) = -\int_{\mathbf{r}_1}^{\mathbf{r}_2} \mathbf{F}(\mathbf{r}) \cdot d\mathbf{r}, \\ &\mathcal{V}(\mathbf{r}) = \text{potential energy} \end{aligned} \tag{3.4}$$

with a conservative and path-independent force field, i.e.,

$$\oint \mathbf{F} \cdot d\mathbf{r} = 0. \tag{3.5}$$

The power of a conservative system, all forces applied to m are derived from potential energy functions, can be written as

$$\frac{d\mathcal{T}}{dt} = \mathbf{F} \cdot \dot{\mathbf{r}} = -\frac{d\mathcal{V}}{dt}, \tag{3.6}$$

which implies that

$$\frac{d}{dt}(\mathcal{T} + \mathcal{V}) = 0 \tag{3.7}$$

or

$$\begin{aligned} &\mathcal{T} + \mathcal{V} = E = \text{constant}, \\ &E = \text{total energy.} \end{aligned}$$

E is called the total energy and is a state function that is path independent.

If the force is a function of time, then the time dependent potential field results in

$$\mathbf{F}(\mathbf{r},t)\cdot d\mathbf{r} = -d\mathcal{V}(\mathbf{r},t), \tag{3.8}$$

and the total energy is modified as

$$\frac{d}{dt}(\mathcal{T}+\mathcal{V}) = \frac{\partial \mathcal{V}}{\partial t}. \tag{3.9}$$

Nonconservative forces can be included by adding the power due to a nonconservative force via

$$\begin{aligned} &\mathbf{F} = \mathbf{F}_c + \mathbf{F}_{NC},\\ &\mathbf{F}_c = \text{conservative force (potential field)},\\ &\mathbf{F}_{NC} = \text{nonconservative force}, \end{aligned} \tag{3.10}$$

which produces

$$\frac{dE}{dt} = \frac{d}{dt}(\mathcal{T}+\mathcal{V}) = \frac{\partial \mathcal{V}}{\partial t} + \mathbf{F}_{NC}\cdot\dot{\mathbf{r}}. \tag{3.11}$$

The focus in this book will be on *natural systems* with no explicit time-dependence [6]

$$\boxed{\frac{dE}{dt} = \frac{d}{dt}(\mathcal{T}+\mathcal{V}) = \mathbf{F}_{NC}\cdot\dot{\mathbf{r}}} \tag{3.12}$$

with nonconservative applied forces that perform work on and flow power into the system (generators) as well as dissipate energy within in the system by frictional forces (dissipators). These systems are described by thermodynamics as adiabatic irreversible work processes. The energy and entropy balances are described by: work is done on the system; no heat or entropy is exchanged with the environment; and irreversible entropy is produced through dissipation or

$$\boxed{\begin{aligned} &\mathcal{E}_2 - \mathcal{E}_1 = \mathcal{W}_{12},\\ &\mathcal{S}_2 - \mathcal{S}_1 = \mathcal{S}_{IRR},\\ &\text{and}\\ &\dot{\Xi} = \dot{W} - T_o\dot{\mathcal{S}}_i. \end{aligned}} \tag{3.13}$$

Note that the power flows of (3.12) include irreversible processes such as frictional effects. This means that the exergy rate equation of (3.13) is directly related to (3.12). In particular, the total energy is the stored energy and stored exergy of the system since potential and kinetic energies can do work. The details of this relationship are given in Sect. 3.5.

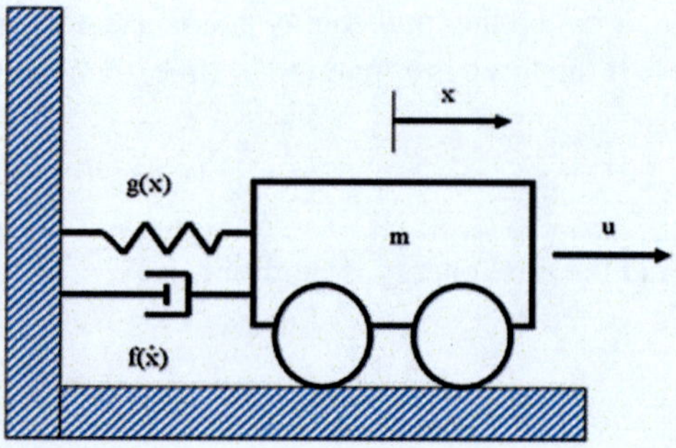

Fig. 3.1 General mass, spring, damper system. Robinett III, R.D. and Wilson, D.G. [46], reprinted by permission of the publisher (Taylor & Francis Ltd, http://www.tandf.co.uk/journals)

3.3 Energy Diagrams and Phase Planes

Energy diagrams and phase planes play an important role in the evaluation of the performance and stability of dynamical systems. The energy surface, which is actually the energy storage surface because it is the sum of the kinetic and potential energies, defines the accessible phase space of the system and can be projected onto the phase plane. Specifically, the trajectory/path of the dynamical system is constrained to the energy storage surface as power flows into and dissipated within the system. A simple example used throughout this book is the nonlinear mass, spring, damper system in Fig. 3.1. It is instructive to begin with the linear version of this example.

$$\begin{aligned}
\mathcal{H} &= \mathcal{T} + \mathcal{V} = \text{Hamiltonian},\\
\mathcal{T} &= \frac{1}{2}m\dot{x}^2 = \text{kinetic energy},\\
\mathcal{V} &= \frac{1}{2}kx^2 = \text{potential energy},\\
g(x) &= kx = \text{spring force},\\
f(\dot{x}) &= c\dot{x} = \text{damping force},\\
u &= \text{control force},\\
F &= u - c\dot{x} = \text{nonconservative forces},\\
m\ddot{x} + kx &= u - c\dot{x} = \text{equation of motion}.
\end{aligned} \tag{3.14}$$

The energy storage surface for a conservative system with

$$\begin{aligned}
u &= c\dot{x} = 0,\\
E &= \mathcal{T} + \mathcal{V} = \text{constant}
\end{aligned} \tag{3.15}$$

is a paraboloid shown in Fig. 3.2 (left). The energy diagrams and phase plane are the projections onto orthogonal planes (see Fig. 3.2, right, and Fig. 3.3).

The system path (see Fig. 3.3) is constrained to the energy storage surface along constant energy orbits that are ellipses in the phase plane given by

$$E = \frac{1}{2}m\dot{x}_0^2 + \frac{1}{2}kx_0^2 \tag{3.16}$$

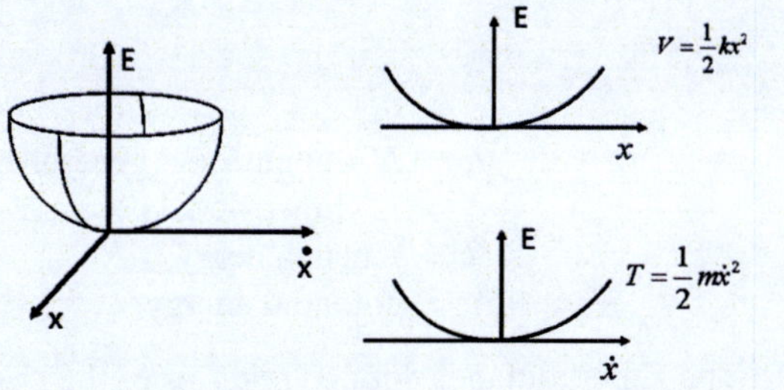

Fig. 3.2 Paraboloid (*left*) and parabolas (*right*)

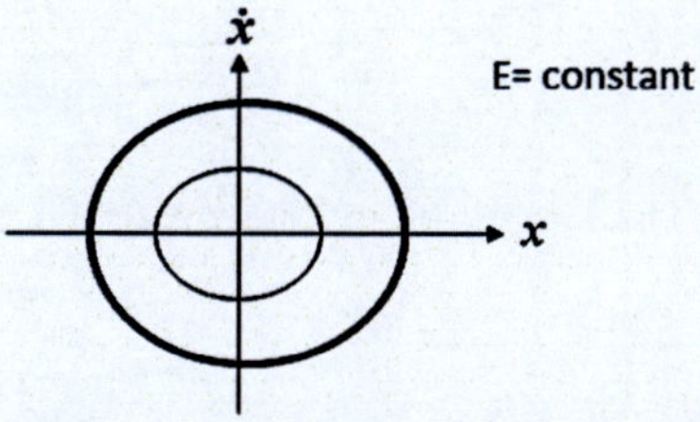

Fig. 3.3 Phase plane

with the initial condition $(x_0, \dot{x}_0)$.

Next, set $u = 0$ and $c > 0$, to create a nonconservative dissipative system which produces the trajectories in Fig. 3.7. The system trajectory/path is constrained to the energy storage surface

$$E = \mathcal{H} = \frac{1}{2}m\dot{x}^2 + \frac{1}{2}kx^2,$$

as it spirals down into the minimum energy state and maximum entropy state since all of the exergy is converted to irreversible entropy

$$\dot{E} = \dot{\mathcal{H}} = -c\dot{x}^2 = -T_0\dot{\mathcal{S}}_i.$$

Finally, set $u = 0$ and $c < 0$, which produces the trajectories in Fig. 3.6. The system trajectory/path is constrained to the energy storage surface as it spirals up to a higher energy state since power is flowing into the system

$$\dot{E} = \dot{\mathcal{H}} = c\dot{x}^2 = \dot{W}.$$

3.4 Hamiltonian Mechanics

This section follows the discussions in Meirovitch [6] and Robinett and Wilson [13]. The derivation of the Hamiltonian [6] begins with the Lagrangian for a system defined as

$$\mathcal{L} = \mathcal{T}(\mathbf{q}, \dot{\mathbf{q}}, t) - \mathcal{V}(\mathbf{q}, t), \tag{3.17}$$

where

t = time,
$\mathbf{q}$ = N-dimensional generalized coordinate vector,
$\dot{\mathbf{q}}$ = N-dimensional generalized velocity vector,
$\mathcal{T}$ = kinetic energy,
$\mathcal{V}$ = potential energy.

The Hamiltonian is defined in terms of the Lagrangian as

$$\mathcal{H} \equiv \sum_{i=1}^{n} \frac{\partial \mathcal{L}}{\partial \dot{q}_i} \dot{q}_i - \mathcal{L}(\mathbf{q}, \dot{\mathbf{q}}, t) = \mathcal{H}(\mathbf{q}, \dot{\mathbf{q}}, t). \tag{3.18}$$

The Hamiltonian in terms of the canonical coordinates $(\mathbf{q}, \mathbf{p})$ is

$$\mathcal{H}(\mathbf{q}, \mathbf{p}, t) = \sum_{i=1}^{n} p_i \dot{q}_i - \mathcal{L}(\mathbf{q}, \dot{\mathbf{q}}, t), \tag{3.19}$$

where the canonical momentum is defined as

$$p_i = \frac{\partial \mathcal{L}}{\partial \dot{q}_i}. \tag{3.20}$$

Then Hamilton's canonical equations of motion become

$$\begin{aligned} \dot{q}_i &= \frac{\partial \mathcal{H}}{\partial p_i}, \\ \dot{p}_i &= -\frac{\partial \mathcal{H}}{\partial q_i} + Q_i, \end{aligned} \tag{3.21}$$

where Q_i is the generalized force vector. Next, taking the time derivative of (3.19) gives

$$\dot{\mathcal{H}} = \sum_{i=1}^{n} \left(\dot{p}_i \dot{q}_i + p_i \ddot{q}_i - \frac{\partial \mathcal{L}}{\partial t} - \frac{\partial \mathcal{L}}{\partial q_i} \dot{q}_i - \frac{\partial \mathcal{L}}{\partial \dot{q}_i} \ddot{q}_i \right). \tag{3.22}$$

Then substituting (3.21) and simplifying gives

$$\dot{\mathcal{H}} = \sum_{i=1}^{n} Q_i \dot{q}_i - \frac{\partial \mathcal{L}}{\partial t}. \tag{3.23}$$

The focus of this book is on *natural systems* that are not explicit functions of time [6, 47], $\partial \mathcal{L}/\partial t = 0$. These natural systems will be referred to as Hamiltonian natural systems. Then for

$$\mathcal{L} = \mathcal{L}(q, \dot{q}), \tag{3.24}$$

the power (work/energy) equation becomes

$$\boxed{\dot{\mathcal{H}}(\mathbf{q},\mathbf{p}) = \sum_{i=1}^{n} Q_i \dot{q}_i.} \tag{3.25}$$

3.5 Connections Between Thermodynamics and Hamiltonian Mechanics

This section summarizes the concepts of conservative mechanical systems as well as reversible and irreversible thermodynamic systems [48]. These summaries provide the basis for the connections between thermodynamics and Hamiltonian mechanics required to support the development of necessary and sufficient conditions for stability of nonlinear Hamiltonian natural systems.

3.5.1 *Conservative Mechanical Systems*

A system is conservative if

$$\dot{\mathcal{H}} = 0 \quad \text{and} \quad \mathcal{H} = \text{constant}.$$

A force is conservative if for any closed path,

$$\oint \mathbf{F}\cdot\mathbf{r} = \oint \mathbf{F}\cdot\dot{\mathbf{r}}\,dt = \oint \sum_j Q_j \dot{q}_j\,dt = 0,$$

where $\mathbf{F}$ is the force vector, $d\mathbf{x}$ the displacement vector, and $\dot{\mathbf{r}}$ the velocity vector. All of the forces can be derived from potential functions. From a physical perspective, they all act as energy storage devices.

3.5.2 *Reversible Thermodynamic Systems*

A thermodynamic system is reversible if

$$dS = \frac{dQ}{T},$$

$$\oint dS = \oint \frac{dQ}{T} = 0,$$

$$\oint dS = \oint [dS_i + dS_e] = \oint [\dot{S}_i + \dot{S}_e]\,dt = 0,$$

which implies that $\dot{S}_e = \dot{Q}/T$ since by definition the second law gives $\dot{S}_i = 0$.

3.5.3 Irreversible Thermodynamic Systems

Since

$$\oint d\mathcal{S} = \oint [\dot{\mathcal{S}}_i + \dot{\mathcal{S}}_e]\, dt = 0,$$

we have $\dot{\mathcal{S}}_e \leq 0$ and $\dot{\mathcal{S}}_i \geq 0$.

3.5.4 Connections

Now the connections between thermodynamics and Hamiltonian mechanics are investigated. Basically, the Hamiltonian is stored exergy since potential and kinetic energies can be converted to work.

1. The irreversible entropy production rate can be expressed as

$$\dot{\mathcal{S}}_i = \sum_k \mathcal{F}_k \dot{\mathcal{X}}_k = \frac{1}{T_0} \sum_k Q_k \dot{q}_k \geq 0, \tag{3.26}$$

   which assumes local equilibrium.
2. The time derivative of the Hamiltonian is equivalent to the exergy rate

$$\begin{aligned} \dot{\mathcal{H}} &= \sum_k Q_k \dot{q}_k = \mathbf{F}_{\mathrm{NC}} \cdot \dot{\mathbf{r}}, \\ \dot{\Xi} &= \dot{W} - T_0 \dot{\mathcal{S}}_i = \sum_{j=1}^{N} Q_j \dot{q}_j - \sum_{l=N+1}^{M+N} Q_l \dot{q}_l, \end{aligned} \tag{3.27}$$

   where N is the number of generators, M the number of dissipators, and $\dot{W} = \sum_j \dot{W}_j$. The following assumptions apply when utilizing the exergy rate equation (2.18) for *Hamiltonian natural systems* (irreversible adiabatic work process):
   a. No substantial heat flow:

$$\dot{Q}_i \approx 0.$$

   b. No substantial exergy flow or assume T_i is only slightly greater than T_0:

$$1 - \frac{T_0}{T_i} \approx 0.$$

   c. No $p_0 \bar{V}$ work on the environment:

$$p_0 \frac{d\bar{V}}{dt} = 0.$$

d. No mass flow rate:

$$\sum_k \dot{m}_k \zeta_k^{\text{flow}} = 0.$$

e. Then define:

$$\dot{W} \geq 0 \text{ power input/generated,}$$
$$T_0\dot{\mathcal{S}}_i \geq 0 \text{ power dissipated.}$$

3. A conservative system is equivalent to a reversible system when

$$\dot{\mathcal{H}} = 0 \quad \text{and} \quad \dot{\mathcal{S}}_e = 0;$$

then

$$\dot{\mathcal{S}}_i = 0 \quad \text{and} \quad \dot{W} = 0.$$

4. A system that "appears to be conservative" but is not reversible is defined as:

$$\begin{aligned} \dot{\mathcal{H}}_{\text{ave}} = 0 &= \frac{1}{\tau_c} \oint [\dot{W} - T_0\dot{\mathcal{S}}_i]\, dt \\ &= (\dot{W})_{\text{ave}} - (T_0\dot{\mathcal{S}}_i)_{\text{ave}} = \frac{1}{\tau_c} \oint \left[\sum_{j=1}^{N} Q_j \dot{q}_j - \sum_{l=N+1}^{M+N} Q_l \dot{q}_l \right] dt \\ &= \text{average power over a cycle,} \end{aligned}$$

where τ_c is the period of the cycle. To be more specific about the average power calculations, the AC power factor [49] provides an excellent example. For the general case of alternating current supplied to a complex impedance, the voltage and current differ in phase by an angle ϕ. The time responses for power, voltage, and current are shown for a general AC circuit in Fig. 3.4 with

$$\begin{aligned} \dot{W} = P = Q\dot{q} = \tilde{V}\tilde{I} &= \sqrt{2}\tilde{v}\cos(\omega t + \phi) \cdot \sqrt{2}\tilde{i}\cos\omega t \\ &= \tilde{v}\tilde{i}[\cos\phi + \cos(2\omega t + \phi)], \end{aligned}$$

where P is power, $\tilde{V}$ is voltage ($\tilde{v}$), $\tilde{I}$ is current ($\tilde{i}$), ϕ is the phase angle, and ω is the frequency. Integrating over a cycle gives

$$(\dot{W})_{\text{ave}} = \tilde{v}\tilde{i}\cos\phi,$$

where for the second term,

$$\oint \cos(2\omega t + \phi)\, dt = 0.$$

This is an important set of conditions that will be used in the next section to find the dynamic stability boundary.

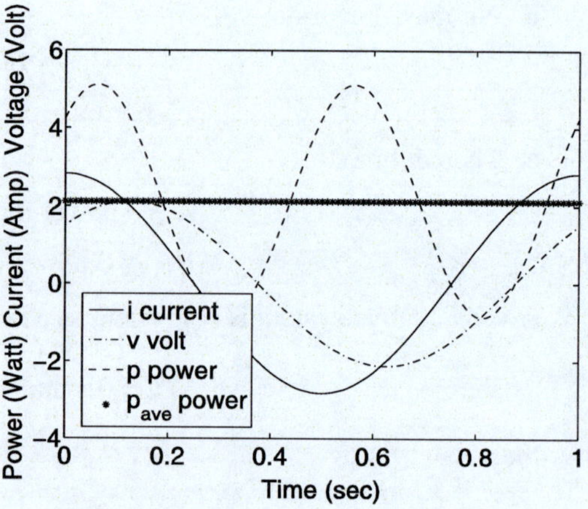

Fig. 3.4 Time response for power in a general AC circuit with $\omega = 2\pi$, $\tilde{v} = 1.5$, $\tilde{i} = 2.0$, and $\theta = \pi/4$. Robinett III, R.D. and Wilson, D.G. [13], reprinted by permission of the publisher (Interscience Publishers)

5. Finally, the power terms are sorted into three categories:
   a. $(\dot{W})_{\text{ave}}$—power generators, $(Q_j\dot{q}_j)_{\text{ave}} > 0$;
   b. $(T_0\dot{S}_i)_{\text{ave}}$—power dissipators, $-(Q_l\dot{q}_l)_{\text{ave}} < 0$;
   c. $(T_0\dot{S}_{\text{rev}})_{\text{ave}}$—reversible/conservative exergy storage terms, $(Q_k\dot{q}_k)_{\text{ave}} = 0$.
   These three categories are fundamental terms in the following design procedures of the next chapter.

3.6 Line Integrals and Limit Cycles

Line integrals are utilized in the previous chapter and some previous sections of this chapter. It is time to more formally describe the concepts of linear and nonlinear limit cycles that result from closed line integrals and balanced power flows. This section follows references [7, 46, 50]

3.6.1 Linear Limit Cycle

This subsection describes the concept of *linear limit cycles*. A linear limit cycle is a strange concept to most people since limit cycles are typically associated with nonlinear systems. A limit cycle is defined by Wikipedia as a closed trajectory in phase space having the property that at least one other trajectory spirals into it either as time approaches infinity or as time approaches minus-infinity. By the end of this section, it will be demonstrated how a center [32] of a second-order system can be interpreted as a linear limit cycle, for example, a goal of power engineering.

The mass–spring–damper system of Fig. 3.1 will be utilized throughout this section by methodically adding more complexity as the section progresses.

The Lagrangian and Hamiltonian of this linear system ($g(x) = kx$, $f(\dot{x}) = c\dot{x}$) are

$$\mathcal{L} = \mathcal{T} - \mathcal{V} = \frac{1}{2}m\dot{x}^2 - \frac{1}{2}kx^2 \tag{3.28}$$

and

$$\mathcal{H} = \mathcal{T} + \mathcal{V} = \frac{1}{2}m\dot{x}^2 + \frac{1}{2}kx^2. \tag{3.29}$$

The equation of motion is derived from Lagrange's equation

$$\frac{d}{dt}\left(\frac{\partial \mathcal{L}}{\partial \dot{q}_i}\right) - \frac{\partial \mathcal{L}}{\partial q_i} = Q_i \tag{3.30}$$

and is determined as

$$m\ddot{x} + kx = -c\dot{x} + u, \tag{3.31}$$

where

$q_i = x =$ generalized coordinate,
$\dot{q}_i = \dot{x} =$ generalized velocity,
$Q_i = -c\dot{x} + u =$ generalized forces, and
$u =$ control input.

The Hamiltonian for natural systems is the stored energy, and the time derivative is the power flow into, dissipated within, and stored in the system and is determined as the first time derivative of (3.29) or

$$\dot{\mathcal{H}} = [m\ddot{x} + kx]\dot{x} = [-c\dot{x} + u]\dot{x}, \tag{3.32}$$

which for a conservative system is

$$\dot{\mathcal{H}} = 0 \quad \Rightarrow \quad \mathcal{H} = \text{constant}$$

and

$$[-c\dot{x} + u] = 0, \tag{3.33}$$

which implies point-by-point cancellation of forces and power flows.

Next (3.32) and (3.33) are investigated in more detail. Equation (3.32) can be rewritten as a line integral of the power flow to produce the work per cycle [26–29]:

$$W_{\text{cyclic}} = \mathcal{H}_{\text{cyclic}} = \oint_\tau \dot{\mathcal{H}}\,dt = \oint_\tau [m\ddot{x} + kx]\dot{x}\,dt = \oint_\tau [-c\dot{x} + u]\dot{x}\,dt, \tag{3.34}$$

where

$\tau =$ period of the limit cycle,
$\oint =$ closed (trajectory) path integral along the Hamiltonian surface.

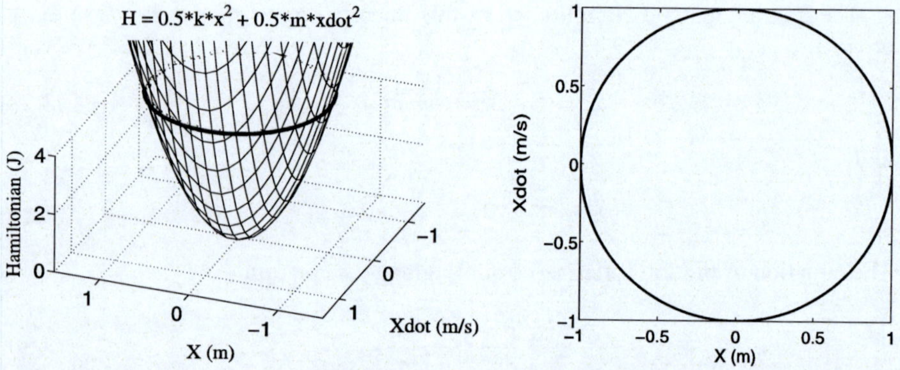

Fig. 3.5 Linear limit cycle: Hamiltonian 3D surface (*left*) and phase plane plot (*right*). Robinett III, R.D. and Wilson, D.G. [46], reprinted by permission of the publisher (Taylor & Francis Ltd, http://www.tandf.co.uk/journals)

Equation (3.34) can be further rewritten as

$$\begin{aligned}\mathcal{H}_{\text{cyclic}} &= \oint [\mathrm{p}(x, y)\,dx + \mathrm{q}(x, y)\,dy] = \oint [\mathcal{H}_x\,dx + \mathcal{H}_y\,dy] \\ &= \oint [kx\,dx + my\,dy] = \oint_\tau [kx\dot{x} + my\dot{y}]\,dt = 0 \end{aligned} \tag{3.35}$$

for $[c\dot{x} - u] = 0$, $y = \dot{x}$, $\mathcal{H}_x = \partial\mathcal{H}/\partial x$, since $\mathcal{H}$ is an exact integral, conservative, and path independent. This result can be confirmed by Green's Theorem

$$\mathcal{H}_{\text{cyclic}} = \oint [\mathrm{p}\,dx + \mathrm{q}\,dy] = \iint_\Omega [\mathrm{q}_x - \mathrm{p}_y]\,dx\,dy = 0$$

since

$$\begin{aligned} \mathrm{p} &= \mathcal{H}_x = kx \quad \Rightarrow \quad \mathrm{p}_y = 0, \\ \mathrm{q} &= \mathcal{H}_y = my \quad \Rightarrow \quad \mathrm{q}_x = 0. \end{aligned}$$

Equation (3.35) describes a center which will be referred to as a linear limit cycle shown in Fig. 3.5 (where $m = 10$ kg, $k = 10$ N/m, $x_0 = 1.0$ m, and $\dot{x}_0 = 0.0$ m/s), which defines a constant energy orbit in phase space, $\mathcal{H} =$ constant, given some initial condition that defines an initial energy state

$$\mathcal{H} = \frac{1}{2}m\dot{x}_0^2 + \frac{1}{2}kx_0^2. \tag{3.36}$$

Equation (3.36) defines the orbit that is an orthogonal cut across the Hamiltonian manifold and projected onto the phase plane. The Hamiltonian surface determines the accessible phase space as a function of energy level. The power flow, $\dot{\mathcal{H}}$, determines the system trajectory across the Hamiltonian surface as a function of time and initial position.

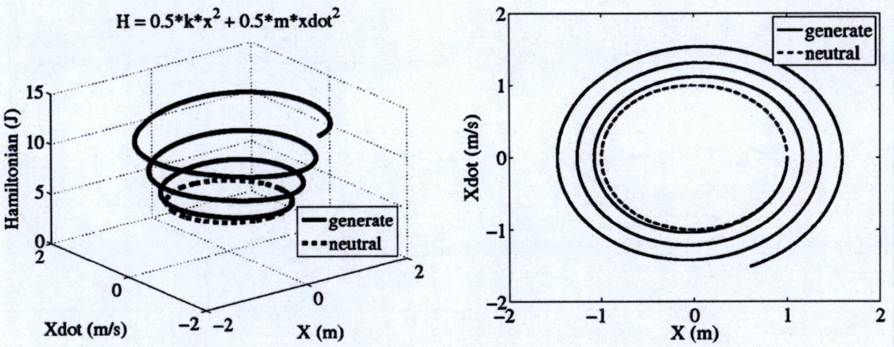

Fig. 3.6 Linear limit cycle mass–spring–damper system rising to higher energy state—generative with 3D Hamiltonian (*left*) and projected 2D phase plane (*right*) plots ($K_p = c = K_D = 0$, $K_I = 0.5$). Robinett III, R.D. and Wilson, D.G. [46], reprinted by permission of the publisher (Taylor & Francis Ltd, http://www.tandf.co.uk/journals)

Now, one can explore (3.33) further by selecting a Proportional-Integral-Derivative (PID) controller

$$u = -K_P x - K_I \int_0^t x\, d\tau - K_D \dot{x} \tag{3.37}$$

and rewriting (3.32) as

$$\dot{\mathcal{H}} = [m\ddot{x} + (k + K_P)x]\dot{x} = \left[-(c + K_D)\dot{x} - K_I \int_0^t x\, d\tau\right]\dot{x}, \tag{3.38}$$

which leads to a rewritten version of (3.33) for $m > 0$, $k > 0$, $K_P > 0$ as

$$\left[-(c + K_D)\dot{x} - K_I \int_0^t x\, d\tau\right] = 0. \tag{3.39}$$

Equation (3.39) was solved in [30] and is presented in Sect. 4.3, where the integral feedback is shown to be a power generator term that balances the power dissipator terms point-for-point which leads to another linear limit cycle like the one in Fig. 3.5. It is as though no forces and power flows are acting on the system. Notice that the balance of power flows determines the stability of the system. Basically, a system is unstable for increasing energy while stable, possibly asymptotically stable, for decreasing energy. If $K_P > 0$, $c > 0$, $K_D > 0$, and

$$K_I \int_0^t x\, d\tau \dot{x} > -(c + K_D)\dot{x}^2,$$

then the system rises to a higher energy state, and the system is unstable which spirals up from the center (see Fig. 3.6). On the other hand, if

$$K_I \int_0^t x\, d\tau\, \dot{x} < -(c + K_D)\dot{x}^2,$$

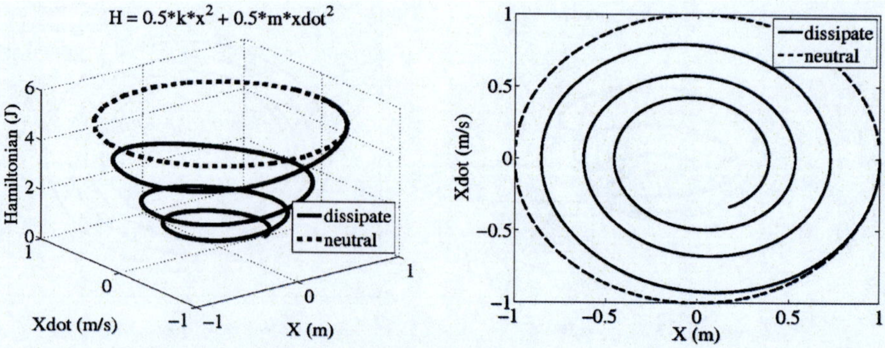

Fig. 3.7 Linear limit cycle mass–spring–damper system falls to lower energy state—dissipative with 3D Hamiltonian (*left*) and projected 2D phase plane (*right*) plots ($K_P = c = K_I = 0$, $K_D = 1.0$). Robinett III, R.D. and Wilson, D.G. [46], reprinted by permission of the publisher (Taylor & Francis Ltd, http://www.tandf.co.uk/journals)

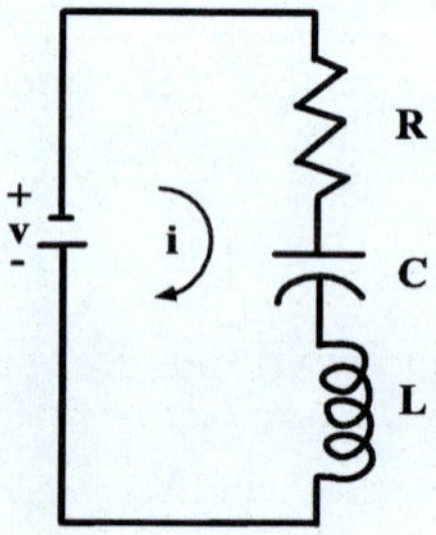

Fig. 3.8 Thevinen equivalent RLC circuit. Robinett III, R.D. and Wilson, D.G. [46], reprinted by permission of the publisher (Taylor & Francis Ltd, http://www.tandf.co.uk/journals)

then the system falls to a lower energy state, and the system is asymptotically stable or spirals down from the center (see Fig. 3.7). The linear limit cycle is a stability boundary. Equation (3.38) defines an eigenvalue problem that will be discussed in Sect. 4.3.5.

Next, a particular example of designing linear limit cycles will be explored for a power engineering application [49]. Returning to (3.31), the control input is a sinusoidal voltage, and the electrical analogy from Fig. 3.8 gives

$$L\ddot{\bar{q}} + \frac{1}{C}\bar{q} = -R\dot{\bar{q}} + v_0 \cos \Omega t, \tag{3.40}$$

where $i = \dot{\bar{q}} =$ current, and

$$t = \text{time},$$

$$\bar{q} = \text{charge},$$

$$L = \text{inductance},$$

$$C = \text{capacitance},$$

$$R = \text{resistance},$$

$$v_0 = \text{voltage},$$

$$\Omega = \text{driving frequency}.$$

The Hamiltonian is

$$\mathcal{H} = \frac{1}{2}L\dot{\bar{q}}^2 + \frac{1}{2}\frac{1}{C}\bar{q}^2. \tag{3.41}$$

The time derivative of the Hamiltonian is

$$\dot{\mathcal{H}} = \left[L\ddot{\bar{q}} + \frac{1}{C}\bar{q}\right]\dot{\bar{q}} = [-R\dot{\bar{q}} + v_0 \cos \Omega t]\dot{\bar{q}}, \tag{3.42}$$

where the two goals are to make $\dot{\mathcal{H}} = 0$ and obtain a power factor of 1 which maximizes the real power to the load.

The first goal occurs when

$$\dot{\mathcal{H}} = 0 = [v_0 \cos \Omega t - R\dot{\bar{q}}]\dot{\bar{q}}, \tag{3.43}$$

which implies

$$v_0 \cos \Omega t = R\dot{\bar{q}}. \tag{3.44}$$

The second goal occurs when

$$\dot{\mathcal{H}} = 0 = \left[L\ddot{\bar{q}} + \frac{1}{C}\bar{q}\right]\dot{\bar{q}}, \tag{3.45}$$

which implies $\omega^2 = 1/LC = \Omega^2$. To verify this, (3.40) is solved for the steady-state solution [25]

$$\bar{q}(t) = \frac{v_0}{[L^2(\omega^2 - \Omega^2)^2 + R^2\Omega^2]^{1/2}} \cos(\Omega t - \delta), \tag{3.46}$$

where

$$\cos \delta = \frac{L(\omega^2 - \Omega^2)}{[L^2(\omega^2 - \Omega^2)^2 + R^2\Omega^2]^{1/2}}.$$

By imposing $\omega^2 = 1/LC = \Omega^2$ we have

$$\cos \delta = 0 \quad \Rightarrow \quad \delta = \pi/2,$$

which implies

$$\bar{q} = \frac{v_0}{R\Omega}\cos\left(\Omega t - \frac{\pi}{2}\right) = \frac{v_0}{R\Omega}\sin \Omega t,$$

$$\dot{\bar{q}} = -\frac{v_0}{R}\sin\left(\Omega t - \frac{\pi}{2}\right) = \frac{v_0}{R}\cos \Omega t,$$

which gives (3.44), and then, for (3.45),

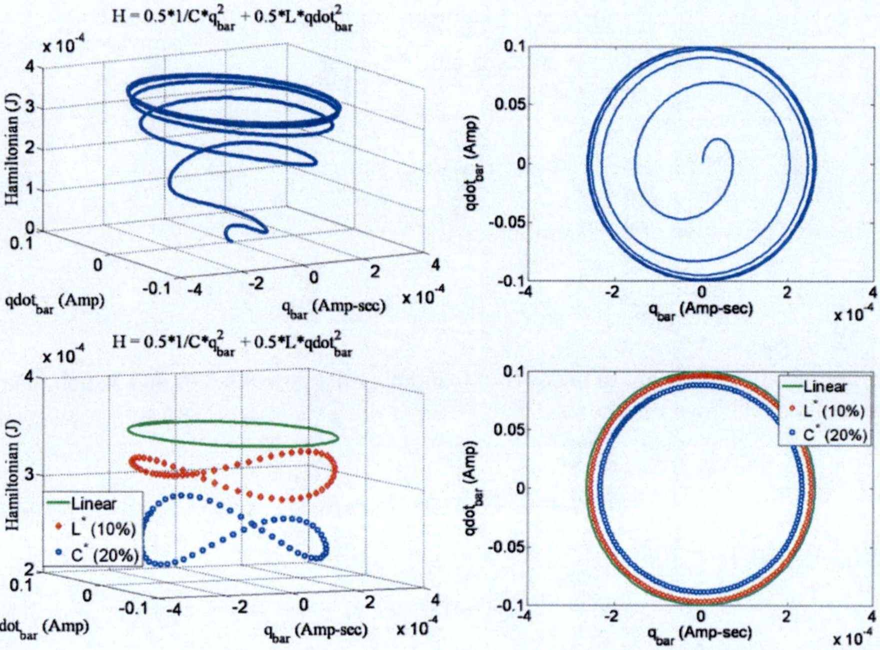

Fig. 3.9 Linear limit cycle Hamiltonian 3D spiral transient (*upper-left*) with corresponding phase plane plot (*upper-right*). The next 3D plots present the linear, 10% variation in inductance, and 20% variation in capacitance, C steady-state responses (*lower-left*) with the corresponding phase plane plots (*lower-right*). Robinett III, R.D. and Wilson, D.G. [46], reprinted by permission of the publisher (Taylor & Francis Ltd, http://www.tandf.co.uk/journals)

$$\begin{aligned}
-L\left(\frac{v_0}{R}\Omega\right)\sin\Omega t + \frac{1}{C}\left(\frac{v_0}{R\Omega}\right)\sin\Omega t &= \left[-L\Omega + \frac{1}{C\Omega}\right]\frac{v_0}{R}\sin\Omega t \\
&= \left[-\Omega^2 + \frac{1}{LC}\right]\frac{L}{\Omega}\frac{v_0}{R}\sin\Omega t \\
&= 0.
\end{aligned}$$

Next, the linear limit cycles for this problem, where $\Omega^2 = 1/LC$ and $\Omega^2 \neq 1/LC$, will be investigated. Figure 3.9 (upper-left 3D Hamiltonian plot and upper-right phase-plane plot) shows the linear limit cycle for a power factor of 1 or $\Omega^2 = 1/LC$ where "at least one other trajectory spirals into it." It appears similar to the response of an undamped, unforced linear system once the system reaches the linear limit cycle. Figure 3.9 (linear, 10% variation in L, and 20% variation in C) are plotted in lower-left along with the phase plane plots lower-right) shows the linear limit cycle for a power factor of less than 1 or $\Omega^2 \neq 1/LC$ (see Table 3.1 for numerical values). The energy level has dropped, and the edges have drooped, which results in the concentric ellipses in the phase plane even though the Hamiltonian surface has been deformed by $\Omega^2 \neq 1/LC$. Part of the power flow is being used for reactive power or VAR support. This ensures a concentric linear limit cycle in the phase

Table 3.1 RLC numerical parameters and results. Robinett III, R.D. and Wilson, D.G. [46], reprinted by permission of the publisher (Taylor & Francis Ltd, http://www.tandf.co.uk/journals)

Case description	L (mH)	C (μF)	R (Ohms)	v_0 (Volts)	ω (Hz)	Ω (Hz)	δ (deg)	$\cos(\frac{\pi}{2}-\delta)$ power factor
Linear	70.362	100.0	10.0	1.0	60.0	60.0	90	1.0
10% variation (L)	63.326	100.0	10.0	1.0	63.25	60.0	75	0.9659
20% variation (C)	70.362	125.0	10.0	1.0	53.67	60.0	118	0.8829

plane at a lower energy level. Notice that

$$\oint_\tau \dot{\mathcal{H}}\,dt = \mathcal{H}_{\text{cyclic}} = 0$$

implies that $\mathcal{H}(t)$ is cyclic and not constant, where $\tau = 2\pi/\Omega$ and $\Omega^2 \neq \omega^2 = 1/LC$. Note that a goal of power engineering is to generate a power factor of 1 which is equivalent to generating a linear limit cycle. This *out of plane* limit cycle behavior is the first hint at how to generalize (3.32) and (3.34) to nonlinear limit cycles.

3.6.2 Nonlinear Limit Cycles

This section extends the concept of linear limit cycles to nonlinear limit cycles. The linear mass–spring–damper system is extended to a nonlinear system by adding nonlinear stiffness, $g(x)$, and damping $f(\dot{x})$ as shown in Fig. 3.1.

The Lagrangian and Hamiltonian for this system are

$$\mathcal{L} = \frac{1}{2}m\dot{x}^2 - \mathcal{V}(x), \tag{3.47}$$

$$\mathcal{H} = \frac{1}{2}m\dot{x}^2 + \mathcal{V}(x) \tag{3.48}$$

with the following equation of motion:

$$m\ddot{x} + g(x) = -f(\dot{x}) + u, \tag{3.49}$$

where for $\mathcal{V}(x) > 0$ when $x \neq 0$, $\mathcal{V}(x) = 0$ when $x = 0$, $\partial\mathcal{V}(x)/\partial x = g(x)$, and $f(\dot{x})$ is the generalized damping force. The time derivative of the Hamiltonian is

$$\dot{\mathcal{H}} = [m\ddot{x} + g(x)]\dot{x} = [-f(\dot{x}) + u]\dot{x}, \tag{3.50}$$

where the goal is to find the nonlinear limit cycles given by

$$\mathcal{H}_{\text{cyclic}} = \oint_\tau \dot{\mathcal{H}}\,dt = \oint_\tau [m\ddot{x} + g(x)]\dot{x}\,dt = \oint_\tau [-f(\dot{x}) + u]\dot{x}\,dt = 0. \tag{3.51}$$

For a conservative system, $\dot{\mathcal{H}} = 0$ and $\mathcal{H} = \text{constant}$, and

$$[-f(\dot{x}) + u] = 0 \quad \text{(for point-by-point)}. \tag{3.52}$$

The more general solution that replaces the point-by-point force balance is a cyclic balance between the power flowing into the system versus the power being dissipated within the system is

$$\mathcal{H}_{\text{cyclic}} = \oint_{\tau} \dot{\mathcal{H}}\, dt = 0 \quad \text{and} \quad \mathcal{H}(t) \text{ is cyclic.} \tag{3.53}$$

A familiar example to begin the discussion involving (3.53) is the van der Pol equation

$$\ddot{x} - \mu(1 - x^2)\dot{x} + x = 0, \tag{3.54}$$

which shows the effect of nonlinear damping on a system with linear mass and stiffness. The Hamiltonian with nonunity mass and stiffness is

$$\mathcal{H} = \frac{1}{2} m\dot{x}^2 + \frac{1}{2} kx^2 \tag{3.55}$$

with the time derivative of

$$\dot{\mathcal{H}} = [m\ddot{x} + kx]\dot{x} = [\mu(1 - x^2)\dot{x}]\dot{x}, \tag{3.56}$$

and the nonlinear limit cycle becomes

$$\mathcal{H}_{\text{cyclic}} = 0 = \oint_{\tau} [m\ddot{x} + kx]\dot{x}\, dt = \oint_{\tau} [\mu(1 - x^2)\dot{x}]\dot{x}\, dt, \tag{3.57}$$

which implies

$$\oint_{\tau} [\mu\dot{x}^2]\, dt = \oint_{\tau} [\mu x^2 \dot{x}^2]\, dt. \tag{3.58}$$

These conditions are simulated numerically with the parameter values given in Table 3.2. For the generative case, the generators are greater than the dissipators, as shown in Fig. 3.10 with the power flow (left) and energy (right) time-domain responses. The neutral case energy balance over a cycle is shown in Fig. 3.11 for the power flow (left) and energy (right) time-domain responses. This is a stable nonlinear limit cycle. Notice that the nonelliptical shape of the nonlinear limit cycle

Table 3.2 Van der Pol model numerical values. Robinett III, R.D. and Wilson, D.G. [46], reprinted by permission of the publisher (Taylor & Francis Ltd, http://www.tandf.co.uk/journals)

Case	Description	x_0 (m)	$\dot{x}_0$ (m/s)	μ (kg/s)	m (kg)	k (kg/s^2)
1	generate	0.1	−0.1	1.5	1.0	1.0
2	neutral	1.0	−1.0	1.5	1.0	1.0
3	dissipate	2.0	−2.0	1.5	1.0	1.0

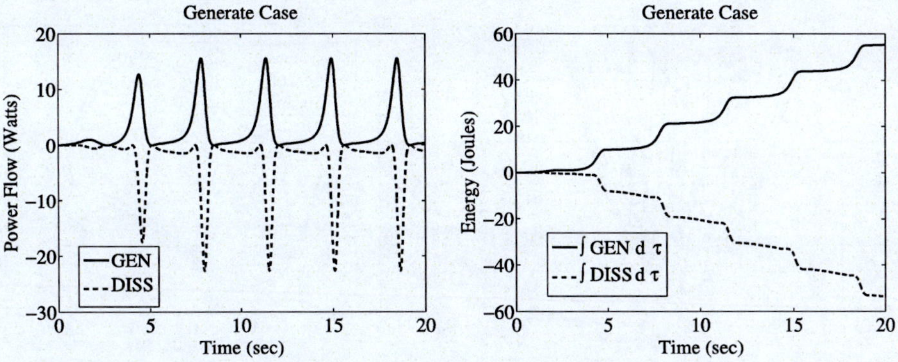

Fig. 3.10 Van der Pol power flow (*left*) and energy (*right*) responses—generative case. Robinett III, R.D. and Wilson, D.G. [46], reprinted by permission of the publisher (Taylor & Francis Ltd, http://www.tandf.co.uk/journals)

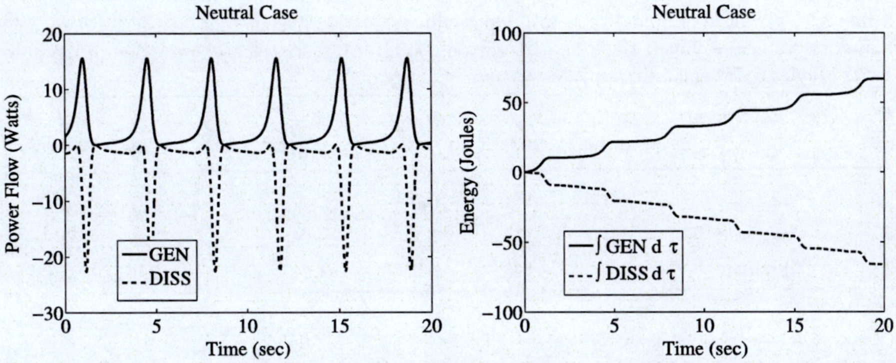

Fig. 3.11 Van der Pol power flow (*left*) and energy (*right*) responses—neutral case. Robinett III, R.D. and Wilson, D.G. [46], reprinted by permission of the publisher (Taylor & Francis Ltd, http://www.tandf.co.uk/journals)

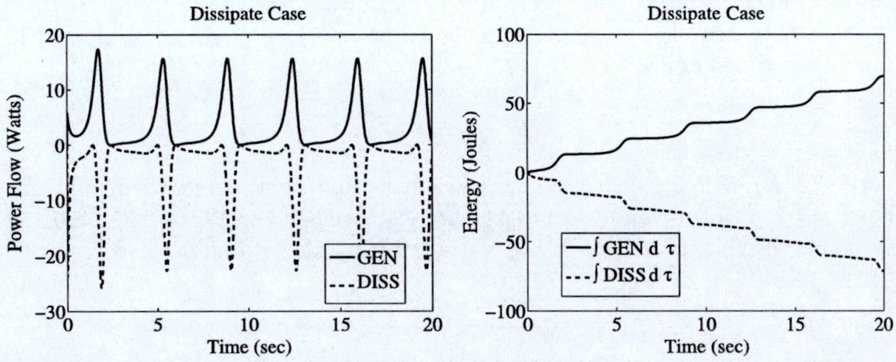

Fig. 3.12 Van der Pol power flow (*left*) and energy (*right*) responses—dissipative case. Robinett III, R.D. and Wilson, D.G. [46], reprinted by permission of the publisher (Taylor & Francis Ltd, http://www.tandf.co.uk/journals)

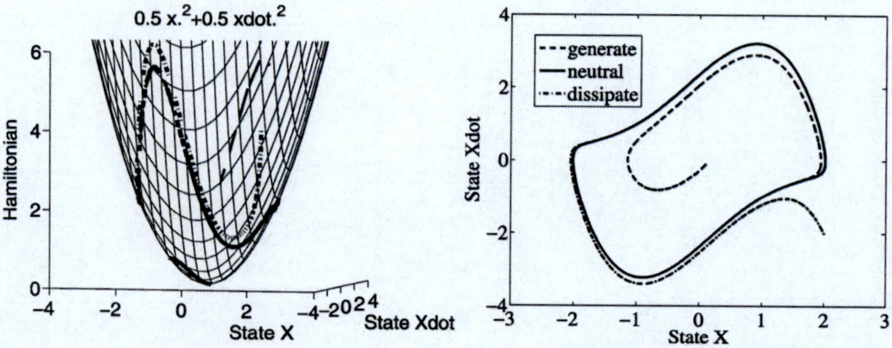

Fig. 3.13 Van der Pol responses: Hamiltonian 3D surface (*left*) and phase plane 2D projection (*right*). Robinett III, R.D. and Wilson, D.G. [46], reprinted by permission of the publisher (Taylor & Francis Ltd, http://www.tandf.co.uk/journals)

Table 3.3 Van der Pol model as control input with generative K_G and dissipative K_D gains and numerical values. Robinett III, R.D. and Wilson, D.G. [46], reprinted by permission of the publisher (Taylor & Francis Ltd, http://www.tandf.co.uk/journals)

Case	Description	x_0 (m)	$\dot{x}_0$ (m/s)	K_G (kg/s)	K_D (kg/(m$^2\cdot$s))	m (kg)	k (kg/s^2)
1	neutral	0.0	−1.0	0.0	0.0	1.0	1.0
2	generate	0.0	−1.0	1.5	0.0	1.0	1.0
3	dissipate	0.0	−1.0	0.0	1.5	1.0	1.0

is a result of the cyclic rise and fall of the energy level of the system, $\mathcal{H}(t)$, and is due to the nonlinear damping. The dissipative case is shown in Fig. 3.12 with the power flow (left) and energy (right) time-domain responses. The results for all three cases are visually shown in Fig. 3.13 as a 3D Hamiltonian plot (left) with the corresponding 2D phase-plane projection (right), respectively.

Also, note that the nonlinear damping can be rewritten as a control input with two gains instead of one, or

$$u = K_G\dot{x} - K_D x^2\dot{x}.$$

For $K_G = K_D = 0$, the system response is a linear limit cycle (as shown in Fig. 3.5). For $K_G = 0$, $K_D > 0$, the system is asymptotically stable. For $K_G > 0$, $K_D = 0$, the system is unstable. For $K_G > 0$, $K_D > 0$, a stable nonlinear limit cycle exists at

$$\oint_\tau K_G\dot{x}^2\,dt = \oint_\tau K_D x^2\dot{x}^2\,dt.$$

These conditions are simulated numerically with the parameter values given in Table 3.3. The generative case is shown in Fig. 3.14 with the power flow (left) and energy (right) time-domain responses. The neutral case energy balance and power

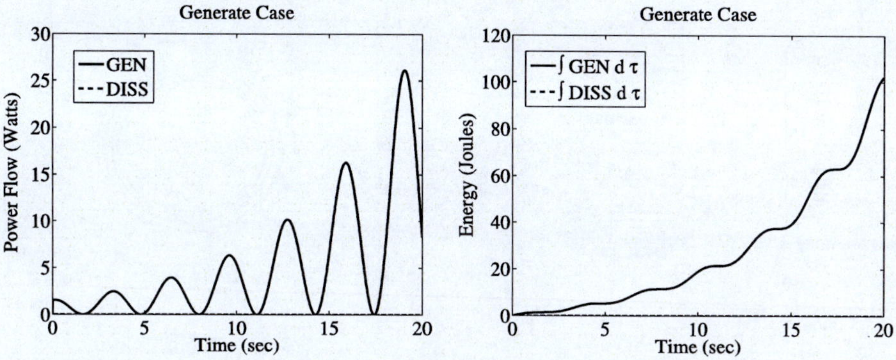

Fig. 3.14 Van der Pol reconstructed as controller inputs: power flow (*left*) and energy (*right*) responses—generative case (Note: DISS and $\int$ DISS $d\tau$ are zero). Robinett III, R.D. and Wilson, D.G. [46], reprinted by permission of the publisher (Taylor & Francis Ltd, http://www.tandf.co.uk/journals)

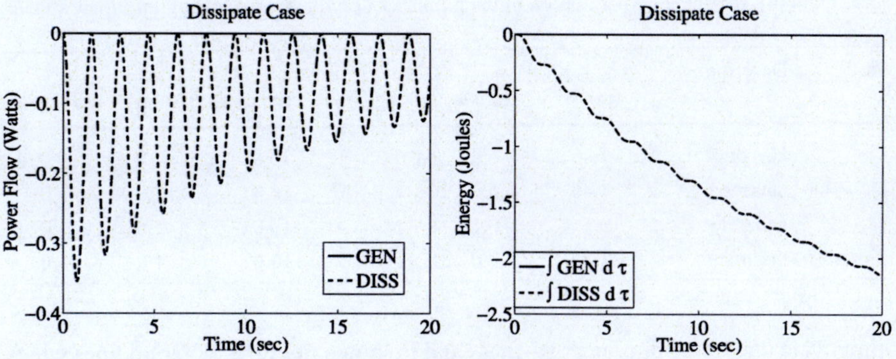

Fig. 3.15 Van der Pol reconstructed as controller inputs: power flow (*left*) and energy (*right*) responses—dissipative case (Note: GEN and $\int$ GEN $d\tau$ are zero). Robinett III, R.D. and Wilson, D.G. [46], reprinted by permission of the publisher (Taylor & Francis Ltd, http://www.tandf.co.uk/journals)

flow are zero over a cycle and are the same as the linear limit cycle case discussed earlier. The dissipative case is shown in Fig. 3.15 with the power flow (left) and energy (right) time-domain responses. The results for all three cases are visually shown in Fig. 3.16 as a 3D Hamiltonian plot (left) with the corresponding 2D phase-plane projection (right), respectively. Notice the slow decay rate due to the nonlinear damping.

The second example combines the van der Pol and Duffing oscillators

$$m\ddot{x} + kx + k_{\mathrm{NL}}x^3 = \mu\left(1 - x^2\right)\dot{x} = c_{\mathrm{NL}}(x)\dot{x}, \tag{3.59}$$

which demonstrates the effect of nonlinear stiffness and nonlinear damping. Setting $\mu = 0$ (see Table 3.4 for all parameters investigated, Case 1) results in a nonlinear

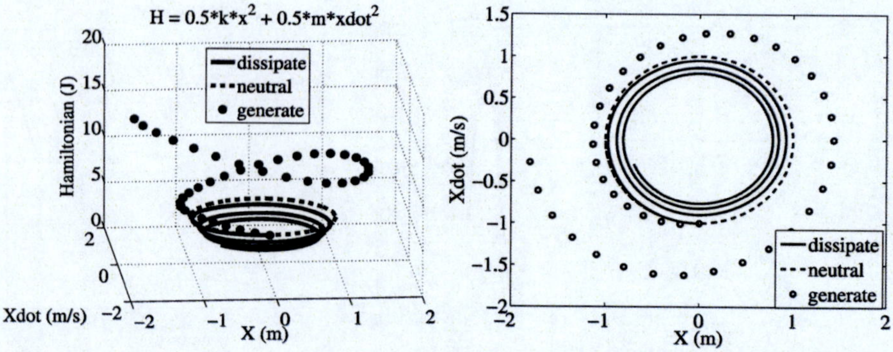

Fig. 3.16 Van der Pol reconstructed as controller inputs responses: Hamiltonian 3D surface (*left*) and phase plane 2D projection (*right*). Robinett III, R.D. and Wilson, D.G. [46], reprinted by permission of the publisher (Taylor & Francis Ltd, http://www.tandf.co.uk/journals)

Table 3.4 Van der Pol and Duffing oscillator model numerical values. Robinett III, R.D. and Wilson, D.G. [46], reprinted by permission of the publisher (Taylor & Francis Ltd, http://www.tandf.co.uk/journals)

Case	Description	x_0 (m)	$\dot{x}_0$ (m/s)	μ (kg/s)	k_{NL} (N/m^3)	m (kg)	k (kg/s^2)
1	k_{NL} only	1.45	−6.0	0.0	10.0	1.0	1.0
2	neutral	1.45	−6.0	1.5	10.0	1.0	1.0
3	generate	0.1	−0.1	1.5	10.0	1.0	1.0
4	dissipate	−1.75	9.0	1.5	10.0	1.0	1.0

limit cycle due to the nonlinear stiffness and is shown in Fig. 3.17 (solid line) which is an orthogonal cut across the Hamiltonian manifold and projected onto the phase plane. Setting $\mu = 1.5$ (Case 2 in Table 3.4) results in a stable nonlinear limit cycle shown in Fig. 3.17 (dashed line) derived from the Hamiltonian

$$\mathcal{H} = \frac{1}{2}m\dot{x}^2 + \frac{1}{2}kx^2 + \frac{1}{4}k_{NL}x^4 \tag{3.60}$$

with a time derivative of

$$\dot{\mathcal{H}} = [m\ddot{x} + kx + k_{NL}x^3]\dot{x} = [\mu(1 - x^2)\dot{x}]\dot{x}, \tag{3.61}$$

which leads to

$$\oint_\tau [\mu\dot{x}^2]\,dt = \oint_\tau [\mu x^2\dot{x}^2]\,dt. \tag{3.62}$$

The shape of the nonlinear limit cycle is different than the van der Pol oscillator due to the nonlinear stiffness which changes the accessible phase space (Hamiltonian surface). Cases 2–4 in Table 3.4 correspond to the neutral, generative, and dissipative cases subject to different initial conditions and are given in the combined plots

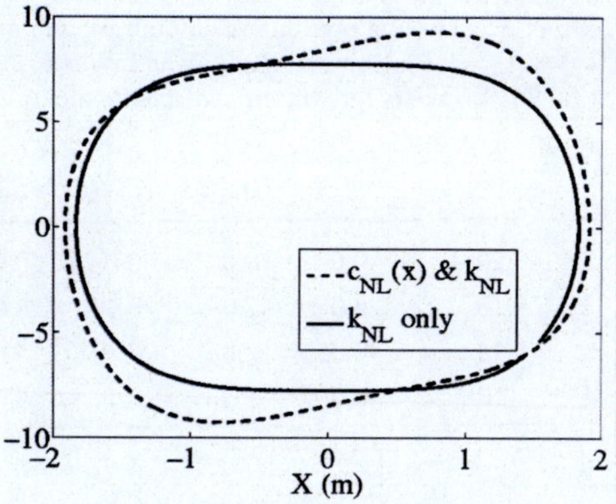

Fig. 3.17 Nonlinear limit cycles with nonlinear stiffness and damping effects. Robinett III, R.D. and Wilson, D.G. [46], reprinted by permission of the publisher (Taylor & Francis Ltd, http://www.tandf.co.uk/journals)

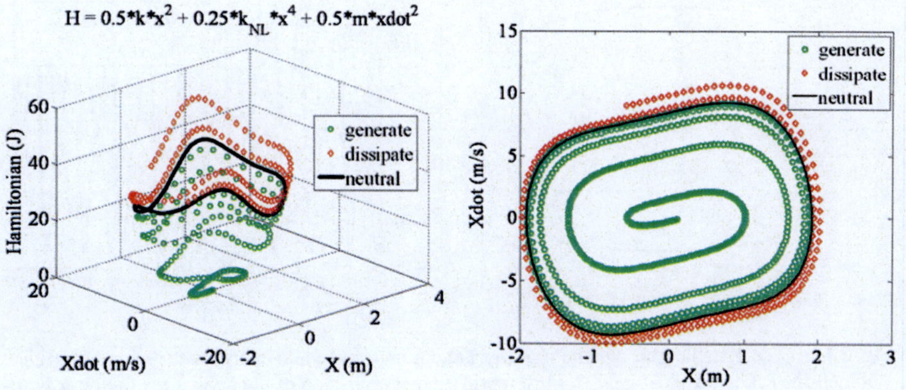

Fig. 3.18 Combined van der Pol and Duffing responses: Hamiltonian 3D surface (*left*) and phase plane 2D projection (*right*). Robinett III, R.D. and Wilson, D.G. [46], reprinted by permission of the publisher (Taylor & Francis Ltd, http://www.tandf.co.uk/journals)

in Fig. 3.18 for the 3D Hamiltonian (left) and projected 2D phase plane (right) plots, respectively. Notice the consistent shape on the phase plane that has been rotated like the van der Pol oscillator and the cyclic rise and fall of the energy level.

The third example is a Duffing oscillator equation with Coulomb friction given by

$$m\ddot{x} + kx + k_{\mathrm{NL}}x^3 = -c\dot{x} - c_{\mathrm{NL}}\mathrm{sign}(\dot{x}) + u \tag{3.63}$$

and controlled with a PID controller

$$u = -K_P x - K_I \int_0^t x\, d\tau - K_D \dot{x}. \tag{3.64}$$

The Hamiltonian is

$$\mathcal{H} = \frac{1}{2}m\dot{x}^2 + \frac{1}{2}(k + K_P)x^2 + \frac{1}{4}k_{\mathrm{NL}}x^4 \tag{3.65}$$

Table 3.5 Duffing oscillator/Coulomb friction model and PID control system gains (Note: for all cases, $\dot{x}_0 = 0.0$). Robinett III, R.D. and Wilson, D.G. [46], reprinted by permission of the publisher (Taylor & Francis Ltd, http://www.tandf.co.uk/journals)

Case No.	K_P (kg/s^2)	K_I (kg/s^3)	K_D (kg/s)	c (kg/s)	C_{NL} (N)	m (kg)	k (N/m)	k_{NL} (N/m^3)	T_f (s)	x_0 (m)	x_r (m)
1	10	0.0	0.0	0.0	0.0	10	10	100.0	10	1	0
2	10	20.0	2.0	0.1	5.0	10	10	100.0	10	1	0
3	10	40.05	2.0	0.1	5.0	10	10	100.0	10	1	0
4	10	80.0	2.0	0.1	5.0	10	10	100.0	10	1	0

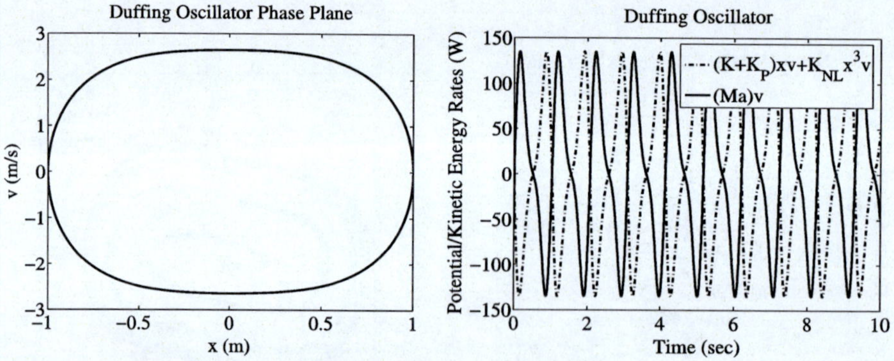

Fig. 3.19 Case 1 stable phase plane plot (*left*) and kinetic/potential energy rate responses (*right*). Robinett III, R.D. and Wilson, D.G. [46], reprinted by permission of the publisher (Taylor & Francis Ltd, http://www.tandf.co.uk/journals)

with a time derivative of

$$\begin{aligned}\dot{\mathcal{H}} &= \left[m\ddot{x} + (k + K_P)x + k_{NL}x^3\right]\dot{x}\\ &= \left[-(c + K_D)\dot{x} - c_{NL}\text{sign}(\dot{x}) - K_I\int_0^t x\,d\tau\right]\dot{x}.\end{aligned} \tag{3.66}$$

The nonlinear limit cycle occurs when

$$\oint_\tau\left[-(c + K_D)\dot{x} - c_{NL}\text{sign}(\dot{x})\right]\dot{x}\,dt = \oint_\tau\left[K_I\int_0^t x\,d\tau\right]\dot{x}\,dt. \tag{3.67}$$

Numerical simulations are performed to demonstrate where the nonlinear stability boundary lies for the Duffing oscillator/Coulomb friction dynamic model subject to PID control. Four separate cases (Cases 1–4) were conducted with the numerical values listed in Table 3.5. For Case 1, all the generative/dissipative terms are set to zero which results in a stable orbit, nonlinear limit cycle, for the nonlinear system (see Fig. 3.19, left). In addition, the sum of the generator/dissipator terms over a

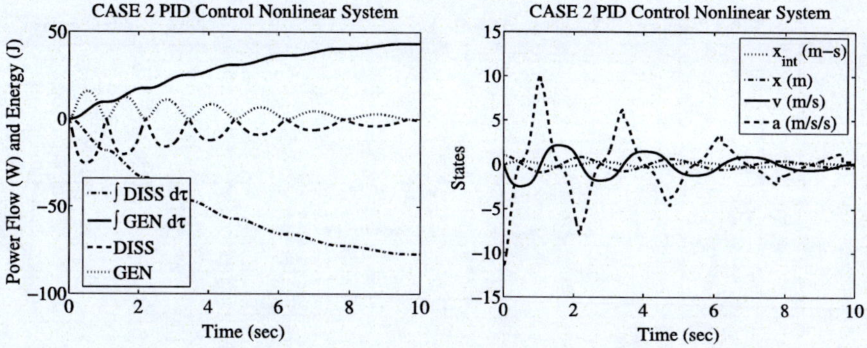

Fig. 3.20 Case 2 power flow and energy responses (*left*) and system responses (*right*). Robinett III, R.D. and Wilson, D.G. [46], reprinted by permission of the publisher (Taylor & Francis Ltd, http://www.tandf.co.uk/journals)

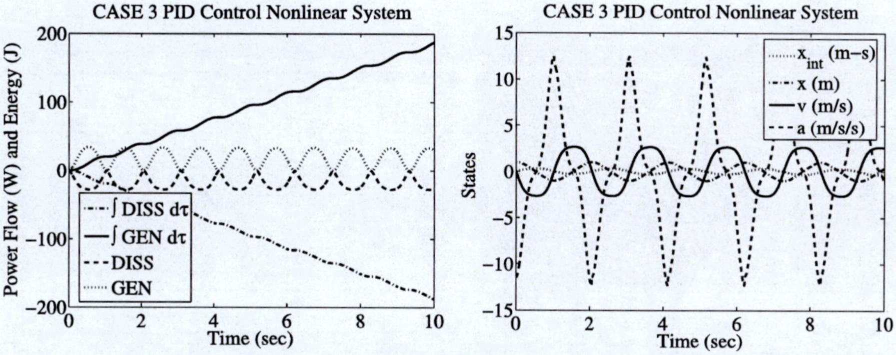

Fig. 3.21 Case 3 power flow and energy responses (*left*) and system responses (*right*). Robinett III, R.D. and Wilson, D.G. [46], reprinted by permission of the publisher (Taylor & Francis Ltd, http://www.tandf.co.uk/journals)

cycle between kinetic and potential is zero as shown in offsetting power flow plots in Fig. 3.19 (right).

For Case 2, the power flow and energy responses (left) along with the integral of position, position, velocity, and acceleration responses (right) are plotted in Fig. 3.20. For this case, the dissipative term is greater than the generative term. This is observed from the decaying system responses.

In Case 3, the power flow and energy responses (left) along with the system responses (right) are shown in Fig. 3.21. In this case, the average energy slopes and integrated power areas for the dissipative and generative terms are equivalent, which demonstrates (3.67): a nonlinear limit cycle. This results in system responses that do not decay, displaying constant nonlinear oscillatory behavior.

In Case 4, the power flow and energy responses (left) along with the system responses (right) are shown in Fig. 3.22. In this case, the dissipative term is less than

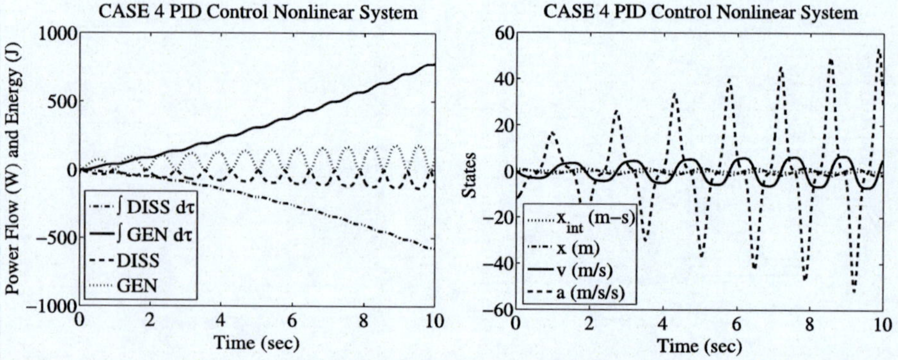

Fig. 3.22 Case 4 power flow and energy responses (*left*) and system responses (*right*). Robinett III, R.D. and Wilson, D.G. [46], reprinted by permission of the publisher (Taylor & Francis Ltd, http://www.tandf.co.uk/journals)

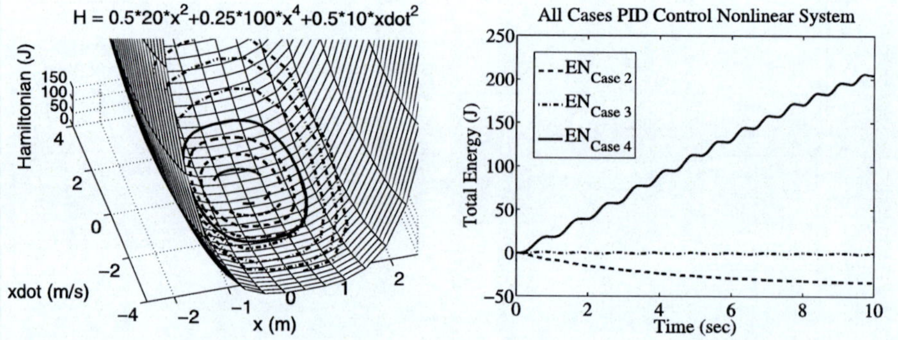

Fig. 3.23 Cases 2–4: mass–spring–damper with Duffing oscillator/Coulomb friction model numerical results: Hamiltonian 3D surface (*left*) and total energy responses (*right*). Robinett III, R.D. and Wilson, D.G. [46], reprinted by permission of the publisher (Taylor & Francis Ltd, http://www.tandf.co.uk/journals)

the generative term which results in a system response with increasing nonlinear oscillatory behavior which rises to another nonlinear limit cycle.

In summary, Fig. 3.23 presents the responses with respect to the Hamiltonian surface with trajectories traversing the surface (left) for each case including the total energy responses (right) for the nonlinear system. For Case 3, the nonlinear stability boundary (or neutral stability) is characteristic of an average zero output for the total energy response or validation of (3.67). For the trajectories on the Hamiltonian surface, Case 2 demonstrates an asymptotically stable decaying response, Case 3 a neutrally stable orbital response, and Case 4 an unstable increasing orbit response.

In addition, the stability boundary condition can be used to identify different operating regions for systems that may need to be gain scheduled. For example, the

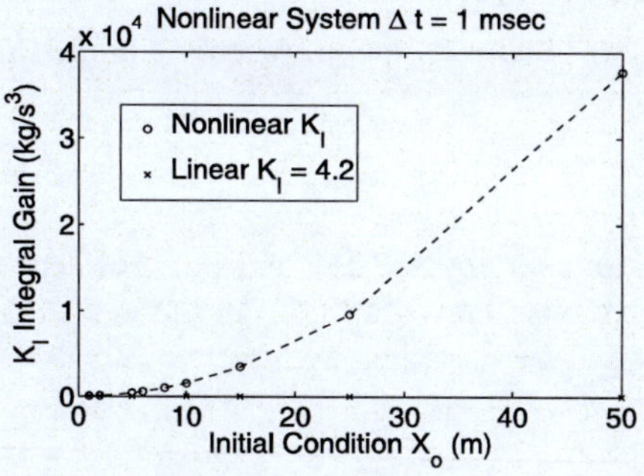

Fig. 3.24 Gain scheduling with the integral gain as a function of initial conditions. Robinett III, R.D. and Wilson, D.G. [46], reprinted by permission of the publisher (Taylor & Francis Ltd, http://www.tandf.co.uk/journals)

integral gain in general may be a function of many different parameters

$$K_I = K_I(x_0, \dot{x}_0, C, C_{\mathrm{NL}}, K_D).$$

For the purpose of illustration, the integral gain was investigated for different initial conditions, x_0, while holding all the other possible parameters constant. The results are shown in Fig. 3.24. For several varying initial conditions for x_0, at the stability boundary condition each K_I was determined. The corresponding linear system resulted in constant K_I values, which is characteristic of a linear system. The gain scheduling is due to the nonlinear spring. It changes the limit cycle behavior as a function of stored energy.

The last example is a generalization of (3.49). Sabatini [23] discusses the uniqueness of limit cycles for a class of plane differential systems defined by

$$\begin{aligned} x' &= \beta(x)[\phi(y) - F(x, y)], \\ y' &= -\alpha(y)g(x). \end{aligned} \tag{3.68}$$

The system given in (3.68) contains the particular cases of Lotka–Volterra systems, the Rayleigh equation, the Liénard equation, the van der Pol equation, and more general second-order equations.

In particular, if one chooses $\beta(x) = \alpha(y) = 1$, $F(x, y) \equiv F(x)$, and $\phi(y) = y$, then the Liénard equation results in

$$x'' + f(x)x' + g(x) = 0, \tag{3.69}$$

where $f(x) = \partial F(x)/\partial x$. The van der Pol equation is produced with $g(x) = x$ and $f(x) = -\mu(1 - x^2)$:

$$x'' - \mu(1 - x^2)x' + x = 0. \tag{3.70}$$

Additionally, if one chooses $\beta(x) = \alpha(y) = 1$, $F(x, y) = F(x)$, $g(x) = x$, then the Rayleigh equation results in

$$y'' - F(-y') + \phi(y) = 0. \tag{3.71}$$

In this example, the Liénard systems,

$$x' = y - F(x), \qquad y' = -g(x), \tag{3.72}$$

will be investigated, where a unique limit cycle exists when

$$\int_0^T F(x)g(x)\,dt = 0 \tag{3.73}$$

for every cycle [23]. The goal is to show that (3.73) is equivalent to (3.53). The first step is to rewrite (3.72) in second-order form:

$$\dot{x} = y - F(x), \qquad \dot{y} = -g(x),$$

$$\ddot{x} = \dot{y} - \frac{d}{dt}\big(F(x)\big) = -g(x) - f(x)\dot{x},$$

and simplifying gives

$$\ddot{x} + f(x)\dot{x} + g(x) = 0. \tag{3.74}$$

The second step is to formulate the Hamiltonian for (3.74):

$$\mathcal{H} = \frac{1}{2}\dot{x}^2 + \mathcal{V}(x), \tag{3.75}$$

where $m = 1$ and $\partial\mathcal{V}(x)/\partial x = g(x)$. Next, take the time derivative of the Hamiltonian in (3.75) or

$$\dot{\mathcal{H}} = [\ddot{x} + g(x)]\dot{x} = -f(x)\dot{x}^2 = -\frac{d}{dt}\big(F(x)\big)\dot{x}, \tag{3.76}$$

which leads to a limit cycle

$$\mathcal{H}_{\text{cyclic}} = 0 = \oint_\tau [\ddot{x} + g(x)]\dot{x}\,dt = \oint_\tau \left[-f(x)\dot{x}^2\right] dt = \oint_\tau \left[-\frac{d}{dt}\big(F(x)\big)\dot{x}\right] dt. \tag{3.77}$$

The third step is to rewrite (3.73) as

$$\begin{aligned}
\int_0^T F(x)g(x)\,dt &= \oint_\tau F(x)g(x)\,dt = \int_0^T [y - \dot{x}]g(x)\,dt \\
&= \int_0^T \left[-y\dot{y} - \frac{\partial\mathcal{V}(x)}{\partial x}\dot{x}\right] dt \\
&= -\frac{1}{2}y^2\big|_0^T - \mathcal{V}(x)\big|_0^T.
\end{aligned} \tag{3.78}$$

The fourth step is to rewrite (3.77) as

$$\begin{aligned}
\int_0^T f(x)\dot{x}^2\,dt &= \int_0^T \frac{d}{dt}\big(F(x)\big)\dot{x}\,dt = \int_0^T \dot{F}[y - F]\,dt = -\frac{1}{2}F^2\Big|_0^T + \int_0^T \dot{F}y\,dt \\
&= -\frac{1}{2}F^2\Big|_0^T + Fy\big|_0^T + \int_0^T Fg\,dt = \left[\frac{1}{2}y^2 - \frac{1}{2}\dot{x}^2\right]_0^T
\end{aligned}$$

$$+\int_0^T F(x)g(x)\,dt, \tag{3.79}$$

which upon substitution of (3.78) gives

$$\oint_\tau f(x)\dot{x}^2\,dt = \int_0^T f(x)\dot{x}^2\,dt = -\left[\frac{1}{2}\dot{x}^2 + \mathcal{V}(x)\right]_0^T = -\mathcal{H}_{\text{cyclic}} = 0. \tag{3.80}$$

This example demonstrates the applicability of the Hamiltonian-based approach to a *large class* of nonlinear systems.

3.6.3 Connection of Line Integrals and Limit Cycles to Thermodynamics

It is time to extend the concepts of cyclic equilibrium thermodynamics to cyclic nonequilibrium thermodynamics to limit cycles of Hamiltonian systems. For cyclic equilibrium thermodynamics, (2.13), then

$$\begin{aligned}\oint dU &= \oint T\,dS - \oint p\,d\bar{V} = 0,\\ \oint dU &= \oint T\,dS - \oint dW = 0,\end{aligned} \tag{3.81}$$

which yields a closed cycle

$$\boxed{\begin{aligned}\oint p\,d\bar{V} &= \oint T\,dS,\\ \oint dW &= \oint T\,dS.\end{aligned}} \tag{3.82}$$

For cyclic nonequilibrium thermodynamics with local equilibrium, (3.27), a limit cycle exists for

$$\oint \dot{\Xi}\,dt = \oint [\dot{W} - T_0\dot{S}_i]\,dt = 0,$$

which implies that

$$\boxed{\oint \dot{W}\,dt = \oint T_0\dot{S}_i\,dt.} \tag{3.83}$$

For Hamiltonian natural systems with applied nonconservative forces,

$$\oint \dot{\mathcal{H}}\,dt = \oint \dot{\Xi}\,dt = \oint [\dot{W} - T_0\dot{S}_i]\,dt$$

$$= \oint \left[\sum_{j=1}^{N} Q_j \dot{q}_j - \sum_{k=N+1}^{M+N} Q_k \dot{q}_k \right] dt = 0,$$

which implies that

$$\boxed{\oint \left[\sum_{j=1}^{N} Q_j \dot{q}_j \right] dt = \oint \left[\sum_{k=N+1}^{M+N} Q_k \dot{q}_k \right] dt.} \tag{3.84}$$

Equation (3.84) is implemented by sorting terms into the following categories from

$$\oint \dot{\mathcal{H}}\, dt = \oint \left[\dot{W} - T_0 \dot{S}_i \right] dt,$$

$$\oint \mathbf{F}_C \cdot \dot{\mathbf{r}}\, dt = \oint \left[(\mathbf{F}_{\mathrm{NC}})_G + (\mathbf{F}_{\mathrm{NC}})_D \right] \cdot \dot{\mathbf{r}}\, dt.$$

1. Conservative terms are

$$\oint \mathbf{F}_C \cdot \dot{\mathbf{r}}\, dt = 0, \quad \text{which implies } \textit{potential functions.}$$

2. Generator terms are

$$\int_C (\mathbf{F}_{\mathrm{NC}})_G \cdot \dot{\mathbf{r}}\, dt = \int_C \mathbf{F}_{\mathrm{NC}} \cdot \dot{\mathbf{r}}\, dt > 0, \quad \text{which implies} \quad \int_C \left[\sum_{j=1}^{N} Q_j \dot{q}_j \right] dt > 0.$$

3. Dissipator terms are

$$\int_C (\mathbf{F}_{\mathrm{NC}})_D \cdot \dot{\mathbf{r}}\, dt = \int_C \mathbf{F}_{\mathrm{NC}} \cdot \dot{\mathbf{r}}\, dt < 0, \quad \text{which implies}$$

$$- \int_C \left[\sum_{k=N+1}^{N+M} Q_k \dot{q}_k \right] dt < 0.$$

In summary, a limit cycle occurs when

$$\boxed{\begin{aligned} &\oint (\mathbf{F}_{\mathrm{NC}})_G \cdot \dot{\mathbf{r}}\, dt = - \oint (\mathbf{F}_{\mathrm{NC}})_D \cdot \dot{\mathbf{r}}\, dt \\ &\text{subject to} \quad \oint \mathbf{F}_C \cdot \dot{\mathbf{r}}\, dt = 0. \end{aligned}} \tag{3.85}$$

These limit cycle results will be used in the next chapter to determine the necessary and sufficient conditions for dynamic stability as well as designing feedback controllers.

3.7 Chapter Summary

The concepts of energy, work, and power were reviewed in this chapter to set the stage to relate thermodynamics to Hamiltonian mechanics. The Hamiltonian was recognized as exergy which provides the basic relationship necessary to apply the Second Law of Thermodynamics to the power flow, the time derivative of the Hamiltonian, to sort terms into three classes: storage, generation, and dissipation. This sorting process defines the dynamic stability boundary, a limit cycle, while the Hamiltonian surface defines the static stability boundary, a rigid body mode. The combination of the static stability and dynamic stability of Hamiltonian natural systems defines the necessary and sufficient conditions for a class of nonlinear systems, adiabatic irreversible work process, which will be developed in the next chapter.

Chapter 4
Stability and Control

4.1 Introduction

HSSPFC is a two-step control law design and analysis process. The first step treats the Hamiltonian system as if it were a conservative system with no externally applied nonconservative forces. This process enables the shaping of the Hamiltonian surface with acceleration feedback and/or proportional feedback to create an isolated minimum (stable) energy state. Static stability is utilized to find the first stability boundary, a rigid body mode (singular stiffness matrix), which defines the point of static neutral stability. The second step analyzes and designs controllers for a Hamiltonian system with externally applied nonconservative forces. This step applies dynamic stability concepts to modify the power flow with dissipation and generation feedback. In the previous chapters, the energy balance equation from the First Law of Thermodynamics combined with the entropy balance equation from the Second Law of Thermodynamics are written in rate equation form to create the exergy rate equation. This is equivalent to power flow which is the first time derivative of the Hamiltonian for natural systems (the work/rate equation). The recognition that the Hamiltonian is stored exergy provides the basic relationship necessary to apply the Second Law of Thermodynamics to the power flow to sort terms into three classes: storage, generation, and dissipation. This sorting process identifies the second stability boundary, a limit cycle, which defines the point of dynamic neutral stability.

It will be explained that dynamically stable systems must be statically stable, but the converse is not true. Therefore, static stability is a necessary condition for stability, while dynamic stability is a sufficient condition for stability and produces a maximum entropy state as well as a minimum energy state. The development of this chapter will combine the writings [51] of Meirovitch [6], Chetayev [14], La Salle and Lefschetz [52], Scanlan and Rosenbaum [53], and Robinett and Wilson [7, 13, 46].

R.D. Robinett III, D.G. Wilson, *Nonlinear Power Flow Control Design*, Understanding Complex Systems,
DOI 10.1007/978-0-85729-823-2_4,

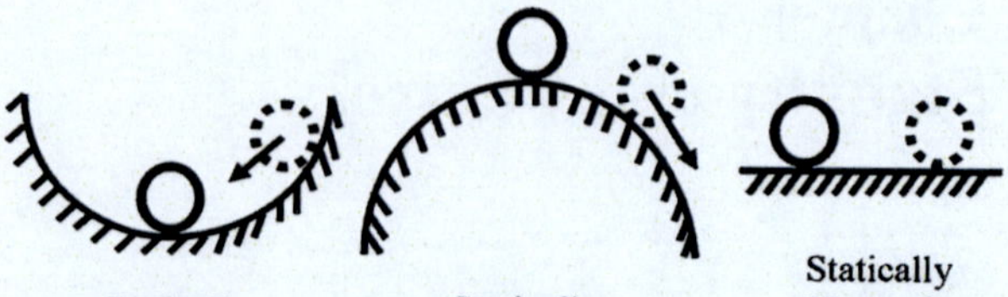

Fig. 4.1 Static stability. Robinett III, R.D. and Wilson, D.G. [56], reprinted by permission of the publisher (©2010 IEEE)

4.2 Static Stability and Dynamic Stability

Basically, the necessary and sufficient conditions for stability of Hamiltonian natural systems, both linear and nonlinear, can be determined from the shape of the Hamiltonian surface and its time derivative, the power flow. The proof of this observation begins with the assistance of the concepts of static stability and dynamic stability.

The concepts of static stability and dynamic stability will be reviewed, built upon, and expanded in this section to provide the basis for a two-step control law design and analysis process referred to as Hamiltonian Surface Shaping and Power Flow Control (HSSPFC). HSSPFC will be formalized in the last section of this chapter.

In aerospace engineering, flight stability of airplanes and missiles is broadly defined by two categories, static stability and dynamic stability, which leads to a two-step design process for implementing the desired handling qualities into an airplane/missile system [31, 54, 55]. To begin the review, the sum of the forces and moments acting on a missile while on a steady, straight flight path must be in static equilibrium. Typically, static stability is considered as a linearization of the system about a prescribed flight path and defined by the following statement:

If the forces and moments on a body caused by a disturbance tend initially to return (move) the body toward (away from) its equilibrium state, the body is statically stable (unstable) [54, 55].

An equilibrium state is an unaccelerated motion wherein the sums of the forces and moments on the body are zero. Static neutral stability occurs when the body is disturbed and the sums of the forces and moments on the body remain zero. This happens when the system has a rigid body mode or zero stiffness in the system. Figure 4.1 graphically presents the concept of static stability. It will become evident in the following description of dynamic stability that a dynamically stable body must always be statically stable, but static stability is not sufficient to ensure dynamic stability [31, 54, 55]. *Therefore, static stability is a necessary condition for stability.*

A specific example of static stability is found in reentry vehicle flight stability [55]. For an axisymmetric reentry vehicle, the static margin determines the static stability. The static margin (SM) is the difference in length between the center-of-mass and the center-of-pressure relative to the nose (see Fig. 4.2) of the reentry

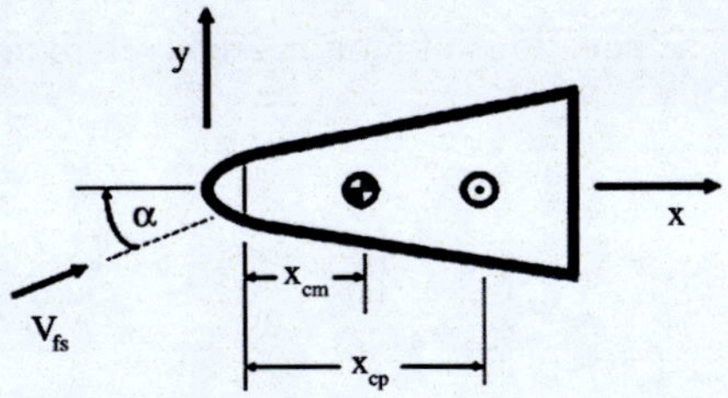

Fig. 4.2 Static margin

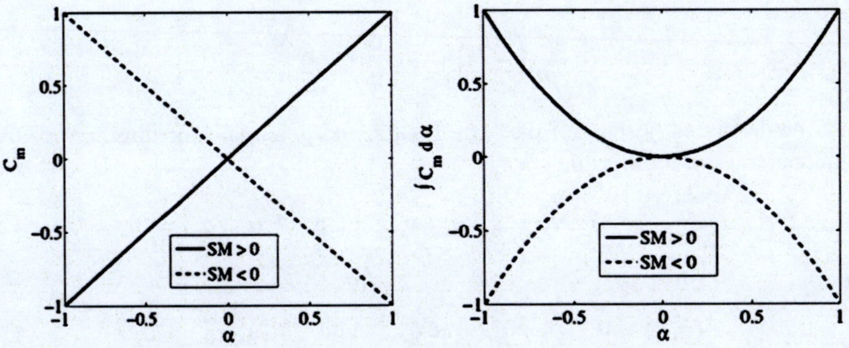

Fig. 4.3 Aerodynamic moment (*left*) and integral of aerodynamic moment with respect to angle of attack (*right*)

vehicle. For

$$SM = x_{cp} - x_{cm}, \tag{4.1}$$

then the following definitions apply:

$$\begin{aligned} x_{cp} &= \text{center-of-pressure}, \\ x_{cm} &= \text{center-of-mass}, \\ \alpha &= \text{angle-of-attack}, \\ V_{fs} &= \text{free stream velocity}. \end{aligned}$$

If $SM < 0$, the reentry vehicle is statically unstable. If $SM > 0$, the reentry vehicle is statically stable. The aerodynamic moment for a reentry vehicle is presented in Fig. 4.3 (left), which upon integration with respect to the angle of attack, α, gives a quadratic potential function (see Fig. 4.3, right). Clearly, static stability for a reentry vehicle is equivalent to the analysis of the conservative forces applied to a body which is defined by the energy storage surface, the Hamiltonian.

To build and expand upon static stability, the system is treated as though it is a conservative natural Hamiltonian system with no externally applied nonconservative forces or moments, without linearization. The Lagrange–Dirichlet Theorem (a state where the potential energy is an isolated minimum is a stable equilibrium state [52]) can be applied at this point to the energy storage surface, Hamiltonian, which is a constant:

$$\mathcal{H} = \mathcal{T} + \mathcal{V} = E = \text{constant}. \tag{4.2}$$

The equations of motion are given in first-order canonical form as

$$\dot{q}_j = \frac{\partial \mathcal{H}}{\partial p_j}, \quad \dot{p}_j = -\frac{\partial \mathcal{H}}{\partial q_j}, \tag{4.3}$$

and in second-order form as

$$\frac{d}{dt}\left(\frac{\partial \mathcal{L}}{\partial \dot{\mathbf{q}}}\right) - \left(\frac{\partial \mathcal{L}}{\partial \mathbf{q}}\right) = \mathbf{0}. \tag{4.4}$$

A system is statically stable if the Hamiltonian and potential functions are positive definite about the equilibrium state:

$$\mathcal{H}(\dot{\mathbf{q}}, \mathbf{q}) > 0 \quad \forall \dot{\mathbf{q}} \neq \dot{\mathbf{q}}_e, \mathbf{q} \neq \mathbf{q}_e, \quad \text{and} \quad \mathcal{H}(\dot{\mathbf{q}}_e, \mathbf{q}_e) = 0 \tag{4.5}$$

for

$$\mathcal{T}(\dot{\mathbf{q}}, \mathbf{q}) > 0 \quad \forall \dot{\mathbf{q}} \neq \dot{\mathbf{q}}_e, \mathbf{q} \neq \mathbf{q}_e, \quad \text{and} \quad \mathcal{T}(\dot{\mathbf{q}}_e, \mathbf{q}_e) = 0, \tag{4.6}$$

and

$$\mathcal{V}(\mathbf{q}) > 0 \quad \forall \mathbf{q} \neq \mathbf{q}_e, \quad \text{and} \quad \mathcal{V}(\mathbf{q}_e) = 0. \tag{4.7}$$

The Converse of the Lagrange–Dirichlet Theorem, Lyapunov's Theorem (at an isolated maximum of the potential energy, the equilibrium state is unstable [51, 52]), can be applied, and the system is statically unstable if

$$\mathcal{V}(\mathbf{q}) < 0 \quad \forall \mathbf{q} \neq \mathbf{q}_e, \quad \text{and} \quad \mathcal{V}(\mathbf{q}_e) = 0. \tag{4.8}$$

The Instability Theorem of Chetayev [14] (if at an equilibrium state the potential energy is not a minimum, then the equilibrium state is unstable [52]) can be applied, and the system is statically neutrally stable, a rigid body mode, if

$$\mathcal{V}(\mathbf{q}) = 0 \quad \forall \mathbf{q}. \tag{4.9}$$

These theorems will be discussed in more detail in Sect. 4.4, which is focused on Lyapunov analysis.

Continuing the review, the sum of the forces and moments acting on a missile while on an unsteady or curved flight path must be in dynamic equilibrium. Typically, dynamic stability is considered as a linearization of the system about a prescribed flight path and defined by the following statement in terms of the time history of the motion of a body after encountering a disturbance:

A body is dynamically stable (unstable) if, out of its own accord, it eventually returns to (deviates from) and remains at (away from) its equilibrium state over a period of time [54, 55].

A dynamically neutral stable body occurs when a limit cycle exists [7, 46]. A dynamically stable body must always be statically stable, but static stability is not sufficient to ensure dynamic stability [31, 54, 55]. *Therefore, static stability is a necessary condition for stability, and dynamic stability is a sufficient condition for stability.*

To build and expand upon dynamic stability, the system is treated as though it is a natural Hamiltonian system with externally applied nonconservative forces and/or moments that is statically stable without linearization. The equations of motion are given in first-order canonical form as

$$\begin{aligned} \dot{q}_j &= \frac{\partial \mathcal{H}}{\partial p_j}, \\ \dot{p}_j &= -\frac{\partial \mathcal{H}}{\partial q_j} + Q_j, \end{aligned} \tag{4.10}$$

and second-order form as

$$\frac{d}{dt}\left(\frac{\partial \mathcal{L}}{\partial \dot{\mathbf{q}}}\right) - \left(\frac{\partial \mathcal{L}}{\partial \mathbf{q}}\right) = \mathbf{Q}. \tag{4.11}$$

The energy and power flow discussions with respect to stability of the previous chapters, in particular Sect. 3.3, are used to determine dynamic stability.

The system path/trajectory traverses a positive definite energy storage surface (statically stable) defined by the Hamiltonian as a result of the power flow. The time derivative of the energy/Hamiltonian surface defines the power flow into, dissipated within, and stored in the system. This determines whether the system is rising to a higher energy state (away from its equilibrium state), dropping to a lower energy state (returning to its equilibrium state), or staying on a closed cyclic path (limit cycle) constrained to the energy/Hamiltonian surface. Average power flow calculations are used because we cannot guarantee that the opposing power flows will cancel or be dominant generators or dissipators point-for-point in time. In fact, limit cycles balance over the cycle (see Chap. 3, Sect. 3.6).

A system is dynamically stable if the power flow on the average drives the perturbed system to a lower energy state which eventually converges to the statically stable equilibrium state,

$$\dot{\mathcal{H}}_{\text{ave}} = \frac{1}{\tau_c}\int_0^{\tau_c} \dot{\mathcal{H}}\, dt < 0. \tag{4.12}$$

A system is dynamically unstable if the power flow on the average drives the perturbed system to a higher energy state which eventually diverges from the statically stable equilibrium state,

$$\dot{\mathcal{H}}_{\text{ave}} = \frac{1}{\tau_c}\int_0^{\tau_c} \dot{\mathcal{H}}\, dt > 0. \tag{4.13}$$

A system is dynamically neutral stable if the power flow on the average drives the perturbed system to a closed cyclic path (limit cycle) constrained to the en-

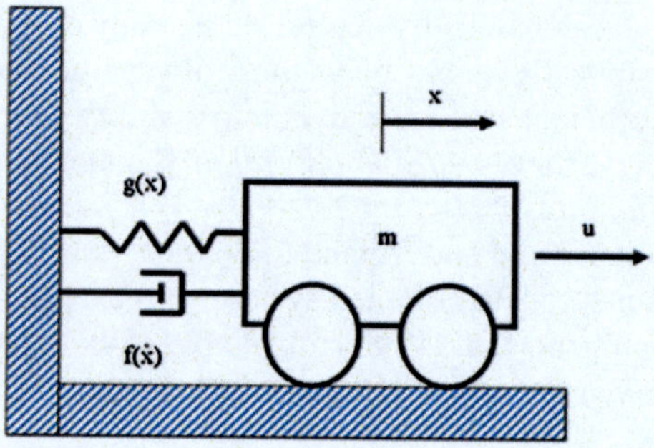

Fig. 4.4 Nonlinear mass, spring, damper system. Robinett III, R.D. and Wilson, D.G. [46], reprinted by permission of the publisher (Taylor & Francis Ltd, http://www.tandf.co.uk/journals)

ergy/Hamiltonian surface which orbits the statically stable equilibrium state,

$$\dot{\mathcal{H}}_{\text{ave}} = \frac{1}{\tau_c} \oint_{\tau_c} \dot{\mathcal{H}}\, dt = 0. \tag{4.14}$$

These concepts will be proved in the next two sections with respect to an extension of eigenanalysis and Lyapunov analysis.

4.3 Eigenanalysis

The goal of this section is to relate and extend eigenanalysis of linear systems to nonlinear systems via the energy storage surface, power flow, and limit cycles. One place to begin achieving this goal is to discuss "self-excited" systems in the context of aeroelasticity and aircraft flutter. Self-excited systems include feedback controllers because the forces acting on the system are functions of the coordinates and their velocities and accelerations. This discussion follows the presentation in Scanlan and Rosenbaum [53] and utilizes the linearized version of the nonlinear mass, spring, damper system in Fig. 4.4. The Hamiltonian is

$$\mathcal{H} = E = \frac{1}{2} m\dot{x}^2 + \mathcal{V}(x) \tag{4.15}$$

with

$$g(x) = \frac{\partial \mathcal{V}(x)}{\partial x}, \tag{4.16}$$

and equation of motion is

$$m\ddot{x} + g(x) = -f(\dot{x}) + u. \tag{4.17}$$

For the linearized system,

$$\begin{aligned} \mathcal{V}(x) &= \frac{1}{2} k x^2 > 0 \quad \forall x \neq 0 \quad \text{and} \quad \mathcal{V}(0) = 0, \\ f(\dot{x}) &= c\dot{x} \quad \text{with } m, c, k > 0, \end{aligned} \tag{4.18}$$

the Hamiltonian becomes

$$\mathcal{H} = \frac{1}{2}m\dot{x}^2 + \frac{1}{2}kx^2 > 0 \quad \forall x \neq 0,\ \dot{x} \neq 0 \quad \text{and} \quad \mathcal{H}(0,0) = 0, \tag{4.19}$$

and the equation of motion is

$$m\ddot{x} + kx = -c\dot{x} + u. \tag{4.20}$$

The time derivative of the Hamiltonian is

$$\dot{\mathcal{H}} = [m\ddot{x} + kx]\dot{x} = [-c\dot{x} + u]\dot{x}. \tag{4.21}$$

Four feedback controller examples will be discussed next.

First, assume that $c = 0$ and $u = -K_P x$, which is proportional feedback and derivable from a potential function. The Hamiltonian and its derivative become

$$\begin{aligned}
\mathcal{H}(x,\dot{x}) &= \frac{1}{2}m\dot{x}^2 + \frac{1}{2}(k + K_P)x^2 \\
&= \mathcal{T} + \mathcal{V} + \mathcal{V}_c > 0 \quad \forall x \neq 0,\ \dot{x} \neq 0, \\
\mathcal{H}(0,0) &= 0, \\
\mathcal{V}(x) + \mathcal{V}_c(x) &> 0 \quad \forall x \neq 0, \\
\mathcal{V}(0) + \mathcal{V}_c(0) &= 0, \\
\mathcal{V}_c &= \frac{1}{2}K_P x^2,
\end{aligned} \tag{4.22}$$

and

$$\dot{\mathcal{H}} = [m\ddot{x} + (k + K_P)x]\dot{x} = 0, \tag{4.23}$$

which produces the eigenvalue problem

$$\begin{gathered}
m\ddot{x} + (k + K_P)x = 0, \\
\omega^2 = \frac{k + K_P}{m},
\end{gathered} \tag{4.24}$$

with the undamped natural frequency (eigenvalue) of a statically stable and dynamically neutral stable system with a second-order center (linear limit cycle) for

$$k + K_P > 0. \tag{4.25}$$

On the other hand, if

$$k + K_P < 0, \tag{4.26}$$

then the system is statically (and dynamically) unstable and exponentially divergent without oscillation. For $k + K_P = 0$, the system is statically neutral stable with a rigid body mode and dynamically unstable.

Second, assume that $c = 0$ and $u = -K_A\ddot{x}$, which is acceleration feedback and derivable from a kinetic energy function. The Hamiltonian and its derivative become

$$\begin{aligned}\mathcal{H} &= \frac{1}{2}(m + K_A)\dot{x}^2 + \frac{1}{2}kx^2\\ &= \mathcal{T} + \mathcal{T}_c + \mathcal{V} > 0 \quad \forall x \neq 0,\ \dot{x} \neq 0, \quad \text{and} \quad \mathcal{H}(0,0) = 0,\\ \mathcal{T}_c &= \frac{1}{2}K_A\dot{x}^2,\\ \dot{\mathcal{H}} &= [(m + K_A)\ddot{x} + kx]\dot{x} = 0,\end{aligned} \tag{4.27}$$

which produces the eigenvalue problem

$$\begin{aligned}(m + K_A)\ddot{x} + kx = 0,\\ \omega^2 = \frac{k}{m + K_A},\end{aligned} \tag{4.28}$$

with the undamped natural frequency (eigenvalue) of a statically stable and dynamically neutral stable system with a second-order center (linear limit cycle) for

$$m + K_A > 0. \tag{4.29}$$

On the other hand, if

$$m + K_A < 0, \tag{4.30}$$

then the system is statically (and dynamically) unstable and exponentially divergent without oscillation. For $m + K_A = 0$, the system is reduced to a kinematic system (see Chap. 7 on Collective Plume Tracing).

Third, assume that $u = -K_D\dot{x}$, which is damping feedback and treated as additional damping from a nonconservative applied force, where

$$\begin{aligned}\mathcal{H} &= \frac{1}{2}m\dot{x}^2 + \frac{1}{2}kx^2 > 0 \quad \forall x \neq 0,\ \dot{x} \neq 0, \quad \text{and} \quad \mathcal{H}(0,0) = 0,\\ \dot{\mathcal{H}} &= [m\ddot{x} + kx]\dot{x} = [-(c + K_D)\dot{x}]\dot{x}, \quad \text{with}\\ m\ddot{x} &+ kx = -(c + K_D)\dot{x},\end{aligned} \tag{4.31}$$

which produces a complex eigenvalue problem with the real part of the eigenvalue defined by the right-hand term in the time derivative of the Hamiltonian where the system is statically stable and dynamically stable if

$$c + K_D > 0, \tag{4.32}$$

dynamically unstable if

$$c + K_D < 0, \tag{4.33}$$

and dynamically neutral stable if

$$c + K_D = 0. \tag{4.34}$$

This damping feedback is either the additional power being dissipated within the system to drive it to the statically stable equilibrium state or additional power flowing into the system that is driving it to a higher energy level diverging from the statically stable equilibrium state. Notice that *the damping feedback has no effect on the static stability of the system*, which will be addressed further in Sect. 4.4.

Fourth, assume that $u = -K_I \int_0^t x\, d\tau$, which is integral feedback and treated as a power generator from a nonconservative applied force, where

$$\begin{aligned} \mathcal{H} &= \frac{1}{2} m\dot{x}^2 + \frac{1}{2} kx^2 > 0 \quad \forall x \neq 0,\ \dot{x} \neq 0, \quad \text{and} \quad \mathcal{H}(0,0) = 0, \\ \dot{\mathcal{H}} &= [m\ddot{x} + kx]\dot{x} = \left[-c\dot{x} - K_I \int_0^t x\, d\tau \right] \dot{x}, \end{aligned} \tag{4.35}$$

which produces a complex eigenvalue problem where the system is statically stable and dynamically stable if

$$\left[-c\dot{x} - K_I \int_0^t x d\tau \right] \dot{x} < 0, \tag{4.36}$$

dynamically unstable if

$$\left[-c\dot{x} - K_I \int_0^t x d\tau \right] \dot{x} > 0, \tag{4.37}$$

and dynamically neutral stable if

$$\left[-c\dot{x} - K_I \int_0^t x\, d\tau \right] \dot{x} = 0. \tag{4.38}$$

The next three subsections will help provide insights on how to determine that the integral control term is a power or exergy generator.

4.3.1 Integral Feedback Is an Exergy Generator—Comparison to a Lag Stabilized System

In this subsection, the idea of phase shifting a feedback control signal is used to explain the effect of integral feedback in an analogous way to time-delayed feedback. First, compare a PID controller

$$M\ddot{x} + C\dot{x} + Kx = u_{\text{PID}} = -K_P x - K_I \int_0^t x\, d\tau - K_D \dot{x} \tag{4.39}$$

with Lag-stabilization [57]

$$M\ddot{x} + Kx = u_{\mathrm{LS}} = K_P x(t-\tau). \tag{4.40}$$

Note that from PID the PD portions contribute to system stiffness and damping. What does the "I" or integral portion contribute to? Negative damping. Since lag-stabilization [57] was shown to phase shift x to $\dot{x}$, some prescribed amount of damping, for the "integration" of x, the phase shift would be proportional to $-\dot{x}$. These insights will be expanded upon in Chap. 5 when discussing fractional calculus control. Recognizing this, we have

$$\alpha \int_0^t x\,d\tau = -\dot{x}. \tag{4.41}$$

Differentiating yields

$$\alpha x = -\ddot{x}, \quad \text{which implies that} \quad \ddot{x} + \alpha x = 0 \tag{4.42}$$

for no net damping (neutral stability), and where

$$\dot{\mathcal{H}} = [M\ddot{x} + (K + K_P)x]\dot{x} = \left[-(C + K_D)\dot{x} - K_I \int_0^t x\,d\tau\right]\dot{x} = 0, \tag{4.43}$$

we have

$$\alpha = \omega^2 = \frac{K + K_P}{M} = \frac{K_I}{C + K_D}. \tag{4.44}$$

4.3.2 *Integral Feedback Is an Exergy Generator—Investigation by Exergy/Entropy Control Stability Boundary*

For the mass–spring–damper system with PID control equation (4.39), the stability boundary is at

$$-T_0\dot{\mathcal{S}}_i + \dot{W} = 0 \tag{4.45}$$

or

$$T_0\dot{\mathcal{S}}_i = \dot{W}, \tag{4.46}$$

where the averages can be removed for a linear system. Substituting the appropriate terms from (4.39) gives

$$[(C + K_D)\dot{x}]\dot{x} = -\left[K_I \int_0^t x\,d\tau\right]\dot{x}; \tag{4.47}$$

then rearranging as

$$\int_0^t x\,d\tau = -\frac{C + K_D}{K_I}\dot{x} \tag{4.48}$$

and differentiating both sides gives

$$x = -\frac{C + K_D}{K_I}\ddot{x}, \tag{4.49}$$

resulting in

$$\ddot{x} + \omega^2 x = 0, \quad \text{which implies that} \quad \omega^2 = \frac{K_I}{(C + K_D)} = \frac{K + K_P}{M}. \tag{4.50}$$

Next, this result is further clarified with a conventional Routh–Hurwitz analysis.

4.3.3 *Integral Feedback Is an Exergy Generator—Routh–Hurwitz Stability Analysis*

First, convert the combined mass–spring–damper PID control system into the following equivalent third-order system:

$$M\ddot{x} + (C + K_D)\dot{x} + (K + K_P)x + K_I \int_0^t x\, d\tau = 0. \tag{4.51}$$

Next, invoking the change of variable $y = \int_0^t x\, d\tau$ gives

$$\dddot{y} + \frac{C + K_D}{M}\ddot{y} + \frac{K + K_P}{M}\dot{y} + \frac{K_I}{M}y = 0, \tag{4.52}$$

and transforming to the s-domain yields

$$s^3 + \frac{C + K_D}{M}s^2 + \frac{K + K_P}{M}s + \frac{K_I}{M} = 0. \tag{4.53}$$

For the third-order system to be stable, the Routh–Hurwitz analysis [32] results in the following necessary and sufficient conditions:

1. All the polynomial coefficients must be positive, and
2. $$\left(\frac{C + K_D}{M}\right)\left(\frac{K + K_P}{M}\right) > \frac{K_I}{M}. \tag{4.54}$$

The specific equality condition

$$\left(\frac{C + K_D}{M}\right)\left(\frac{K + K_P}{M}\right) = \frac{K_I}{M} \tag{4.55}$$

implies that

$$\frac{K_I}{C + K_D} = \frac{K + K_P}{M} = \omega^2.$$

Table 4.1 Rotary mass–spring–damper model and PID control system gains. Robinett III, R.D. and Wilson, D.G. [13], reprinted by permission of the publisher (Interscience Publishers)

Case No.	K_P (N m)	K_I (N m/s)	K_D (N m/s)	J (kg/m^2)	K (N m)	C (N m/s)
1A	10.0	1.0	0.0	10.0	10.0	0.0
1	10.0	1.0	2.0	10.0	10.0	0.1
2	10.0	1.0	0.4	10.0	10.0	0.1
3	10.0	1.0	0.0	10.0	10.0	0.1

This result is the marginal or neutral stability boundary which agrees with the previous lag-stabilized and exergy/entropy control results:

$$\alpha = \omega^2.$$

Also note that for PID control numerical simulation results (presented in the next subsection) that Case 1, (4.36) is asymptotically stable, Case 2, (4.38) is neutrally stable, and Case 3, (4.37) is unstable, which matches the analysis results.

4.3.4 PID Control Design Numerical Example

The PID control law is partitioned into terms of exergy dissipation, exergy generation, and exergy storage. By applying exergy/entropy control design, the time derivative of the Hamiltonian yields for $\theta = x$ (rotary mass–spring–damper analogy)

$$\begin{aligned} T_0\dot{\mathcal{S}}_i &= (C + K_D)\dot{\theta} \cdot \dot{\theta}, \\ \dot{W} &= -K_I \int_0^t \theta d\tau \cdot \dot{\theta}, \\ \left(T_0\dot{\mathcal{S}}_{\text{rev}}\right)_{\text{ave}} &= \left(J\ddot{\theta} \cdot \dot{\theta} + (K + K_P)\theta \cdot \dot{\theta}\right)_{\text{ave}} = 0. \end{aligned} \tag{4.56}$$

Once again, the first row in (4.56) is identified as a dissipative term composed of the derivative control term along with the damping term. The second row in (4.56) is the integral action and is identified as the generative input. Note that the proportional control term is added to the system stiffness term and contributes to the reversible portion of the third row in (4.56).

For PID control, the stability boundary is determined as

$$\dot{\mathcal{H}} = 0 = \dot{W} - T_0\dot{\mathcal{S}}_i \quad \forall t_0 \le t \le t_{\text{final}}$$

and is also considered a special case of the average power for linear systems. In general for nonlinear systems, the average power terms will need to be taken into account.

Numerical simulations are performed for four separate PID control regulator cases with the numerical values listed in Table 4.1. The rotary mass–spring–damper system is subject to an initial condition of $\theta_0 = 1.0$ rad and $\dot{\theta}_0 = 0.0$ rad/s. For

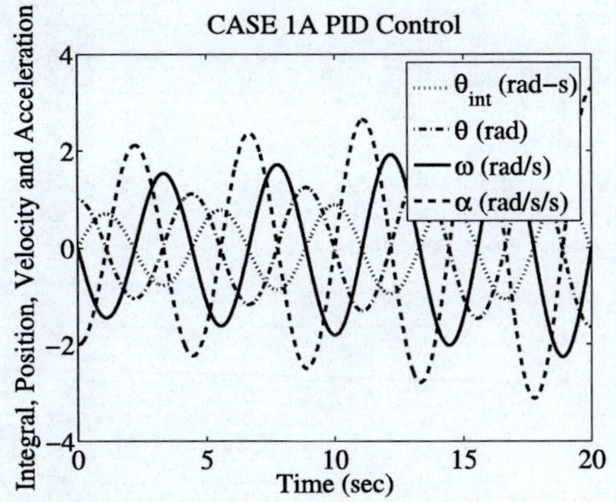

Fig. 4.5 Case 1A: rotary mass–spring–damper model with PID control numerical transient responses. Robinett III, R.D. and Wilson, D.G. [13], reprinted by permission of the publisher (Interscience Publishers)

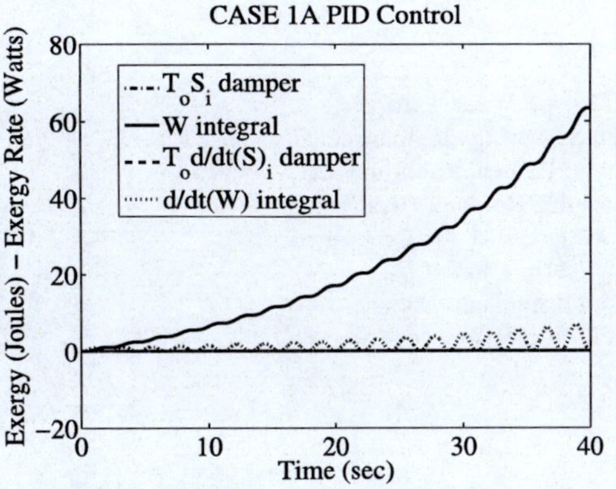

Fig. 4.6 Case 1A: rotary mass–spring–damper model with PID control numerical results. Robinett III, R.D. and Wilson, D.G. [13], reprinted by permission of the publisher (Interscience Publishers)

Case 1A, pure generative input is demonstrated where growing state oscillations (see Fig. 4.5) and positively increasing values are observed in the responses for exergy and exergy rate (see Fig. 4.6). The system has experienced pure generative input without any dissipation mechanisms to damp out the exergy coming into the system. For Case 1, the integral of position, position, velocity, and acceleration responses along with the exergy and exergy rate responses are plotted in Figs. 4.7 and 4.8, respectively. For this case, the dissipative term is greater than the generative term. This is observed from the decaying system responses. In Case 2, the system responses along with the exergy and exergy rate responses are shown in Figs. 4.9 and 4.10, respectively. In this case, the dissipative term is equal to the generative term. This results in system responses that do not decay or have constant oscillatory behavior. Further investigation of the exergy and exergy rate responses show that the mirror images for dissipation and generation will cancel each other out. For Case 2, the potential and kinetic energy rate responses (see Fig. 4.11) are shown to have zero contribution or are reversible over each cycle. In Case 3, the system responses along with the exergy and exergy rate responses are shown in Figs. 4.12 and 4.13, respectively. In this case, the dissipative term is less than the generative

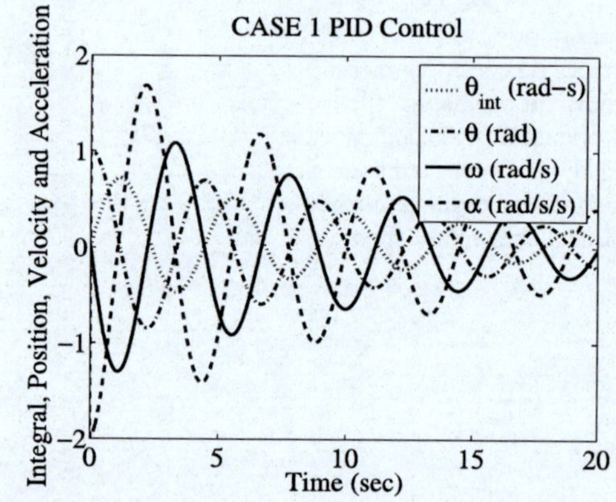

Fig. 4.7 Case 1: rotary mass–spring–damper model with PID control numerical transient responses. Robinett III, R.D. and Wilson, D.G. [13], reprinted by permission of the publisher (Interscience Publishers)

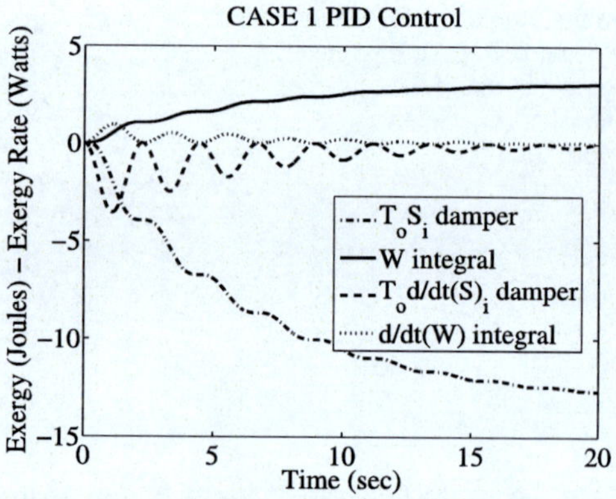

Fig. 4.8 Case 1: rotary mass–spring–damper model with PID control numerical results. Robinett III, R.D. and Wilson, D.G. [13], reprinted by permission of the publisher (Interscience Publishers)

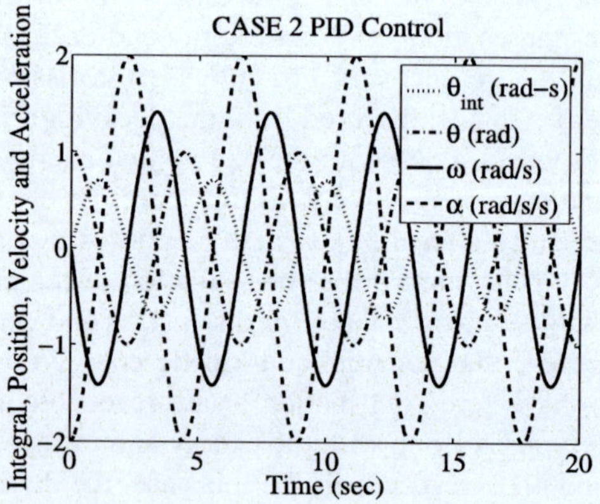

Fig. 4.9 Case 2: rotary mass–spring–damper model with PID control numerical transient responses. Robinett III, R.D. and Wilson, D.G. [13], reprinted by permission of the publisher (Interscience Publishers)

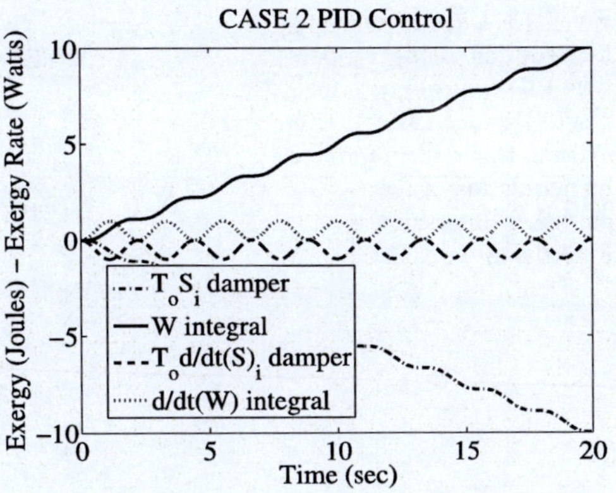

Fig. 4.10 Case 2: rotary mass–spring–damper model with PID control numerical results. Robinett III, R.D. and Wilson, D.G. [13], reprinted by permission of the publisher (Interscience Publishers)

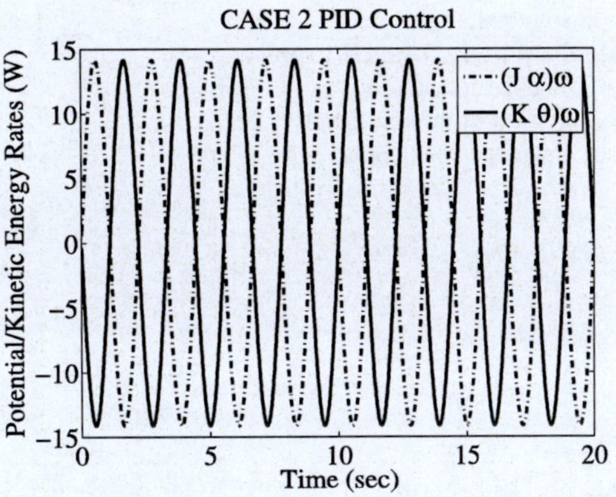

Fig. 4.11 Case 2: rotary mass–spring–damper model spring restoring and inertial effects numerical results. Robinett III, R.D. and Wilson, D.G. [13], reprinted by permission of the publisher (Interscience Publishers)

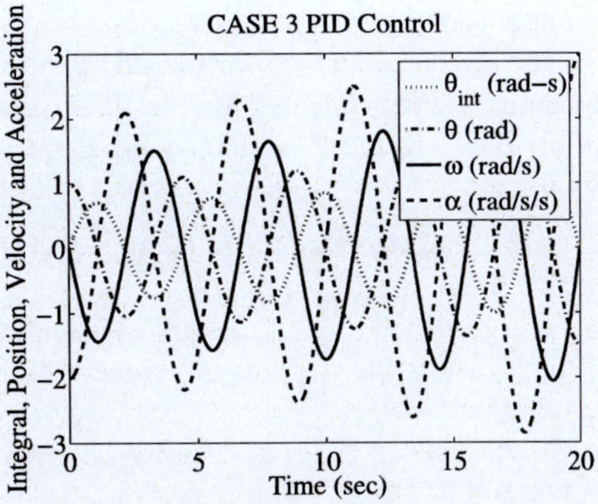

Fig. 4.12 Case 3: rotary mass–spring–damper model with PID control numerical transient responses. Robinett III, R.D. and Wilson, D.G. [13], reprinted by permission of the publisher (Interscience Publishers)

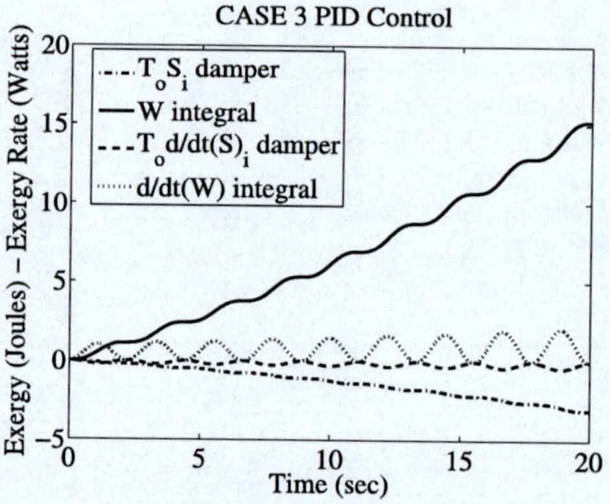

Fig. 4.13 Case 3: rotary mass–spring–damper model with PID control numerical results. Robinett III, R.D. and Wilson, D.G. [13], reprinted by permission of the publisher (Interscience Publishers)

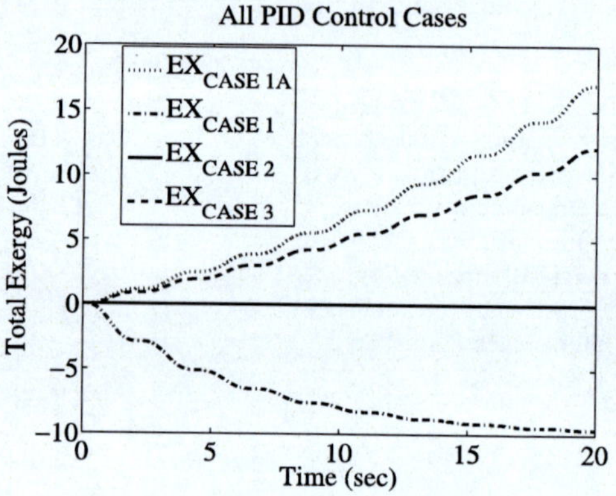

Fig. 4.14 All cases: rotary mass–spring–damper model with PID control exergy transient responses. Robinett III, R.D. and Wilson, D.G. [13], reprinted by permission of the publisher (Interscience Publishers)

term. This results in system responses with increasing oscillatory behavior. In addition, the sum of generative and dissipative terms is increasingly positive, hence exergy is being pumped into the system at a greater rate than can be dissipated. In conclusion, Fig. 4.14 shows the responses for the total exergy, and Fig. 4.15 for exergy rates with respect to each case. Again, for Case 2, a balance or boundary (neutral stability) can be observed.

The three cases (1–3) for the PID control regulator rotary mass–spring–damper dynamic system demonstrate the three subcases in Sect. 3.6.1 and [13]: Given a priori $T_0\dot{S}_i > 0$ and $\dot{W} > 0$, the inertial-spring-damper with PID control showed the following:

- Case 1 yielded $(T_0\dot{S}_i)_{\text{ave}} > (\dot{W})_{\text{ave}}$, asymptotic stability, damped stable response, and demonstration of (4.36).
- Case 2 yielded $(T_0\dot{S}_i)_{\text{ave}} = (\dot{W})_{\text{ave}}$, neutral stability, and demonstration of (4.38). This case is the dividing line where derivative and integral action cancel each other out.
- Case 3 yielded $(T_0\dot{S}_i)_{\text{ave}} < (\dot{W})_{\text{ave}}$, increasing unstable system, and demonstration of (4.37).

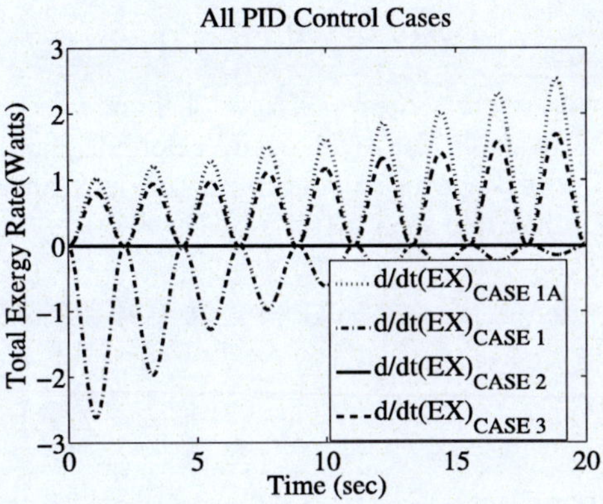

Fig. 4.15 All cases: rotary mass–spring–damper model with PID control exergy rate transient responses. Robinett III, R.D. and Wilson, D.G. [13], reprinted by permission of the publisher (Interscience Publishers)

4.3.5 *The Power Flow Principle of Stability for Nonlinear Systems*

It is now appropriate to extend these linear eigenanalysis results to nonlinear systems by following the concepts in [46]. The following nonlinear extension to the previous eigenvalue problems can be discussed with respect to (4.14) and (4.17):

$$\oint_{\tau}[m\ddot{x}+g(x)]\dot{x}\,dt=\oint_{\tau}[-f(\dot{x})\dot{x}]\,dt=0$$

with $u=0$. The first term describes the nonlinear frequency content [32] of the undamped/undriven system

$$m\ddot{x}+g(x)=0,$$

which effectively extends the undamped natural frequency, (4.24) to nonlinear systems and defines the Hamiltonian surface and the static stability of the system with $g(x)=\partial\mathcal{V}/\partial x$ and $\mathcal{V}(x)>0$ when $x\neq 0$ and $\mathcal{V}(x)=0$ when $x=0$.

The second term describes the dynamic stability of the nonlinear system and the existence of a limit cycle (equations (4.12)–(4.14))

$$\oint_{\tau}[-f(\dot{x})\dot{x}]\,dt \gtrless 0 \quad \text{and} \quad \oint_{\tau}[-f(\dot{x})\dot{x}]\,dt=0,$$

which is an extension of the real part of the eigenvalue, such as equations (4.36)–(4.38), to nonlinear systems. For example, (4.72) shows a nonlinear frequency content (amplitude dependent) and a resulting coupling into the effective damping which produces initial-condition-dependent limit cycle behavior (refer to Figs. 3.22 and 3.24).

An example of a nonlinear Hamiltonian is a cubic nonlinear spring with the following Hamiltonian:

$$\begin{aligned}\mathcal{H}&=\mathcal{T}+\mathcal{T}_c+\mathcal{V}+\mathcal{V}_c\\&=\frac{1}{2}(m+K_A)\dot{x}^2+\frac{1}{2}kx^2+\frac{1}{4}(k_{\mathrm{NL}}+K_{P_{\mathrm{NL}}})x^4>0 \quad \forall x\neq 0,\ \dot{x}\neq 0,\end{aligned}$$

$$\mathcal{H}(0,0) = 0, \tag{4.57}$$

where $u = -K_A\ddot{x} - K_{P_{NL}}x^3 + \hat{u}$ and $m, k, k_{NL}, K_A, K_{P_{NL}} > 0$. This system is statically stable due to the positive definite Hamiltonian surface with the associated nonlinear proportional and linear acceleration feedback and dynamically neutral stable (limit cycle see [46]) if

$$\begin{aligned}\dot{\mathcal{H}}_{\text{ave}} &= \frac{1}{\tau_c}\oint_{\tau_c} [(m + K_A)\ddot{x} + g(x) + g_P(x)]\dot{x}\,dt \\ &= \frac{1}{\tau_c}\oint_{\tau_c} [\hat{u} - f(\dot{x})]\dot{x}\,dt = 0.\end{aligned} \tag{4.58}$$

The nonlinear frequency content of the statically stable system is given by

$$(m + K_A)\ddot{x} + g(x) + g_P(x) = (m + K_A)\ddot{x} + kx + (K_{NL} + K_{P_{NL}})x^3 = 0 \tag{4.59}$$

and is dynamically stable if

$$\dot{\mathcal{H}}_{\text{ave}} = \frac{1}{\tau_c}\int_0^{\tau_c} [\hat{u} - f(\dot{x})]\dot{x}\,dt < 0$$

and dynamically unstable if

$$\dot{\mathcal{H}}_{\text{ave}} = \frac{1}{\tau_c}\int_0^{\tau_c} [\hat{u} - f(\dot{x})]\dot{x}\,dt > 0.$$

This extended analysis is equivalent to the concepts of "flight stability" [31, 58] presented in Sect. 4.2, where systems are analyzed and designed based on static stability and dynamic stability. In summary, a system is dynamically stable only if it is also statically stable; a system can be statically stable and dynamically unstable.

4.4 Lyapunov Analysis

In this section, Lyapunov analysis is utilized to prove the necessary and sufficient conditions for stability of a class of nonlinear systems, Hamiltonian natural systems, based on the concepts of static stability and dynamic stability with respect to the energy storage surface and power flow. In particular, the Hamiltonian surface and its time derivative determine the necessary and sufficient conditions for stability.

- *Necessary*:

$$\mathcal{H}(\dot{\mathbf{q}}, \mathbf{q}) = \mathcal{T}(\dot{\mathbf{q}}, \mathbf{q}) + \mathcal{V}(\mathbf{q})$$

  with (4.5)–(4.9).
- *Sufficient*:

$$\dot{\mathcal{H}}(\dot{\mathbf{q}}, \mathbf{q})$$

  with (4.12)–(4.14).

The proofs will be based on theorems and lemmas from references [6, 14, 52, 59] and [13, 46]. This development is the basis of the HSSPFC control law design and analysis procedure that will be presented and demonstrated in the next section.

Lyapunov analysis is based on the straightforward idea that if the state of a system is near an equilibrium state and the energy of the system is decreasing, then the equilibrium state is stable, possibly asymptotically stable. However, if the energy is increasing near an equilibrium state, then the equilibrium state is unstable. Lyapunov stability, theorems, and functions are an extension and generalization of the energy concept that has been utilized in the previous discussions of stability. The Lyapunov function is the energy storage surface (Hamiltonian) which determines the accessible phase space of the system (constraint surface/manifold). The time derivative of the Lyapunov function is the power flow (work/rate) which determines the path/trajectory of the system as it traverses the Hamiltonian given some initial condition. It will be demonstrated that the design of the Lyapunov function, Hamiltonian Surface Shaping, is equivalent to static stability, which is a necessary condition for stability. The design of the time derivative of the Lyapunov function, Power Flow Control, is equivalent to dynamic stability, which is a sufficient condition for stability for Hamiltonian natural systems.

The proof begins with the stability analysis of conservative dynamical systems which is basically a restatement of the analysis done for static stability within the context of Lyapunov analysis. The system is a conservative natural Hamiltonian system with no externally applied nonconservative forces or moments. The Lagrange–Dirichlet Theorem (a state where the potential energy is an isolated minimum is a stable equilibrium state [52]), which is equivalent to the Lyapunov Stability Theorem [51] (if in some neighborhood Ω of the origin, there exists a Lyapunov function $V(x)$ (where $V(0)=0$, $V(x)>0\ \forall x\neq 0$, and $\dot{V}(x)\leq 0$), then the origin is stable [52]) can be applied at this point to the energy storage surface Lyapunov candidate function, which is a constant, or

$$V(\mathbf{q},\dot{\mathbf{q}})=\mathcal{H}=\mathcal{T}+\mathcal{V}=E=\text{constant}. \tag{4.60}$$

The equations of motion are given in first-order canonical form as

$$\begin{aligned}\dot{q}_j&=\frac{\partial\mathcal{H}}{\partial p_j},\\ \dot{p}_j&=-\frac{\partial\mathcal{H}}{\partial q_j},\end{aligned} \tag{4.61}$$

and second-order form as

$$\frac{d}{dt}\left(\frac{\partial\mathcal{L}}{\partial\dot{\mathbf{q}}}\right)-\left(\frac{\partial\mathcal{L}}{\partial\mathbf{q}}\right)=\mathbf{0}. \tag{4.62}$$

A conservative dynamical system is (statically) stable if the Lyapunov function (Hamiltonian and potential functions) is positive definite about the equilibrium state

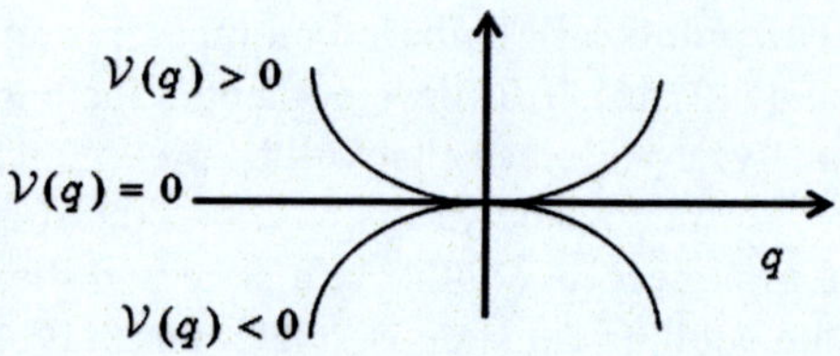

Fig. 4.16 Potential energy function. Robinett III, R.D. and Wilson, D.G. [56], reprinted by permission of the publisher (©2010 IEEE)

and its time derivative is zero,

$$\begin{aligned} V(\mathbf{q}, \dot{\mathbf{q}}) &= \mathcal{H}(\dot{\mathbf{q}}, \mathbf{q}) = \mathcal{T}(\dot{\mathbf{q}}, \mathbf{q}) + \mathcal{V}(\mathbf{q}) \\ &= E = \text{constant} > 0 \quad \forall \mathbf{q} \neq \mathbf{q}_e,\ \dot{\mathbf{q}} \neq \dot{\mathbf{q}}_e, \\ V(\mathbf{q}_e, \dot{\mathbf{q}}_e) &= 0, \end{aligned} \tag{4.63}$$

and

$$\begin{aligned} &\mathcal{V}(\mathbf{q}) > 0 \quad \forall \mathbf{q} \neq \mathbf{q}_e, \\ &\mathcal{V}(\mathbf{q}_e) = 0, \end{aligned} \tag{4.64}$$

and

$$\dot{V} = 0.$$

The Converse of the Lagrange–Dirichlet Theorem, Lyapunov's Theorem (at an isolated maximum of the potential energy, the equilibrium state is unstable [52]), can be applied, and the system is (statically) unstable if

$$\mathcal{V}(\mathbf{q}) < 0 \quad \forall \mathbf{q} \neq \mathbf{q}_e \quad \text{and} \quad \mathcal{V}(\mathbf{q}_e) = 0, \tag{4.65}$$

where the Lyapunov candidate function can be chosen as the Lagrangian

$$\begin{aligned} &V(\mathbf{q}, \dot{\mathbf{q}}) = \mathcal{L} = \mathcal{T} - \mathcal{V} > 0 \quad \forall \mathbf{q} \neq \mathbf{q}_e, \dot{\mathbf{q}} \neq \dot{\mathbf{q}}_e, \\ &V(\mathbf{q}_e, \dot{\mathbf{q}}_e) = 0, \quad \text{and} \\ &\dot{V} > 0. \end{aligned} \tag{4.66}$$

The Instability Theorem of Chetayev [14] or *Extended Lyapunov's Theorem* (if at an equilibrium state the potential energy is not a minimum, then the equilibrium state is unstable [52]) can be applied, and the system is unstable (statically neutrally stable) if

$$\mathcal{V}(\mathbf{q}) = 0 \quad \forall \mathbf{q}.$$

Notice that a conservative dynamical system transitions from stable to unstable as the potential energy function is deformed from a positive definite function to a zero function to a negative definite function. The onset of instability occurs at the point where the potential energy function loses its positive definite convexity. This is presented in Fig. 4.16. Also, a conservative dynamical system is precluded from being asymptotically stable since no external damping forces can exist.

Now, it is necessary to determine what effect power flow, damping and generation, have on conservative dynamical systems. The system is modeled as a natural Hamiltonian system with externally applied nonconservative forces and/or moments. The equations of motion are given in first-order canonical form as

$$\begin{aligned} \dot{q}_j &= \frac{\partial \mathcal{H}}{\partial p_j}, \\ \dot{p}_j &= -\frac{\partial \mathcal{H}}{\partial q_j} + Q_j, \end{aligned} \tag{4.67}$$

and in second-order form as

$$\frac{d}{dt}\left(\frac{\partial \mathcal{L}}{\partial \dot{\mathbf{q}}}\right) - \left(\frac{\partial \mathcal{L}}{\partial \mathbf{q}}\right) = \mathbf{Q}. \tag{4.68}$$

Chetayev [14] and Meirovitch [6] have investigated this situation in detail for complete damping and pervasive damping. Complete damping occurs when

$$\dot{\mathcal{H}}(\dot{\mathbf{q}}, \mathbf{q}) = \sum_{j=1}^{N} Q_j \dot{q}_j$$

is a negative definite function of the generalized velocities $\dot{q}_j$. Pervasive damping occurs when $\dot{\mathcal{H}}$ is a negative semi-definite function of $\dot{q}_j$ and the set of points where $\dot{\mathcal{H}} = 0$ contain no nontrivial positive half-trajectory of the system.

Chetayev [14] and Meirovitch [6] prove: Dissipative forces do not disturb stability of the equilibrium state of a conservative dynamical system in a meaningful way (*static stability is a necessary condition for stability*).

1. If the equilibrium state is stable with potential forces, it becomes asymptotically stable with the addition of dissipative forces with complete damping.
2. An equilibrium state which is unstable with potential forces *cannot* be stabilized by dissipative forces.

Two general theorems follow from these investigations:

Theorem 4.1 [6] *If for the system of equations* (4.67) *and* (4.68), *the Hamiltonian is positive definite and if the system possesses pervasive damping* (*complete damping as a subset*), *then the equilibrium state is asymptotically stable.*

Theorem 4.2 [6] *If for the system of equations* (4.67) *and* (4.68), *the Hamiltonian can assume negative values in the neighborhood of the origin and if the system possesses pervasive damping* (*complete damping as a subset*), *then the equilibrium state is unstable.*

Actually, if the potential energy function is $\mathcal{V}(\mathbf{q}) = 0 \ \forall \mathbf{q}$, then the system is unstable with complete damping since the system has no preferred orientation, a rigid body mode and singular stiffness matrix. The way to solve this problem is with a two-step control law design process that first utilizes proportional and/or acceleration feedback to shape the Hamiltonian surface to ensure static stability. The second step ensures dynamic stability via damping and/or generation feedback. This two-step process, HSSPFC, will be discussed in more detail in the next section.

Returning to the discussion of the stability of the system of equations (4.67) and (4.68), the system path/trajectory traverses a positive definite energy storage surface (statically stable) defined by the Hamiltonian as a result of the power flow relative to an initial condition. The time derivative of the energy storage surface defines the power flow into, dissipated within, and stored in the system. This determines whether the system is rising to a higher energy state (away from its equilibrium state), dropping to a lower energy state (returning to its equilibrium state), or staying on a closed cyclic path (limit cycle) constrained to the energy storage surface. Average power flow calculations are used because one cannot guarantee that the opposing power flows will cancel or be dominant generators or dissipators point-for-point in time [46]. In fact, limit cycles balance over the cycle as described by [6]: A state of stationary motion is achieved in which the system gains energy during part of the cycle and dissipates energy during the remaining part, so that at the end of each cycle the net energy exchange is zero. If the system were linear, then the power flows would cancel or be dominant generators or dissipators point-for-point in time [46]. Consequently, feedback linearization has some attractive features when designing feedback controllers.

The Lyapunov analysis of the system of equations (4.67) and (4.68) provides the sufficient conditions for (dynamic) stability. The system is asymptotically (dynamically) stable if [13]

$$\begin{aligned} V(\mathbf{q},\dot{\mathbf{q}}) &= \mathcal{H} > 0 \quad \forall \mathbf{q} \neq \mathbf{q}_e,\ \dot{\mathbf{q}} \neq \dot{\mathbf{q}}_e, \\ V(\mathbf{q}_e,\dot{\mathbf{q}}_e) &= 0, \\ \left(\dot{V}\right)_{\text{ave}} &= \frac{1}{\tau_c}\int_0^{\tau_c}\left[\sum_{j=1}^{N} Q_j\dot{q}_j\right]\dot{q}_j\,dt < 0. \end{aligned}$$

The system is (dynamically) unstable if [13]

$$\begin{aligned} V(\mathbf{q},\dot{\mathbf{q}}) &= \mathcal{H} > 0 \quad \forall \mathbf{q} \neq \mathbf{q}_e,\ \dot{\mathbf{q}} \neq \dot{\mathbf{q}}_e, \\ V(\mathbf{q}_e,\dot{\mathbf{q}}_e) &= 0, \\ \left(\dot{V}\right)_{\text{ave}} &= \frac{1}{\tau_c}\int_0^{\tau_c}\left[\sum_{j=1}^{N} Q_j\dot{q}_j\right]\dot{q}_j\,dt > 0. \end{aligned}$$

The limit cycle defines the stability boundary between asymptotically (dynamically) stable and (dynamically) unstable [46] and occurs when

$$\begin{aligned} V(\mathbf{q},\dot{\mathbf{q}}) &= \mathcal{H} > 0 \quad \forall \mathbf{q} \neq \mathbf{q}_e,\ \dot{\mathbf{q}} \neq \dot{\mathbf{q}}_e, \\ V(\mathbf{q}_e,\dot{\mathbf{q}}_e) &= 0, \\ \left(\dot{V}\right)_{\text{ave}} &= \frac{1}{\tau_c}\int_0^{\tau_c}\left[\sum_{j=1}^{N} Q_j\dot{q}_j\right]\dot{q}_j\,dt = 0. \end{aligned}$$

Notice that the (static) stability boundary for a conservative dynamical system is a rigid body mode where the potential energy function loses its positive definite convexity, while the (dynamic) stability boundary for the system of equations (4.67) and (4.68) is a limit cycle or a second-order center for a linear system. The (static) stability of a conservative dynamical system is a necessary condition for the (dynamic) stability of the system of equations (4.67) and (4.68), which provides the sufficient condition for stability. The necessary and sufficient conditions for stability of a class of nonlinear systems, Hamiltonian natural systems are the basis for a two-step control law design and analysis process called Hamiltonian Surface Shaping and Power Flow Control (HSSPFC).

4.5 Energy Storage Surface and Power Flow: HSSPFC

The goal of this section is to demonstrate how to design and analyze nonlinear controllers for a class of nonlinear systems, Hamiltonian natural systems from mechanics and adiabatic irreversible work processes from thermodynamics, with a two-step process called Hamiltonian Surface Shaping and Power Flow Control (HSSPFC).

The first example presents the Hamiltonian Surface Shaping of a conservative linear and cubic nonlinear spring system. The bifurcated potential energy surface is given as

$$\mathcal{V}(x) = -\frac{1}{2}kx^2 + \frac{1}{4}k_{\mathrm{NL}}x^4$$

for $k, k_{\mathrm{NL}} > 0$, and the proportional controller is

$$u_P = -K_P x$$

that leads to the Hamiltonian

$$\begin{aligned}\mathcal{H} &= \mathcal{T} + \mathcal{V} + \mathcal{V}_c \\ &= \frac{1}{2}m\dot{x}^2 + \frac{1}{2}[K_P - k]x^2 + \frac{1}{4}k_{\mathrm{NL}}x^4 > 0 \quad \forall x \neq 0,\ \dot{x} \neq 0, \quad \mathcal{H}(0,0) = 0,\end{aligned}$$

which is statically stable if $K_P - k > 0$. The equations of motion become

$$\begin{aligned}m\ddot{x} - kx + k_{\mathrm{NL}}x^3 &= u_P = -K_P x, \\ m\ddot{x} + [K_P - k]x + k_{\mathrm{NL}}x^3 &= 0,\end{aligned}$$

and the Hamiltonian surface is presented in Figs. 4.17 and 4.18, respectively.

To determine the effect that the proportional controller gain K_P has on the system, Hamiltonian phase plane plots are generated. By investigating a system with negative stiffness and by adding enough K_P to result in an overall positive net stiffness, the shape of the Hamiltonian surface changes from a saddle-point surface (see Fig. 4.17) to a positive bowl surface (see Fig. 4.18). A two-dimensional cross-section of the Hamiltonian versus the position presents the characteristics of the

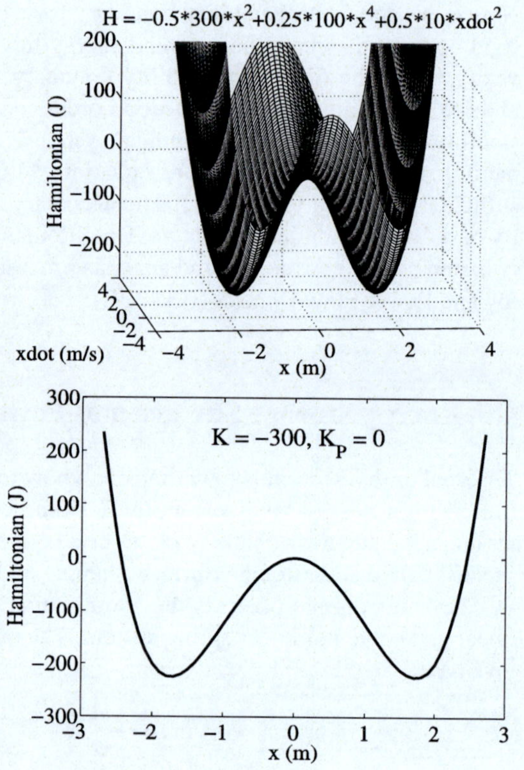

Fig. 4.17 Three-dimensional (*top*) Hamiltonian phase plane plot negative stiffness produces a saddle surface. The two-dimensional cross-section plot (*bottom*) is at $\dot{x} = 0$. Robinett III, R.D. and Wilson, D.G. [60], reprinted by permission of the publisher (©2006 IEEE)

overall storage or potential functions. The operating point at $(\mathcal{H}, \dot{x}, x) = (0, 0, 0)$ changes from being unstable to stable for small values of $|x| > 0$, when enough additional K_P is added, a net positive stiffness for the system results.

In example two, Hamiltonian Surface Shaping of a conservative nonlinear inverted pendulum with an externally applied conservative torque is investigated. The potential energy for the nonlinear pendulum hanging straight down is given as

$$\mathcal{V}(\theta) = mgl[1 - \cos\theta]$$

for $m, g, l > 0$ that leads to the Hamiltonian

$$\mathcal{H} = \frac{1}{2}ml^2\dot{\theta}^2 + mgl[1 - \cos\theta],$$

which leads to the equation of motion

$$ml^2\ddot{\theta} + mgl\sin\theta = u.$$

The nonlinear inverted pendulum model can be obtained by a transformation of coordinates as $\theta = \pi + \beta$, which leads to

$$\mathcal{V}(\beta) = mgl[1 + \cos\beta]$$

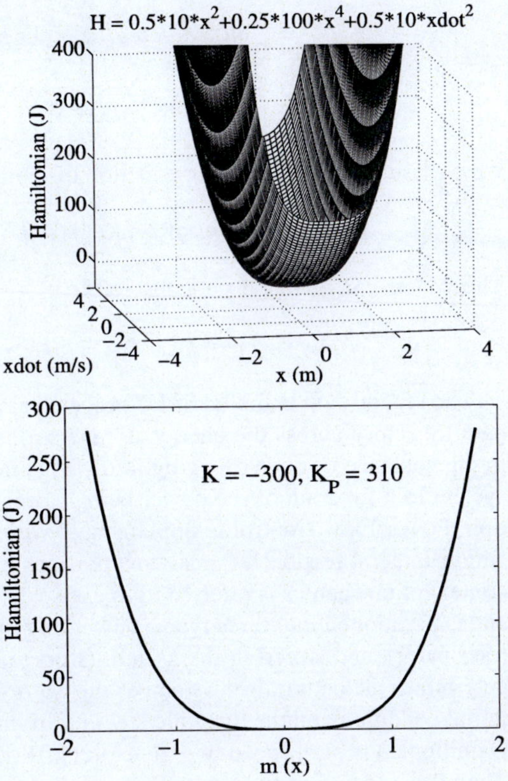

Fig. 4.18 Three-dimensional (*top*) Hamiltonian phase plane plot where the net positive stiffness produces a positive bowl surface. The two-dimensional cross-section plot (*bottom*) is at $\dot{x} = 0$. Robinett III, R.D. and Wilson, D.G. [60], reprinted by permission of the publisher (©2006 IEEE)

and the Hamiltonian as

$$\mathcal{H} = \frac{1}{2}ml^2\dot{\beta}^2 + mgl[1 + \cos\beta],$$

and the equation of motion becomes

$$ml^2\ddot{\beta} - mgl\sin\beta = u.$$

This is a repulsive potential that is statically unstable. A nonlinear proportional feedback controller of the form

$$u_c = -K_P \sin\beta$$

that derives from a potential of the form

$$\mathcal{V}_c(\beta) = K_P[1 - \cos\beta];$$

then the Hamiltonian becomes

$$\mathcal{H} = \mathcal{T} + \mathcal{V} + \mathcal{V}_c$$

$$= \frac{1}{2}ml^2\dot{\beta}^2 + mgl[1+\cos\beta] + K_P[1-\cos\beta]$$

$$= \frac{1}{2}ml^2\dot{\beta}^2 + [mgl + K_P] - [K_P - mgl]\cos\beta,$$

which results in the following equation of motion:

$$ml^2\ddot{\beta} + (K_P - mgl)\sin\beta = u.$$

The system can be made statically stable if

$$K_P - mgl > 0.$$

The second step in the HSSPFC process is Power Flow Control which shapes the path/trajectory across the energy storage surface. This is accomplished by designing the balance of power flowing into versus the power being dissipated within the system as a function of the power being stored in the system and the initial condition. Power Flow Control is implemented with the derivative feedback (dissipator) and/or integral feedback (generator) portions of the control law. Power flow in the context of mechanics is referred to as exergy flow in the context of thermodynamics and is partitioned into three types: power flowing into (generator), dissipated within (dissipator), and stored in the system (storage). The balance of these power flows determines the path/trajectory across the energy storage surface as a function of the initial condition and the dynamic stability of the system. The time derivative of the Hamiltonian is partitioned into generators, $\dot{W}$, dissipators, $T_0\dot{S}_i$, and storage terms, $\dot{\mathcal{H}}$, in order to design the power flow balance defined by

$$\dot{\mathcal{H}} = \dot{W} - T_0\dot{S}_i.$$

Derivative feedback is a dissipator that creates irreversible entropy,

$$u = -K_D\dot{x}, \quad \text{which implies that} \quad T_0\dot{S}_i = K_D\dot{x}^2.$$

Integral feedback is a generator that flows power into the system,

$$u = -K_I\int_0^t x\,d\tau, \quad \text{which implies that} \quad \dot{W} = -K_I\int_0^t x\,d\tau\,\dot{x}.$$

A few examples are presented to demonstrate Power Flow Control.

Example one is the power flow control of the linearized mass, spring, damper system in Fig. 4.4 with numerical simulation results presented in Figs. 4.5–4.14 where the effects of integral feedback are demonstrated in detail. This system is statically stable while being dynamically stable, neutral stable, and unstable depending upon the feedback gains.

The second example is a Duffing oscillator equation with Coulomb friction given by

$$m\ddot{x} + kx + k_{\mathrm{NL}}x^3 = -c\dot{x} - c_{\mathrm{NL}}\mathrm{sign}(\dot{x}) + u \tag{4.69}$$

Table 4.2 Duffing oscillator/Coulomb friction model and PID control system gains (Note: for all cases, $\dot{x}_0 = 0.0$) [7]

Case No.	K_P (kg/s^2)	K_I (kg/s^3)	K_D (kg/s)	c (kg/s)	C_{NL} (N)	m (kg)	k (N/m)	k_{NL} (N/m^3)	T_f (s)	x_0 (m)	x_r (m)
1	10	0.0	0.0	0.0	0.0	10	10	100.0	10	1	0
2	10	20.0	2.0	0.1	5.0	10	10	100.0	10	1	0
3	10	40.05	2.0	0.1	5.0	10	10	100.0	10	1	0
4	10	80.0	2.0	0.1	5.0	10	10	100.0	10	1	0
5	−200	20.0	2.0	0.1	5.0	10	10	100.0	36	1	0
6	−200	10.0	2.0	0.1	5.0	10	10	100.0	36	1	0
7	−200	1.0	2.0	0.1	5.0	10	10	100.0	36	1	0
8	−200	1.0	2.0	0.1	5.0	10	10	100.0	100	1	0
9	−200	50.0	2.0	0.1	5.0	10	10	100.0	15	1	0
10	−200	30.0	2.0	0.1	5.0	10	10	100.0	36	−1	0
11	10	20.0	2.0	0.1	5.0	10	−200	100.0	50	0	1
12	10	20.0	2.0	0.1	5.0	10	−200	100.0	50	0	−1

and controlled with a PID controller

$$u = -K_P x - K_I \int x\,dt - K_D \dot{x}. \tag{4.70}$$

The Hamiltonian is

$$\mathcal{H} = \frac{1}{2} m \dot{x}^2 + \frac{1}{2}(k + K_P) x^2 + \frac{1}{4} k_{\mathrm{NL}} x^4 \tag{4.71}$$

with time derivative

$$\begin{aligned} \dot{\mathcal{H}} &= \left[m\ddot{x} + (k + K_P)x + k_{\mathrm{NL}} x^3\right]\dot{x} \\ &= \left[-(c + K_D)\dot{x} - c_{\mathrm{NL}}\operatorname{sign}(\dot{x}) - K_I \int x\,dt\right]\dot{x}. \end{aligned} \tag{4.72}$$

The nonlinear limit cycle occurs when

$$\oint_\tau [-(c + K_D)\dot{x} - c_{\mathrm{NL}}\operatorname{sign}(\dot{x})]\dot{x}\,dt = \oint_\tau \left[K_I \int x dt\right]\dot{x}\,dt. \tag{4.73}$$

Numerical simulations were performed to demonstrate where the nonlinear stability boundary lies for the Duffing oscillator/Coulomb friction dynamic model subject to PID control. Twelve separate cases (Cases 1–12) were conducted with the numerical values listed in Table 4.2. Cases 1–4 and gain-scheduling of this example are discussed in Sect. 3.6.2 of Chap. 3.

It is instructive to present what happens when the statically stable Hamiltonian surface about the equilibrium point $(0, 0)$ is bifurcated into two equilibrium points near the origin, which makes the system statically unstable at the origin. This situation happens when the linear stiffness is negative, $K_P + k < 0$, and is presented

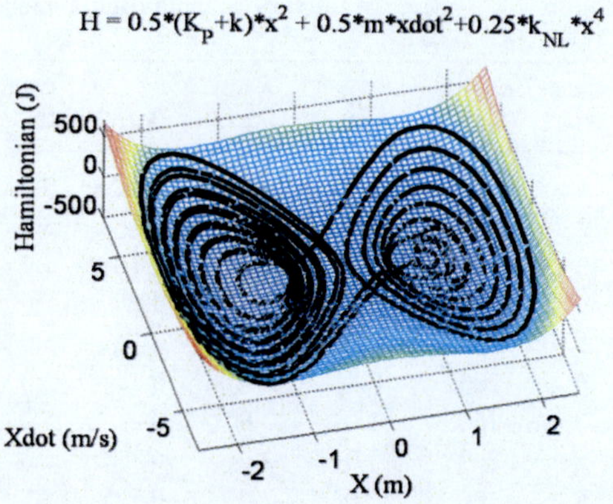

Fig. 4.19 Case 5 Hamiltonian 3D surface plots [7]

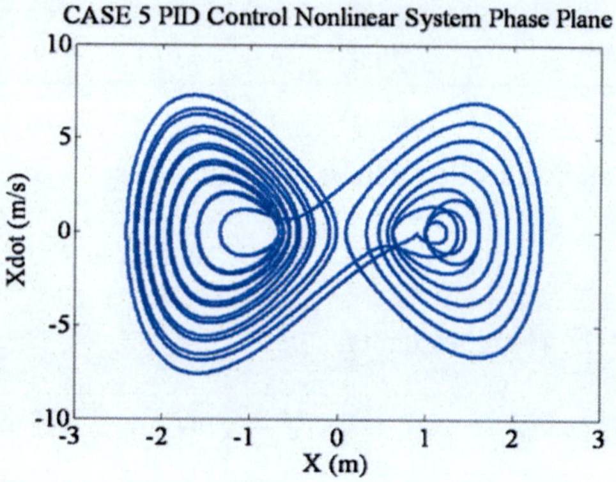

Fig. 4.20 Case 5 2D phase plane plots [7]

in Cases 5–12 in Table 4.2. For Cases 5–10, the system is under regulator control, at $(x, \dot{x}) = (0, 0)$ with positive proportional feedback. This causes the point $(x, \dot{x}) = (0, 0)$ to be an unstable static node, and for the following scenarios with $K_P + k < 0$, the system is forced away from the regulator point.

In Case 5, the system builds up enough energy to transition out of the right well but overshoots into the left well, where the process starts again (see Figs. 4.19, 4.20, 4.21, and 4.22). Eventually, the system achieves a balanced equilibrium between both wells, a nonlinear limit cycle. A similar response results in Case 6 (see Figs. 4.23, 4.24, 4.25, and 4.26). Note that for a reduction in K_I, the responses are slower, in comparison to the previous Case 5.

In Case 7 (see Figs. 4.27 and 4.28), the K_I again is reduced resulting in the appearance that the system decays down to a point in the right well. Also note that in Fig. 4.29 the dissipative term is greater than the generator term with corresponding decaying responses in Fig. 4.30 for both the position and velocity, respectively. Case 7 is building up slower than previous cases, since K_I is reduced. However, in Cases 7 and 8, given enough simulation time ($t_f = 100$ sec), the generator term eventually builds up enough energy to move out of the right well, but again overshoots (see Figs. 4.31 and 4.32) the (0, 0) regulator point and spirals down into the

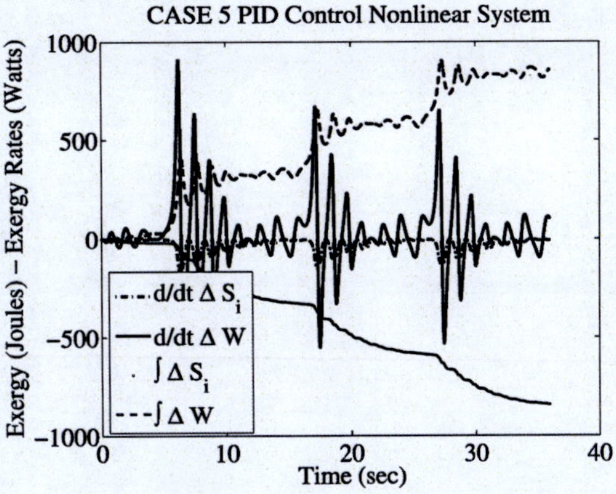

Fig. 4.21 Case 5 exergy and exergy-rate responses [7]

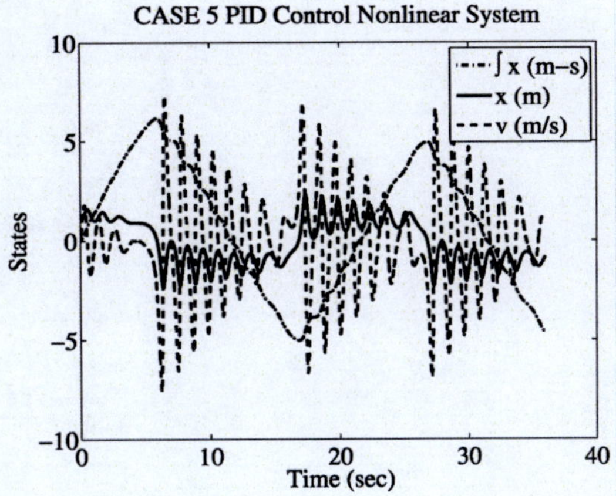

Fig. 4.22 Case 5 system transient responses [7]

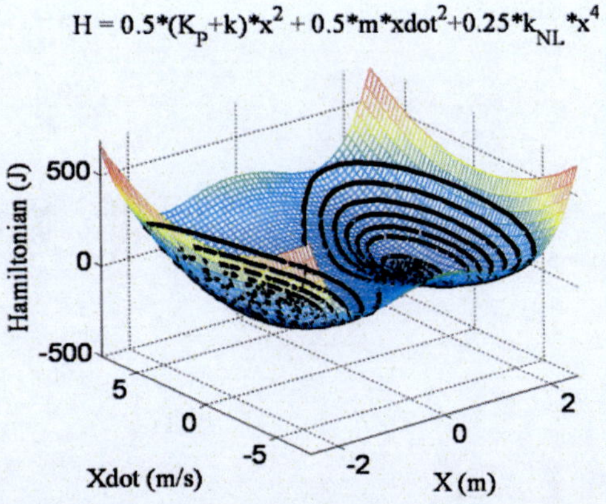

Fig. 4.23 Case 6 Hamiltonian 3D surface plots [7]

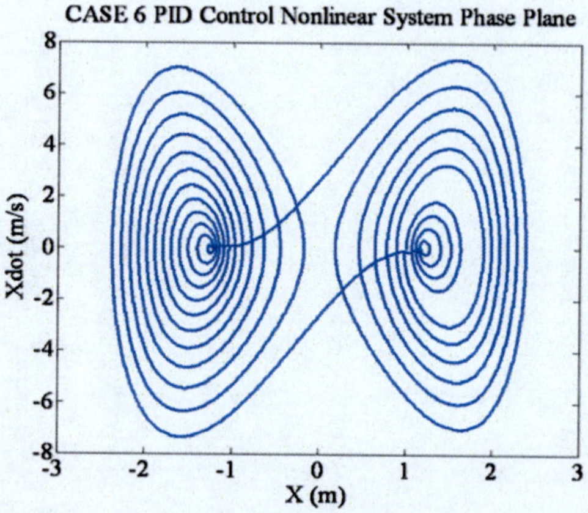

Fig. 4.24 Case 6 2D phase plane plots [7]

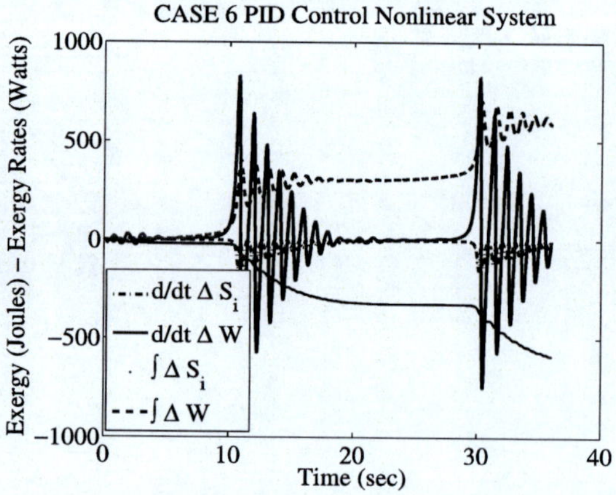

Fig. 4.25 Case 6 exergy and exergy-rate responses [7]

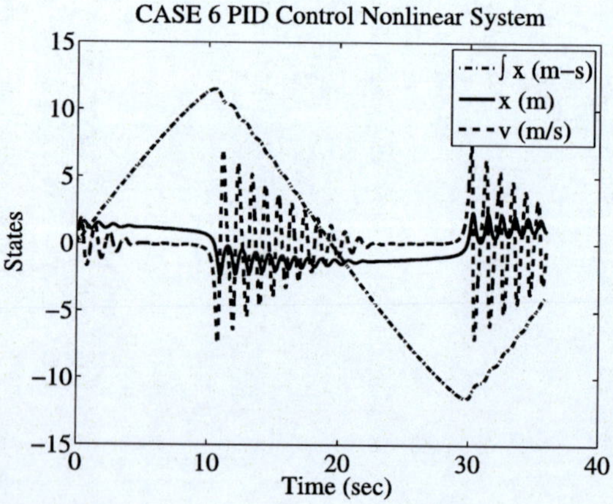

Fig. 4.26 Case 6 system transient responses [7]

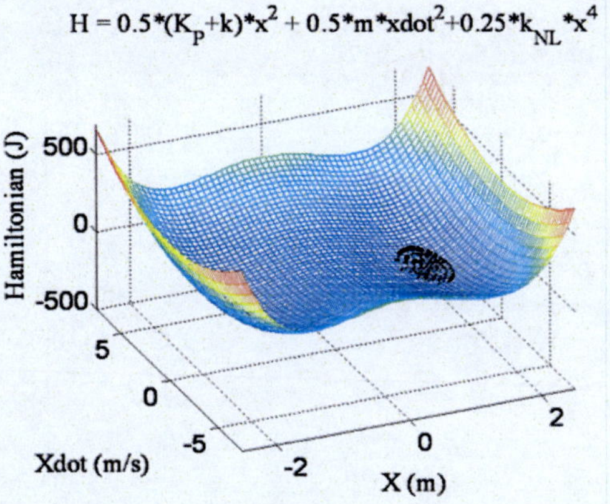

Fig. 4.27 Case 7 Hamiltonian 3D surface plots [7]

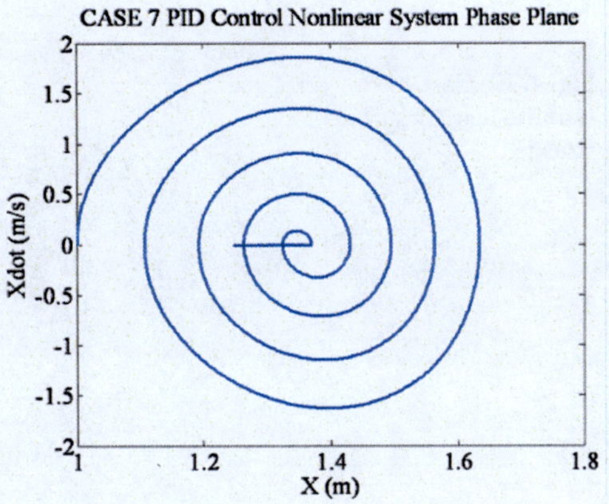

Fig. 4.28 Case 7 2D phase plane plots [7]

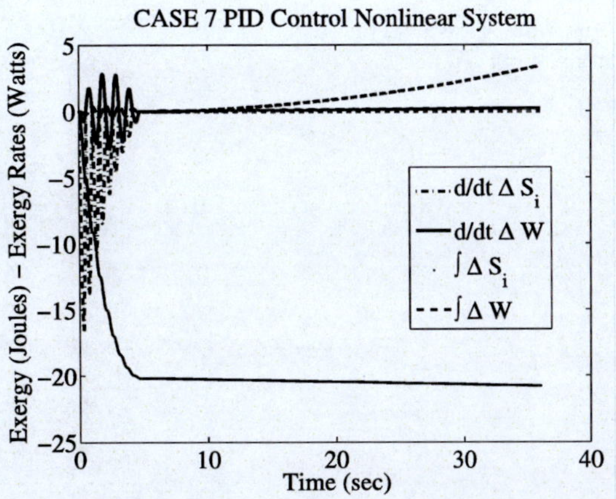

Fig. 4.29 Case 7 exergy and exergy-rate responses [7]

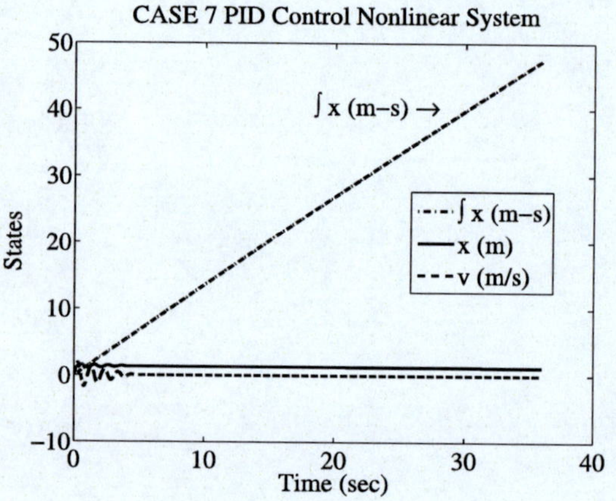

Fig. 4.30 Case 7 system transient responses [7]

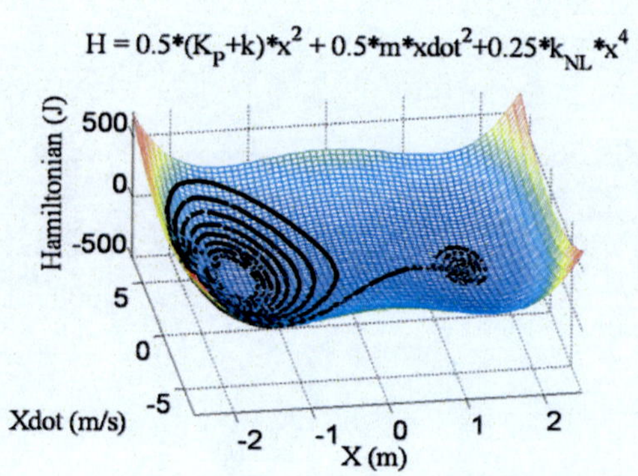

Fig. 4.31 Case 8 Hamiltonian 3D surface plots [7]

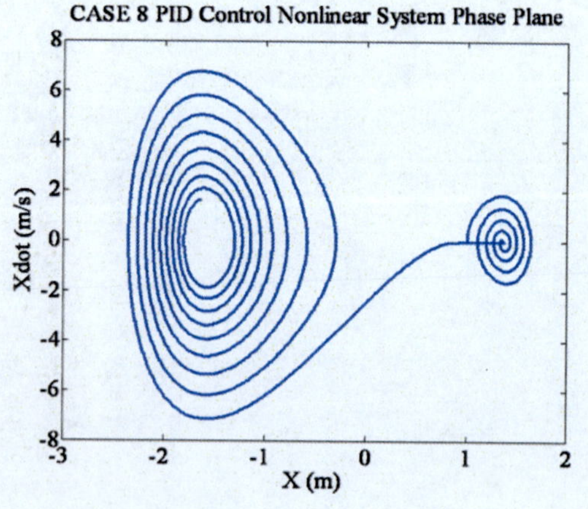

Fig. 4.32 Case 8 2D phase plane plots [7]

Fig. 4.33 Case 8 exergy and exergy-rate responses [7]

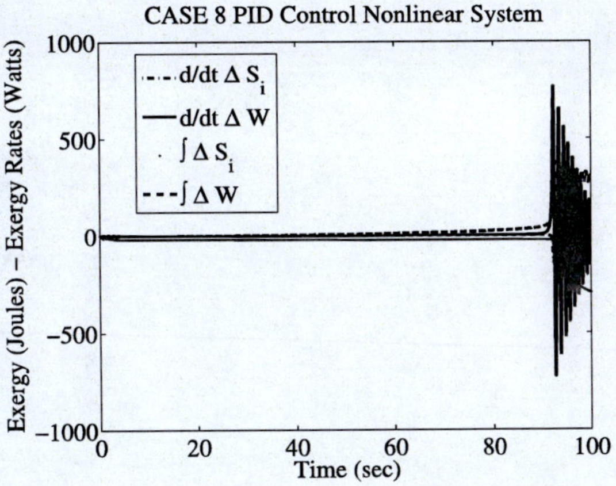

Fig. 4.34 Case 8 system transient responses [7]

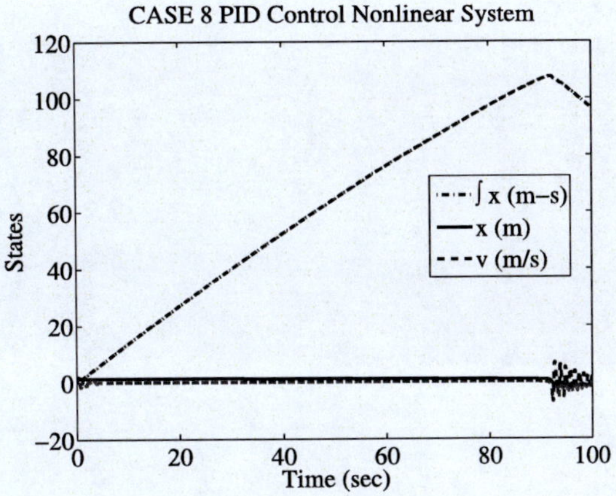

left well. The exergy and exergy-rate responses are given in Fig. 4.33 along with the corresponding state responses in Fig. 4.34.

For Case 9, K_I is increased such that the system traverses around both left and right wells and approaches another higher energy level, nonlinear limit cycle (see Figs. 4.35 and 4.36 along with Figs. 4.37 and 4.38). In this case the generator term maintains a balance with the opposing damping terms subject to the nonlinear spring effect.

Case 10 demonstrates a reduction in K_I and also starts in the left well, builds up enough energy to move over to the right well, and eventually comes back to another stable energy state, nonlinear limit cycle (see Figs. 4.39, 4.40, 4.41, and 4.42).

For the last two cases (Cases 11 and 12), the PID regulator is converted to a tracking controller for which

$$u = -K_P(x - x_r) - K_I \int (x - x_r)\,d\tau - K_D(\dot{x} - \dot{x}_r), \tag{4.74}$$

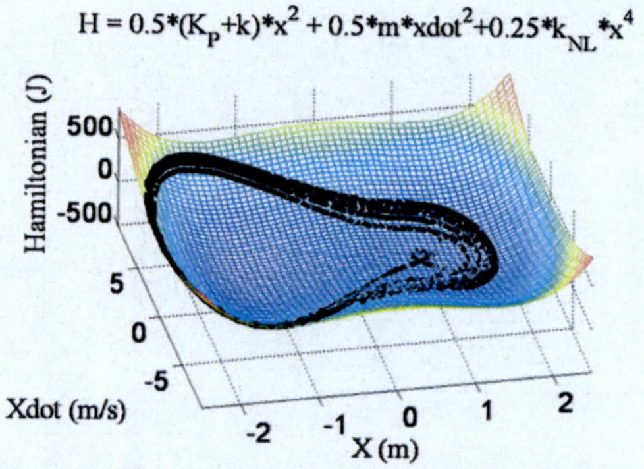

Fig. 4.35 Case 9 Hamiltonian 3D surface plots [7]

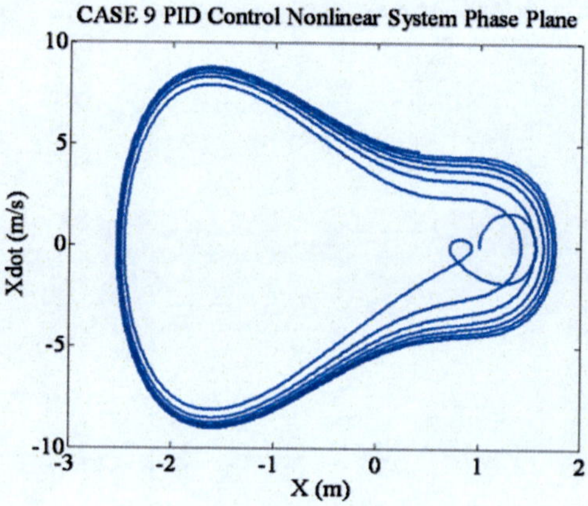

Fig. 4.36 Case 9 2D phase plane plots [7]

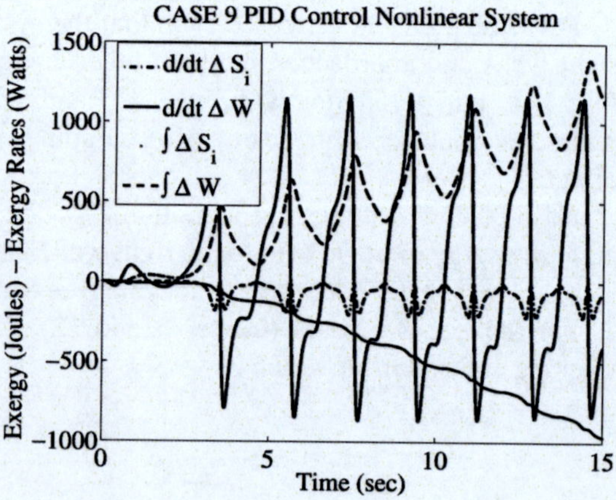

Fig. 4.37 Case 9 exergy and exergy-rate responses [7]

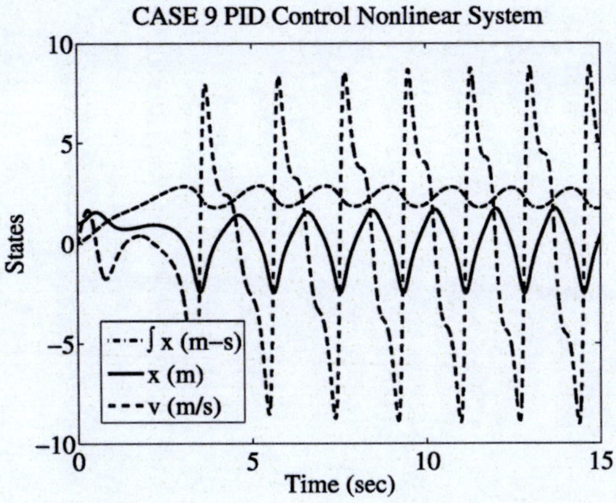

Fig. 4.38 Case 9 system transient responses [7]

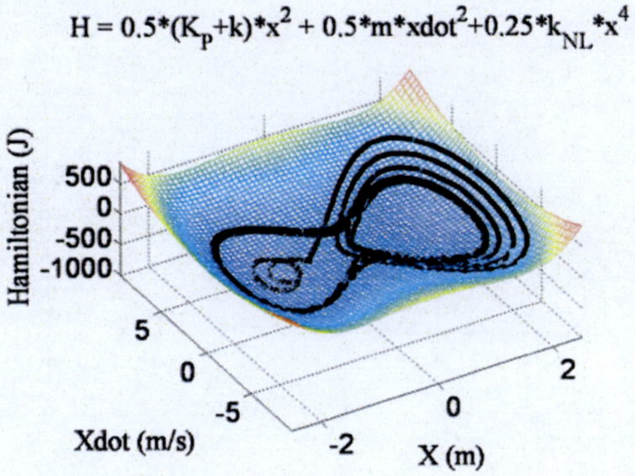

Fig. 4.39 Case 10 Hamiltonian 3D surface plots [7]

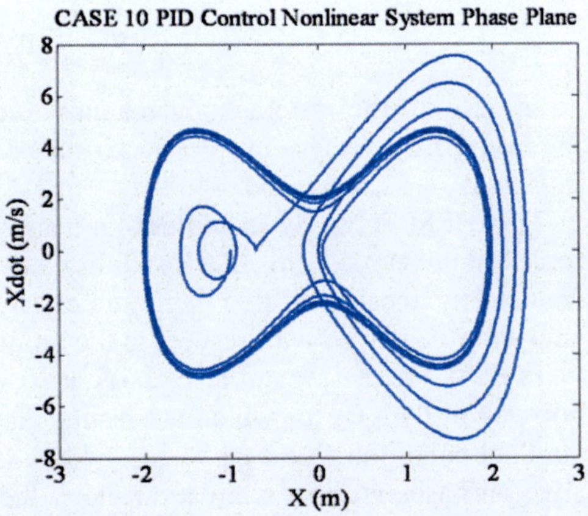

Fig. 4.40 Case 10 2D phase plane plots [7]

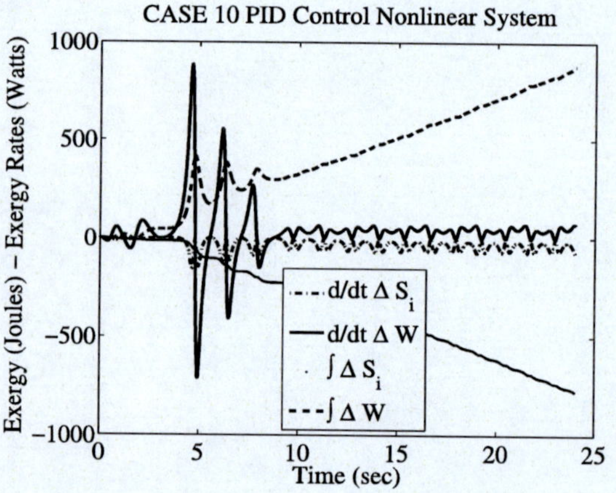

Fig. 4.41 Case 10 exergy and exergy-rate responses [7]

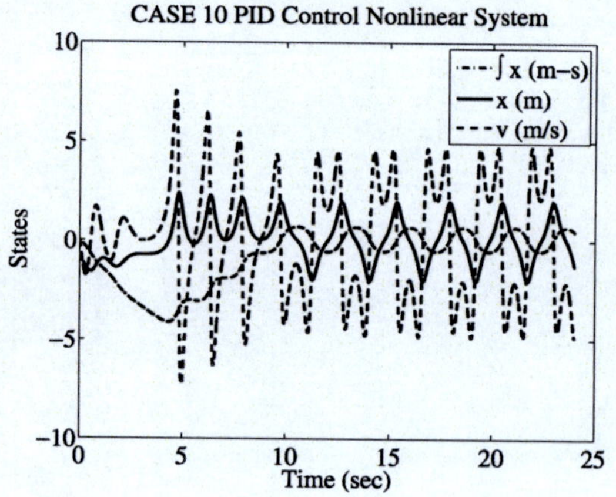

Fig. 4.42 Case 10 system transient responses [7]

where for this discussion, x_r is a reference step input, and the reference velocity is $\dot{x}_r = 0$. The corresponding Hamiltonian becomes

$$\mathcal{H} = \frac{1}{2}m\dot{x}^2 + \frac{1}{4}k_{\mathrm{NL}}x^4 + \frac{1}{2}kx^2 + \frac{1}{2}K_P(x - x_r)^2. \tag{4.75}$$

In Case 11, the system is given a unity reference step input for which after moving from $(0, 0)$ the system spirals into and converges into the right well at $(1, 0)$ (see Figs. 4.43, 4.44, 4.45, and 4.46).

This HSSPFC analysis and design has provided insight for the investigation of forced nonlinear systems. Both stability and performance can be further characterized and synthesized with a better understanding of limit cycles and their relationship and role played with respect to the nonlinear system.

In Case 12, the system also starts at $(0, 0)$ and is commanded to step into the left well at $(-1, 0)$ for which the results also spiral and converge to the appropriate final condition (see Figs. 4.47, 4.48, 4.49, and 4.50). In addition, for comparative purposes without simulation plots, the reference operating point was set to

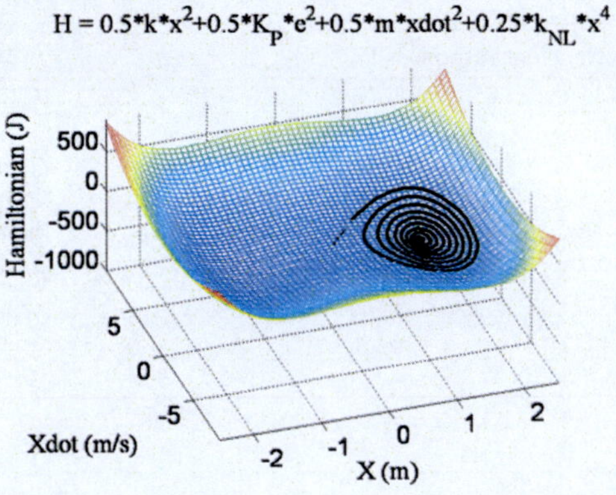

Fig. 4.43 Case 11 Hamiltonian 3D surface plots [7]

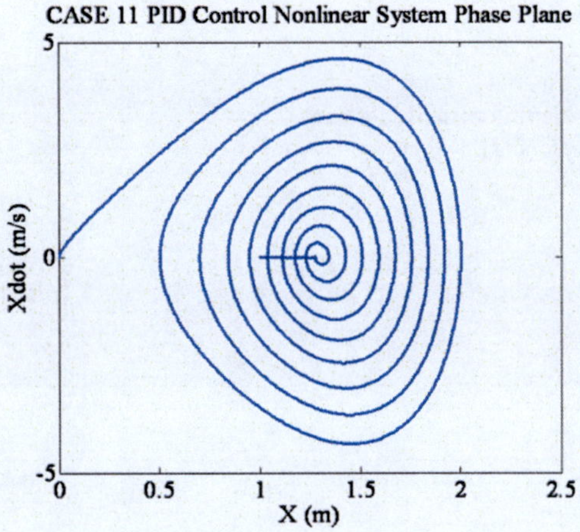

Fig. 4.44 Case 11 2D phase plane plots [7]

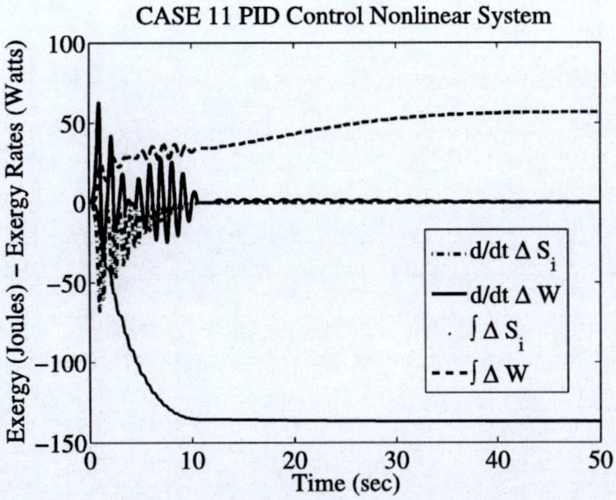

Fig. 4.45 Case 11 exergy and exergy-rate responses [7]

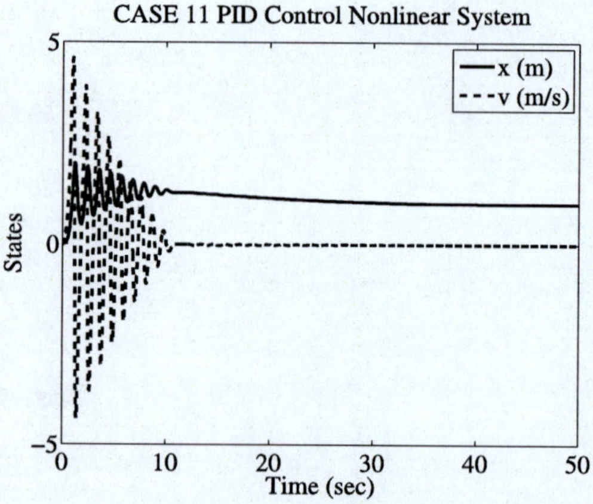

Fig. 4.46 Case 11 system transient responses [7]

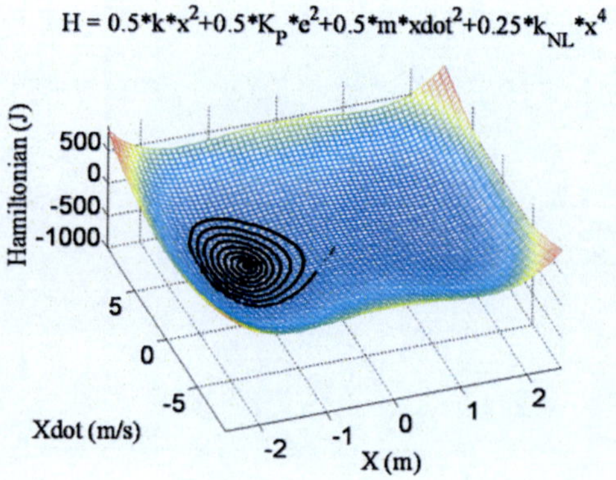

Fig. 4.47 Case 12 Hamiltonian 3D surface plots [7]

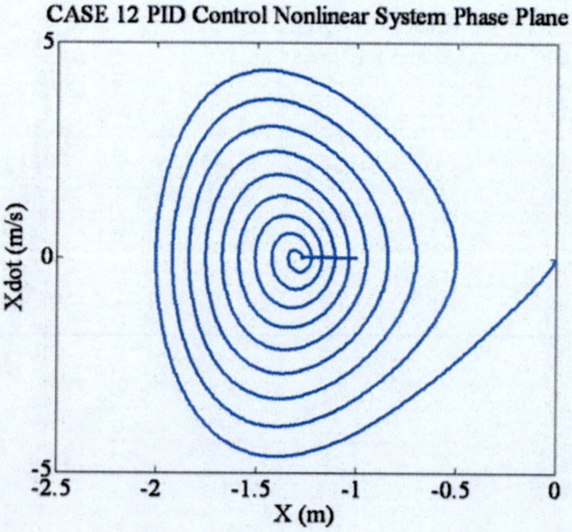

Fig. 4.48 Case 12 2D phase plane plots [7]

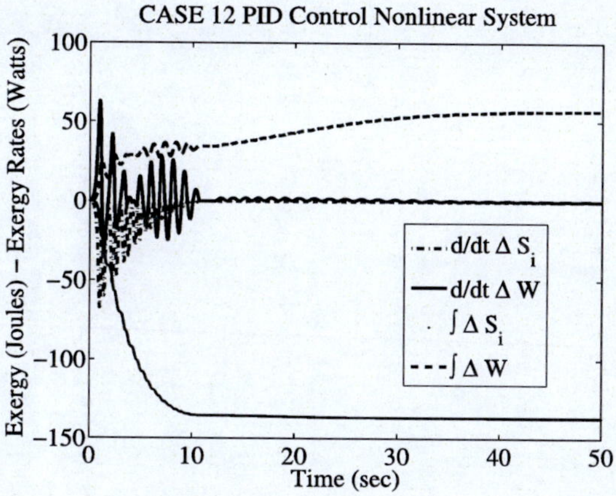

Fig. 4.49 Case 12 exergy and exergy-rate responses [7]

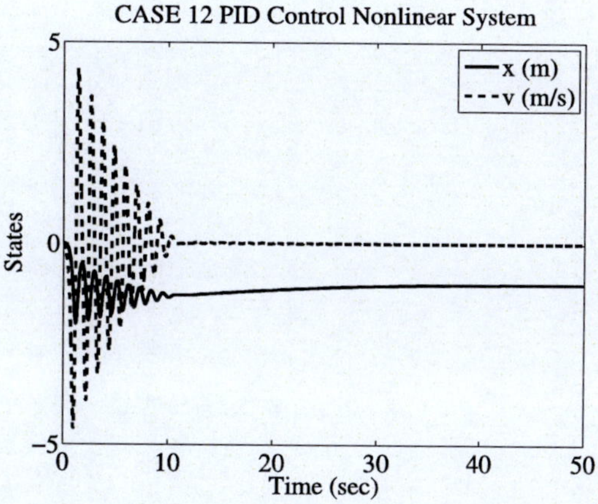

Fig. 4.50 Case 12 system transient responses [7]

$(x_r, \dot{x}_r) = (0, 0)$ (the unstable node), and for nonzero initial conditions, the tracking controller still eventually overshoots the set point with very similar results to the previous cases.

4.6 Chapter Summary

Chapter 4 presented the concepts of static stability, dynamic stability, eigenanalysis, and Lyapunov analysis that were used to develop the necessary and sufficient conditions for stability of a class of nonlinear systems, Hamiltonian natural systems. The combination of static stability of conservative systems and dynamic stability of adiabatic irreversible work processes in the form of exergy rate equations were used to develop Hamiltonian Surface Shaping and Power Flow Control (HSSPFC). Several examples demonstrated the HSSPFC controller design process that will be applied to a series of case studies in Part II.

Chapter 5
Advanced Control Design

5.1 Introduction

The design of HSSPFC controllers for distributed parameter systems is a straightforward extension of the results of the previous chapter by applying calculus of variations to the Extended Hamilton's Principle [6, 26, 61]. Fractional calculus control provides a nice mathematical construct to understand the sorting of power flows into generators, dissipators, and storage based on the order of the derivative. The fractional calculus feedback controller provides a method to couple the independent power flows by using derivatives to fractional order and describes why average power flow calculations are very useful in control designs. Optimal HSSPF controllers are designed for both open-loop and closed-loop implementation. Homotopy is combined with dynamic programming [62] to design open-loop optimal power flow controllers that are nonlinear even for linear system models. The closed-loop controllers are inspired by a discussion and homework problems in [32]. Robust HSSPFC tracking controllers are developed by determining the uncertainty bounds on the system parameters and assessing how these constraints affect the stability boundaries and performance of the system. If the system performance including power consumption is being limited, an adaptive HSSPFC tracking controller is designed by adding the appropriate information potential functions to the Hamiltonian.

5.2 Distributed Parameters/PDEs

A vibrating string is chosen as an example of a distributed parameter system that is modeled with partial differential equations (PDEs). The development follows a homework problem given in Weinstock [26]. Additional details of this approach are given in [63]. The model is derived from the following system presented in Fig. 5.1

R.D. Robinett III, D.G. Wilson, *Nonlinear Power Flow Control Design*,
Understanding Complex Systems,
DOI 10.1007/978-0-85729-823-2_5,

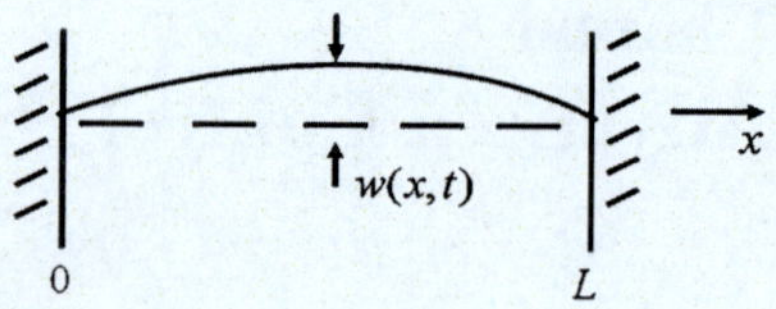

Fig. 5.1 Vibrating string model

with the following definitions and boundary conditions:

$$\begin{aligned}
\tau &= \text{tension in string}, \\
\sigma(x) &= \text{mass/unit length}, \\
\dot{w} &= \frac{\partial w}{\partial t}, \\
w_x &= \frac{\partial w}{\partial x}, \\
w(0,t) &= w(L,t) = 0,
\end{aligned}$$

which leads to the following kinetic and potential energies:

$$\begin{aligned}
\mathcal{T} &= \frac{1}{2}\int_0^L \sigma(x)\dot{w}^2\,dx, \\
\mathcal{V} &= \frac{1}{2}\tau\int_0^L w_x^2\,dx.
\end{aligned}$$

These kinetic and potential energies lead to the following Hamiltonian:

$$\mathcal{H} = \mathcal{T} + \mathcal{V} = \frac{1}{2}\int_0^L \sigma(x)\dot{w}^2\,dx + \frac{1}{2}\tau\int_0^L w_x^2\,dx.$$

The work done by the nonconservative applied distributed force is

$$W = \int_0^L F(x,t)w\,dx = \text{work}.$$

The Extended Hamilton's Principle is given as

$$\begin{aligned}
\bar{I} &= \int_0^t [\mathcal{T} - \mathcal{V} + W]\,dt_1 \\
&= \int_0^t \int_0^L \left[\frac{1}{2}(\sigma\dot{w}^2 - \tau w_x^2) + Fw\right] dx\,dt_1 \\
&= \int_0^t \int_0^L f\,dx\,dt_1,
\end{aligned}$$

where

$$\frac{\partial f}{\partial w} - \frac{\partial}{\partial x}\left(\frac{\partial f}{\partial w_x}\right) - \frac{\partial}{\partial t}\left(\frac{\partial f}{\partial w_t}\right) = 0$$

and

$$F + \tau w_{xx} - \sigma \ddot{w} = 0.$$

The resulting equation of motion from the Euler–Lagrange equation with boundary conditions is

$$\boxed{\begin{array}{l} \sigma \ddot{w} - \tau w_{xx} = F, \\ w(0,t) = w(L,t) = 0. \end{array}} \tag{5.1}$$

This is the PDE that will be used to design an HSSPFC control law. The first step is to evaluate the Hamiltonian surface and the power flow to determine if the uncontrolled system is statically stable and dynamically stable. The static stability of the system is defined by the Hamiltonian surface as

$$\mathcal{H} = \mathcal{T} + \mathcal{V} = \int_0^L \frac{1}{2}\left(\sigma \dot{w}^2 + \tau w_x^2\right) dx,$$

which is positive definite for σ, $\tau > 0$, so the system is statically stable. The dynamic stability of the system is defined by the time derivative of the Hamiltonian as

$$\begin{aligned} \dot{\mathcal{H}} &= \int_0^L [\sigma \dot{w} \ddot{w} + \tau w_x \dot{w}_x]\, dx \\ &= \int_0^L [\sigma \ddot{w} - \tau w_{xx}] \dot{w}\, dx \\ &= \int_0^L F \dot{w}\, dx, \end{aligned}$$

which leads to dynamically neutral stable if

$$\dot{\mathcal{H}}_{\text{ave}} = \frac{1}{\tau_c} \int_0^{\tau_c} \left[\int_0^L F \dot{w}\, dx \right] dt = 0 \quad \text{a limit cycle,}$$

dynamically stable if

$$\dot{\mathcal{H}}_{\text{ave}} = \frac{1}{\tau_c} \int_0^{\tau_c} \left[\int_0^L F \dot{w}\, dx \right] dt < 0,$$

and dynamically unstable if

$$\dot{\mathcal{H}}_{\text{ave}} = \frac{1}{\tau_c} \int_0^{\tau_c} \left[\int_0^L F \dot{w}\, dx \right] dt > 0.$$

The HSSPFC control law is chosen as a distributed PID controller as

$$F = K_P w_{xx} - K_I \int_0^t w\, dt_1 - K_D \dot{w}.$$

Notice that the proportional feedback is positive position feedback in this case to meet the requirements of static stability. This proportional feedback term in the distributed control is generated from a controller potential function as

$$V_c = \frac{1}{2} K_P \int_0^L w_x^2 \, dx,$$

which leads to the following shaped Hamiltonian surface:

$$\mathcal{H} = T + V + V_c = \int_0^T \frac{1}{2} [\sigma \dot{w}^2 + (\tau + K_P) w_x^2] \, dx.$$

This is statically stable for $(\tau + K_P) > 0$, and the equation of motion becomes

$$\sigma \ddot{w} - (\tau + K_P) w_{xx} = -K_I \int_0^t w \, d\tau_1 - K_D \dot{w}$$

and the corresponding power flow is

$$\begin{aligned}
\dot{\mathcal{H}} &= \int_0^L [\sigma \dot{w} \ddot{w} + (\tau + K_P) w_x \dot{w}_x] \, dx \\
&= \int_0^L [\sigma \ddot{w} - (\tau + K_P) w_{xx}] \dot{w} \, dx \\
&= \int_0^L \left[-K_I \int_0^t w \, dt_1 - K_D \dot{w} \right] \dot{w} \, dx.
\end{aligned}$$

The system is dynamically stable if

$$\begin{aligned}
\dot{\mathcal{H}}_{\text{ave}} &= \frac{1}{\tau_c} \int_0^{\tau_c} \left[\int_0^L [\sigma \ddot{w} - (\tau + K_P) w_{xx}] \dot{w} \, dx \right] dt \\
&= \frac{1}{\tau_c} \int_0^{\tau_c} \left[\int_0^L \left[-K_I \int_0^t w \, dt_1 - K_D \dot{w} \right] \dot{w} \, dx \right] dt < 0,
\end{aligned}$$

dynamically unstable if

$$\dot{\mathcal{H}}_{\text{ave}} = \frac{1}{\tau_c} \int_0^{\tau_c} \left[\int_0^L \left[-K_I \int_0^t w \, dt_1 - K_D \dot{w} \right] \dot{w} \, dx \right] dt > 0,$$

and dynamically neutral stable if

$$\dot{\mathcal{H}}_{\text{ave}} = \frac{1}{\tau_c} \int_0^{\tau_c} \left[\int_0^L \left[-K_I \int_0^t w \, dt_1 - K_D \dot{w} \right] \dot{w} \, dx \right] dt = 0.$$

5.3 Fractional Calculus

This section utilizes fractional calculus to clarify and generalize power flow control design for Hamiltonian natural systems in terms of coupled static stability and dynamic stability. The Hamiltonian is the energy storage surface and determines the static stability, while the time derivative is the power flow and determines dynamic stability. These power flow terms can be partitioned into storage, dissipation, and generation based upon the integer derivative order. The differintegral generalizes sorting of these power flow terms as well as the coupling of these terms via fractional derivative orders. The differintegral provides a unique way to sort terms and create hybrid terms with a single formula. The differential hybrid terms are compared to the lag-stabilized hybrid terms with a fractional differential equation formulation of a mass–spring damper system.

The differintegral is a derivative and integral to arbitrary order, and we use the definition given by Oldham and Spanier [64] as

$$\frac{d^q f}{[d(x-a)]^q} = \lim_{N\to\infty}\left\{\frac{[\frac{x-a}{N}]^{-q}}{\Gamma(-q)}\sum_{j=0}^{N-1}\frac{\Gamma(j-q)}{\Gamma(j+1)} f\left(x - j\left[\frac{x-a}{N}\right]\right)\right\}, \tag{5.2}$$

where q is an arbitrary real number, and Γ is the Gamma function. Another definition of the differintegral is given by Podlubny [65] that is useful in developing difference equations for differential equations containing differintegrals:

$$_aD_t^q f(t) = \lim_{\substack{h\to 0\\ nh=t-a}} h^{-q}\sum_{r=0}^{n}(-1)^r\binom{q}{r} f(t-rh), \tag{5.3}$$

where

$$\binom{q}{r} = \frac{q(q-1)(q-2)\cdots(q-r+1)}{r!},$$

and h is a small change in t. Equation (5.3) naturally leads to a first-order fractional difference approximation of the qth derivative as

$$_o\tilde{D}_t^q f(t) = h^{-q}\sum_{k=0}^{[t/h]} w_k^{(q)} f(t-kh), \tag{5.4}$$

where

$$w_k^{(q)} = (-1)^k\binom{q}{k} \quad \text{for } k = 0, 1, 2, \ldots,$$

which are calculated using the recurrence relationships

$$w_o^{(q)} = 1 \quad \text{and} \quad w_k^{(q)} = \left(1 - \frac{q+1}{k}\right) w_{k-1}^{(q)} \quad \text{for } k = 1, 2, 3, \ldots. \tag{5.5}$$

Since the primary interest is in time-domain analyses of power flow, the remainder of the section will focus on the inhomogeneous Bagley–Torvik equation [65]

$$\begin{aligned} &A\ddot{y}(t) + B\,{}_0D_t^{3/2}y(t) + Cy(t) = u(t), \quad t > 0, \\ &y(0) = 0, \quad \dot{y}(0) = 0. \end{aligned} \tag{5.6}$$

Next, the Bagley–Torvik equation (5.6) is generalized to

$$\begin{aligned} &M\ddot{x}(t) + K_q\,{}_0D_t^{q}x(t) + Kx(t) = u(t), \quad t > 0, \\ &x(0) = 0, \quad \dot{x}(0) = 0. \end{aligned} \tag{5.7}$$

From a first-order approximation the numerical solution algorithm becomes

$$\begin{aligned} x_0 &= 0, \qquad x_1 = 0, \\ x_m &= \frac{h^2(u_m - Kx_{m-1}) + M(2x_{m-1} - x_{m-2})}{M + K_q h^{(2-q)}} \\ &\quad + \frac{-K_q h^{(2-q)} \sum_{j=1}^{m} w_j^{(q)} x_{m-j}}{M + K_q h^{(2-q)}} \quad \text{for } m = 2, 3, \ldots, \end{aligned} \tag{5.8}$$

which will be used to sort power terms with respect to the derivative/integral order.

5.3.1 *Sorting Power Terms: Generators, Storage, Dissipators*

As given in [13] and Sect. 4.3, the PID controller contains all three types of terms: proportional feedback is storage, integral feedback is a generator, and derivative feedback is a dissipator. The mass–spring–damper system in Fig. 4.4 provides an illustrative example. The linear equations of motion are given as

$$M\ddot{x} + C\dot{x} + Kx = u$$

and with the PID control given as

$$u = -K_P x - K_I \int_0^t x\,d\tau - K_D\dot{x}.$$

The Hamiltonian is defined as

$$\mathcal{H} = \frac{1}{2}M\dot{x}^2 + \frac{1}{2}Kx^2 + \frac{1}{2}K_P x^2,$$

and taking the derivative yields

$$\dot{\mathcal{H}} = [M\ddot{x} + (K + K_P)x]\dot{x}$$

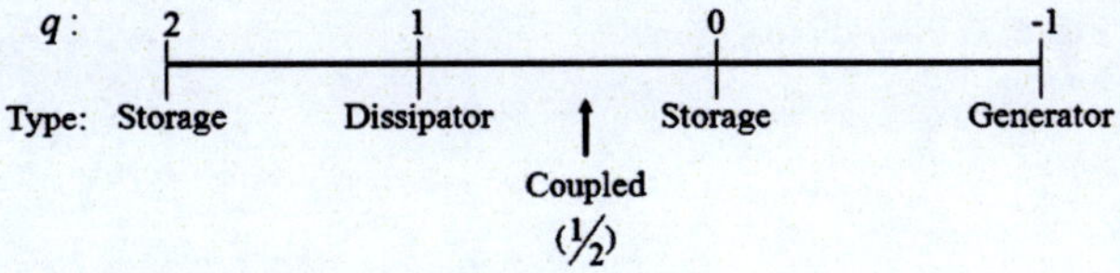

Fig. 5.2 Sorting power terms according to derivative order

$$= \left[-K_I \int_0^t x \, d\tau - K_D \dot{x} - C\dot{x} \right] \dot{x},$$

where the power terms are sorted as

$$\text{Generator:} \quad \dot{W} = -K_I \int_0^t x \, d\tau \dot{x},$$

$$\text{Dissipator:} \quad T_o \dot{S}_i = [K_D + C] \dot{x}^2,$$

$$\text{Storage:} \quad \left(T_o \dot{S}_{\text{rev}} \right)_{\text{ave}} = \int_0^{\tau_c} [M\ddot{x} + (K + K_P)x] \dot{x} \, dt = 0.$$

In terms of a differintegral with $K_q > 0$, from (5.7) $q = 0$ is proportional feedback, $q = -1$ is integral feedback, and $q = 1$ is derivative feedback. Also, the term $M\ddot{x}$ is a storage term with $q = 2$.

$$\text{Generator:} \quad \frac{d^{-1}x}{dt^{-1}},$$

$$\text{Dissipator:} \quad \frac{d^1 x}{dt^1},$$

$$\text{Storage:} \quad \frac{d^2 x}{dt^2} \quad \text{and} \quad \frac{d^0 x}{dt^0}.$$

These results are presented in the Fig. 5.2, where the terms are graphically sorted. This diagram clarifies how the three power flow terms are separated by integer derivative order. The noninteger-order derivative terms couple the power flow terms. This demonstrates why a line integral can be used to understand what the power flow term will be on the average even though these terms come from a linear differential equation. This insight provides a tool that can be extended to nonlinear systems to determine what a nonlinear power flow term is on the average.

The responses to each of these q values employ (5.8) and are plotted and compared to a Runge–Kutta solution from MATLAB in Figs. 5.3–5.17. The results of both methods compare very well as expected. Figure 5.18 shows phase plane plots of these three cases. This implies that integer-order differintegrals can be sorted into the three categories.

What happens when fractional-order differintegrals are used? A $q = 1/2$ produces a storage/dissipator hybrid which is similar to a PD controller with one term instead of two. A $q = -1/2$ produces a storage/generator hybrid which is similar

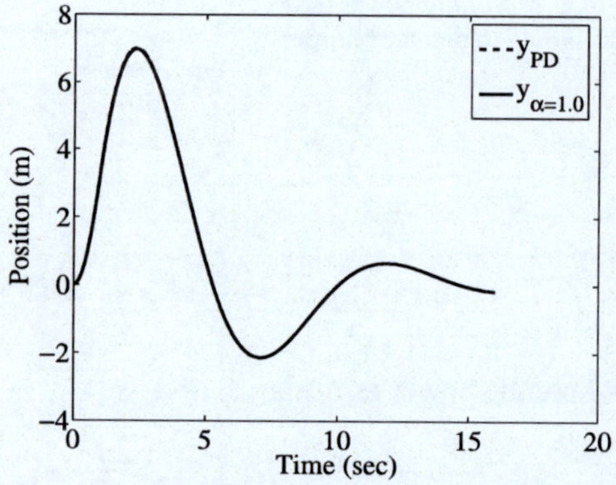

Fig. 5.3 Position responses for $\alpha = 1.0$

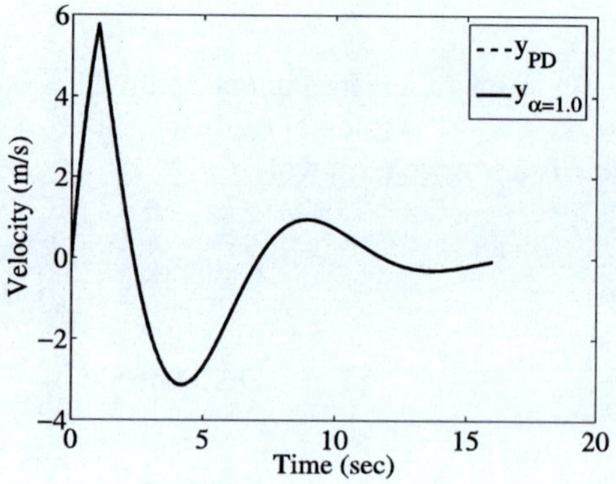

Fig. 5.4 Velocity responses for $\alpha = 1.0$

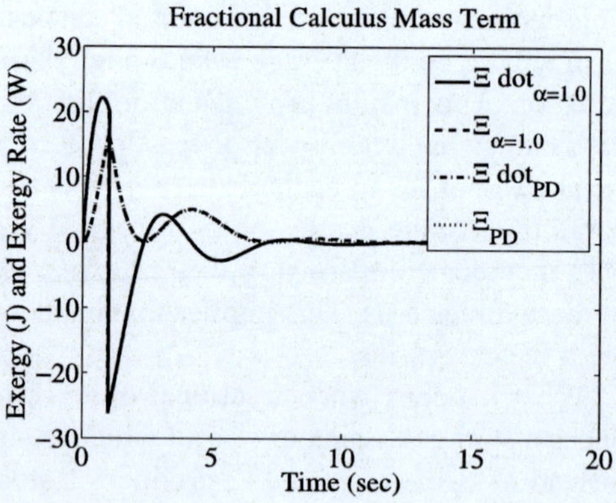

Fig. 5.5 Fractional calculus mass term responses for $\alpha = 1.0$

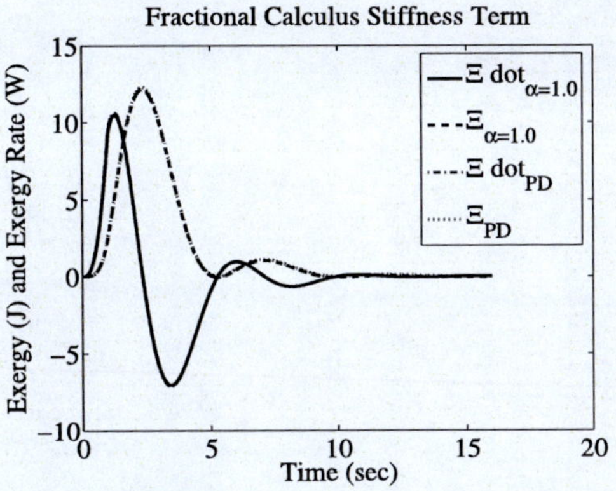

Fig. 5.6 Fractional calculus stiffness term responses for $\alpha = 1.0$

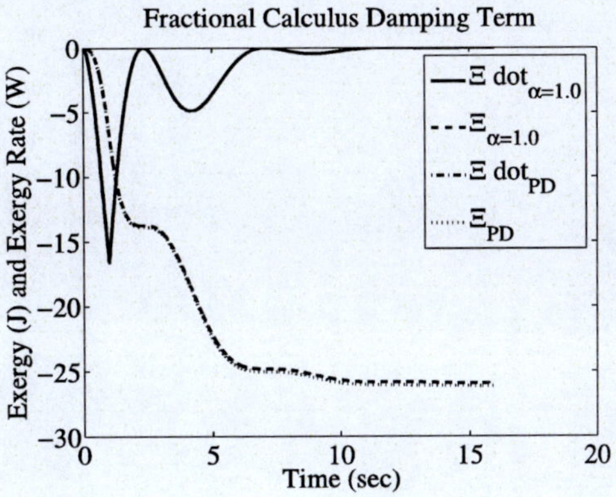

Fig. 5.7 Fractional calculus damping term responses for $\alpha = 1.0$

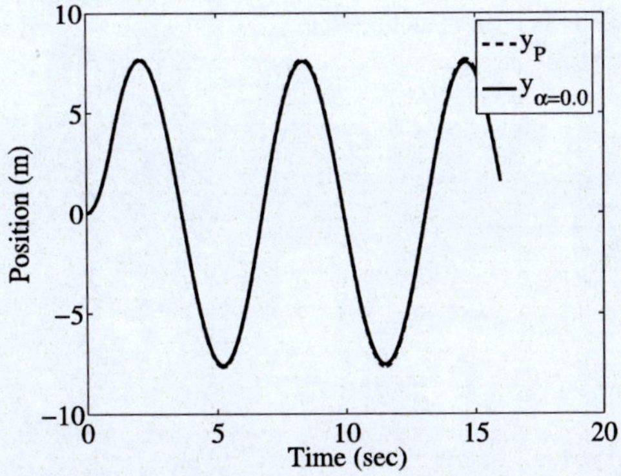

Fig. 5.8 Position responses for $\alpha = 0.0$

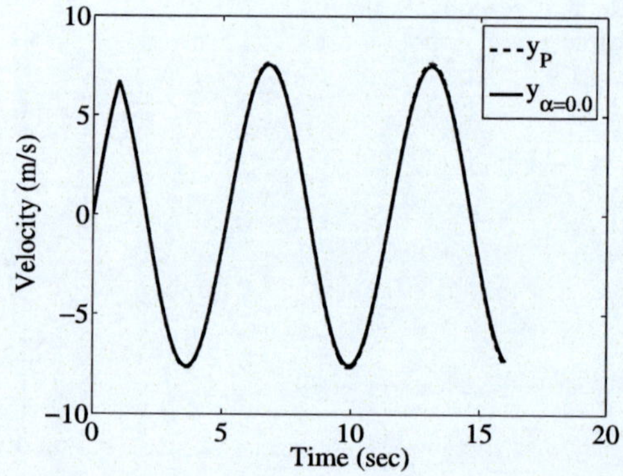

Fig. 5.9 Velocity responses for $\alpha = 0.0$

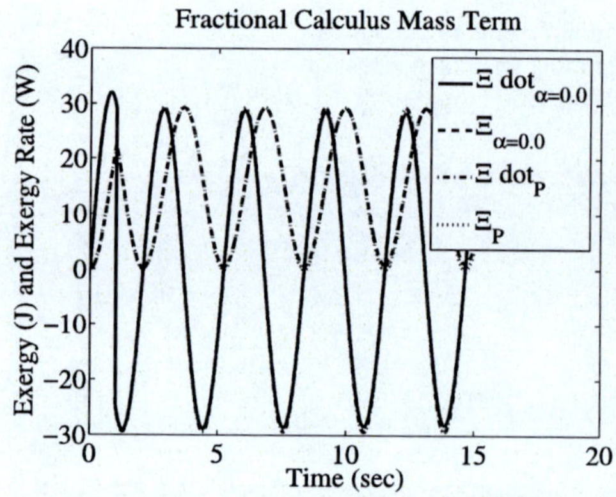

Fig. 5.10 Fractional calculus mass term responses for $\alpha = 0.0$

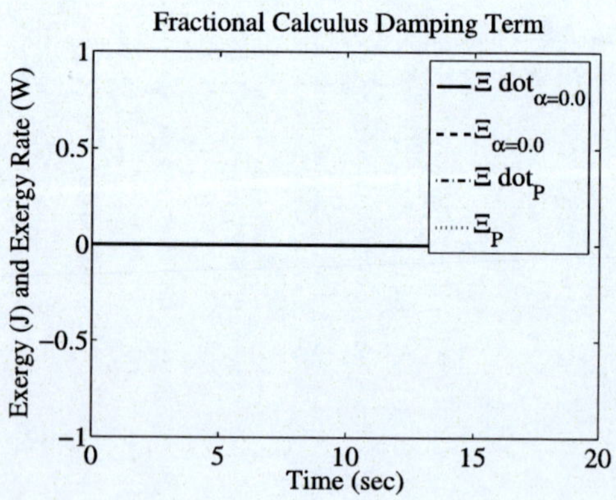

Fig. 5.11 Fractional calculus stiffness term responses for $\alpha = 0.0$

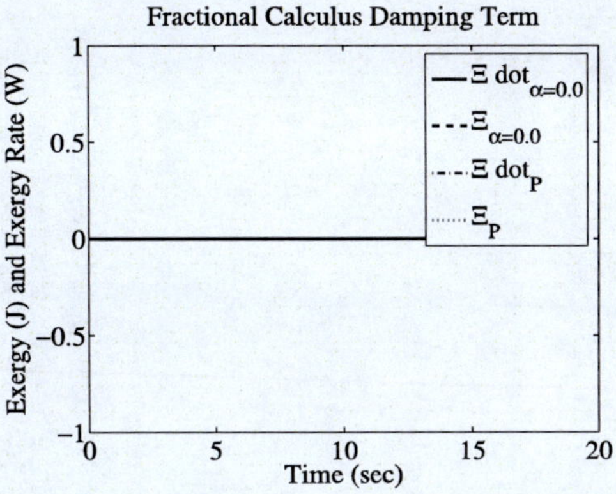

Fig. 5.12 Fractional calculus damping term responses for $\alpha = 0.0$

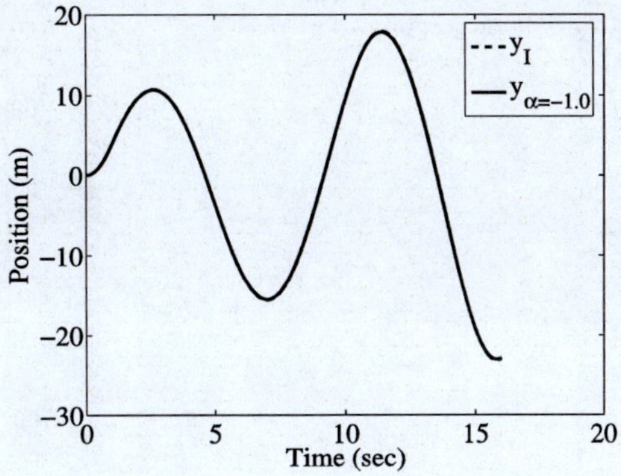

Fig. 5.13 Position responses for $\alpha = -1.0$

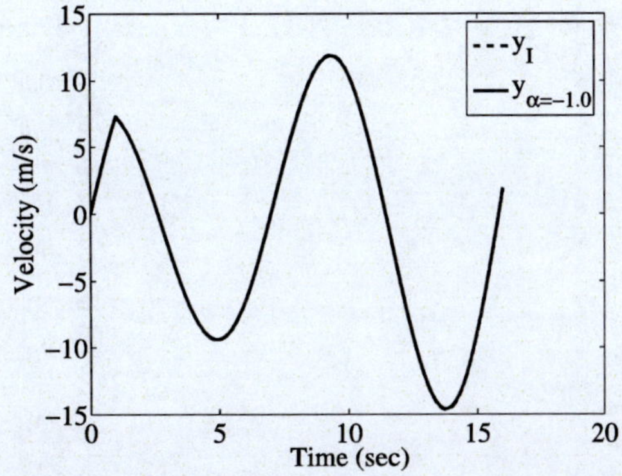

Fig. 5.14 Velocity responses for $\alpha = -1.0$

Fig. 5.15 Fractional calculus mass term responses for $\alpha = -1.0$

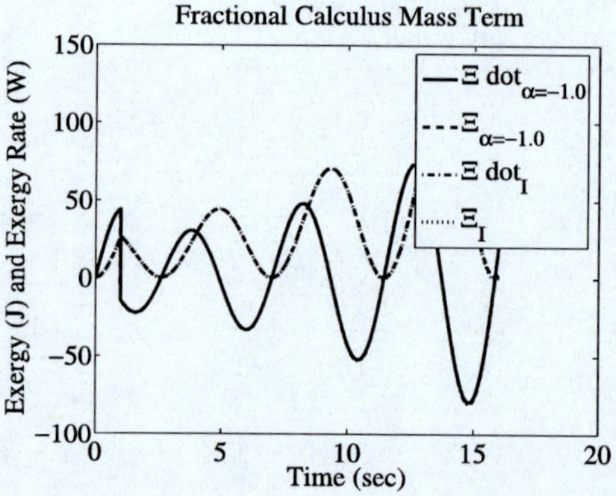

Fig. 5.16 Fractional calculus stiffness term responses for $\alpha = -1.0$

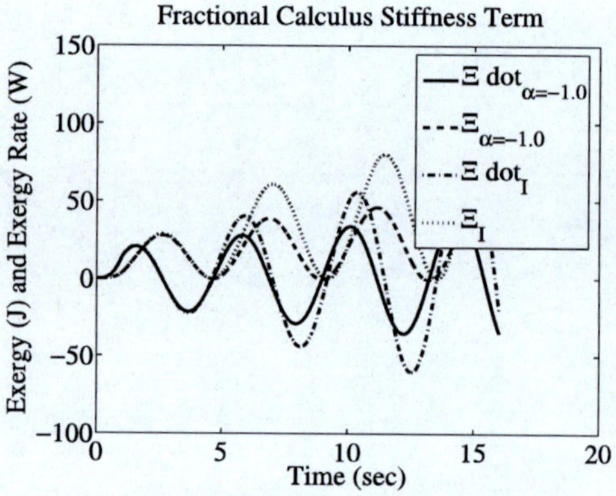

Fig. 5.17 Fractional calculus damping term responses for $\alpha = -1.0$

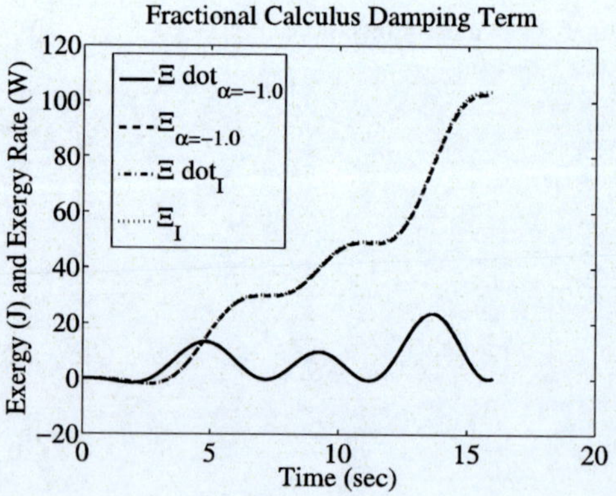

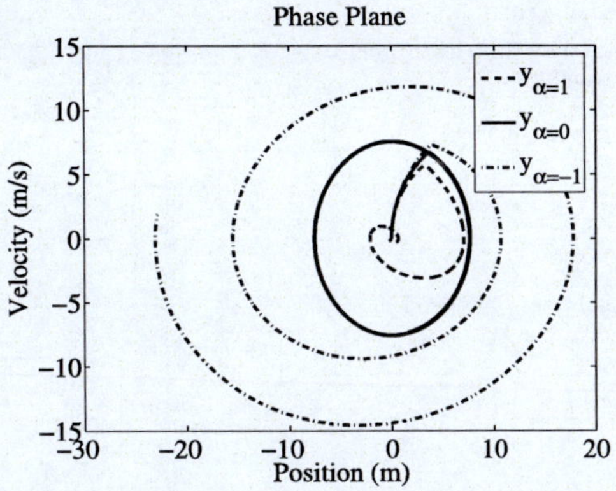

Fig. 5.18 Fractional calculus phase plane responses for $\alpha = 1, 0, -1$

to a PI controller. A $q = 3/2$ produces a storage/dissipator hybrid which is similar to acceleration plus derivative feedback controller. Now it is time to compare performances.

5.3.2 Compare Performance: PID, Lag-Stabilized, Fractional

A good place to look for a hybrid power term is with a lag-stabilized controller [57]. In [57], delayed positive position feedback was shown to stabilize a second-order system given as

$$M\ddot{x} + Kx = u(t) = K_P x(t - \tau) \quad \text{for } K_p > 0,\ \tau > 0.$$

Basically, the delay is phase-shifting the sinusoidal position to produce positive position and negative velocity feedback. This controller generates positive position feedback which reduces the overall system stiffness that shows up in Fig. 5.19 as a lower frequency for the varying time lag scenario (left plot). This controller also generates negative damping as can be seen in the decaying response. For both scenarios presented in Fig. 5.19 and Table 5.1, varying time lag and varying gain, the lag-stabilized study involved a baseline neutrally stable mass–spring (no damping or feedback control) response for comparison. In each scenario, three cases were investigated relative to the baseline case by varying either the time lag value or the gain value to give neutral-dissipative-neutral-generative numerical responses. The mass–spring parameters were set to $M = 1.0$ kg and $K = 0.5$ N/m with the other numerical values used for each scenario given in Table 5.1. The numerical simulation results for the lag-stabilized control are illustrated for all three cases and for both varying time lag in Fig. 5.19 and varying gain in Fig. 5.20 from neutral-dissipative-neutral-generative, respectively. Figure 5.19 presents the effects of varying the lag while fixing the gain at a statically stable value. The positive position feedback becomes negative position feedback as the delay is increased as can be seen by the increasing frequency. The negative velocity feedback becomes positive

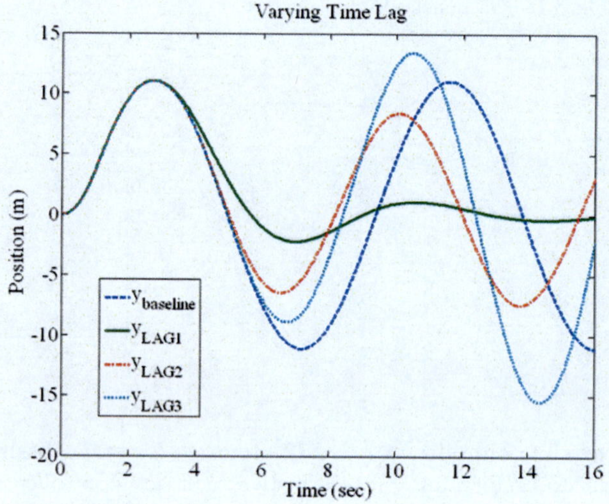

Fig. 5.19 Lag stabilized: varying time lag numerical results

Table 5.1 Numerical values given for each scenario and cases investigated

Scenario description	Case No.	K_P (N/m)	τ (s)
Varying time lag	1	0.25	2.22
Varying time lag	2	0.25	3.60
Varying time lag	3	0.25	4.44
Varying gain	1	0.25	2.22
Varying gain	2	0.50	2.22
Varying gain	3	0.55	2.22

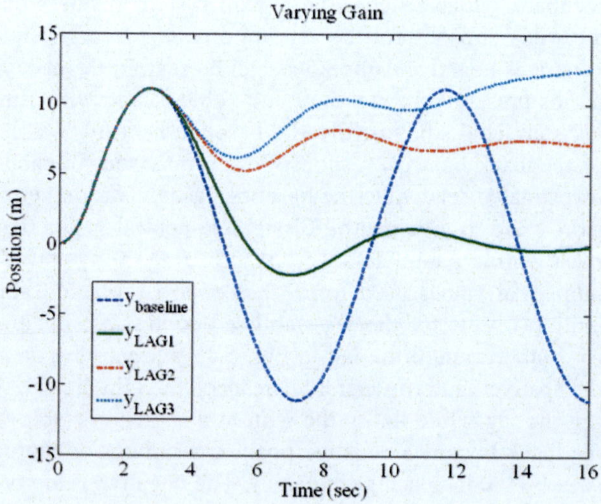

Fig. 5.20 Lag stabilized: varying gain numerical results

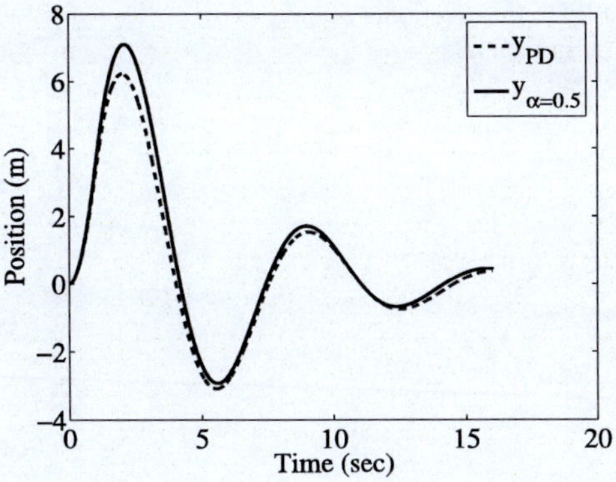

Fig. 5.21 Position responses for $\alpha = 1/2$

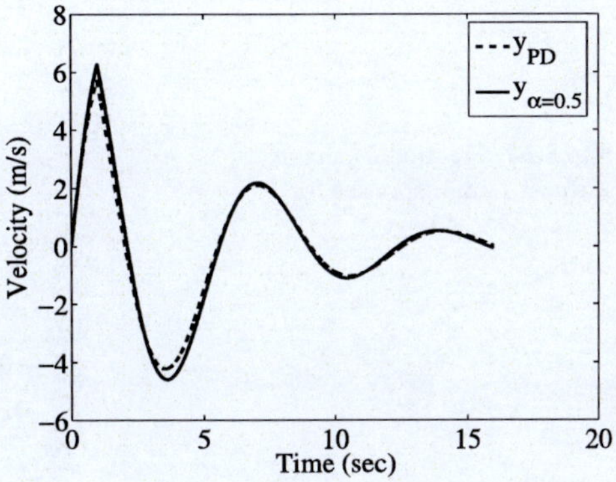

Fig. 5.22 Velocity responses for $\alpha = 1/2$

velocity feedback as the response becomes dynamically unstable. Figure 5.20 illustrates the effect of varying the gain while fixing the lag to produce positive position feedback. The system becomes statically unstable as the gain is increased due to the stiffness becoming negative. The oscillation about a fixed bias is due to the lag in the feedback signal.

The differintegral feedback term provides a "tunable" hybrid term. Setting $q = 1/2$ produces a type of PD controller as can be seen in Figs. 5.21–5.25. The match is not perfect since the stiffness and damping are coupled together [65]. However, it provides a good idea of the amounts of stiffness and damping as well as how much the Hamiltonian surface is deformed due to the differintegral feedback.

Setting $q = -1/2$ produces a type of PI controller as can be viewed in Figs. 5.26–5.30. Again, the match is not perfect, but it provides some valuable insight into the overall system performance due to differintegral feedback. The differintegral provides a unique ability to describe the power flow terms including hybrid terms in a single formula.

One final point, the operation of the lag-stabilized controller is fundamentally different than the fractional controller. The lag-stabilized controller generates positive position and negative velocity feedback. The fractional controller generates both negative position and negative velocity feedback. Consequently, the fractional

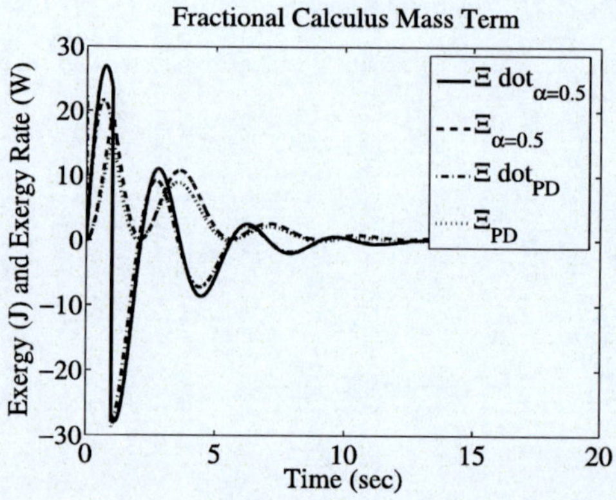

Fig. 5.23 Fractional calculus mass term responses for $\alpha = 1/2$

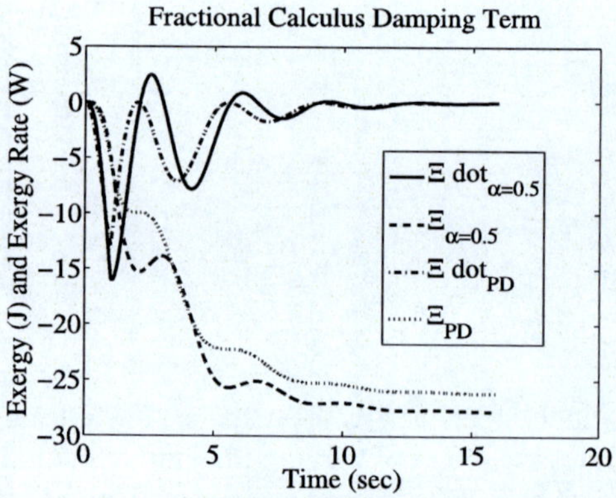

Fig. 5.24 Fractional calculus stiffness term responses for $\alpha = 1/2$

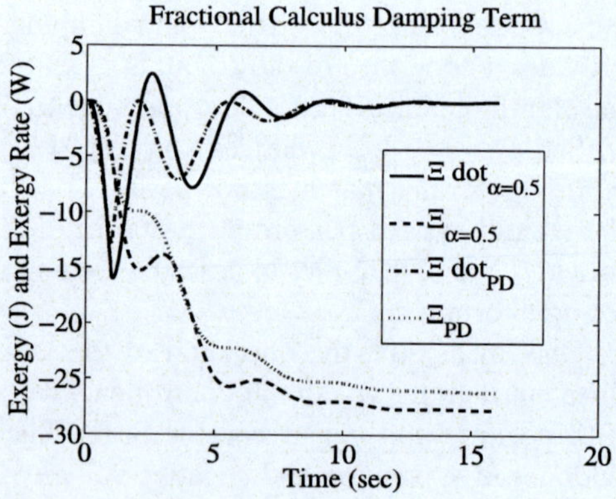

Fig. 5.25 Fractional calculus damping term responses for $\alpha = 1/2$

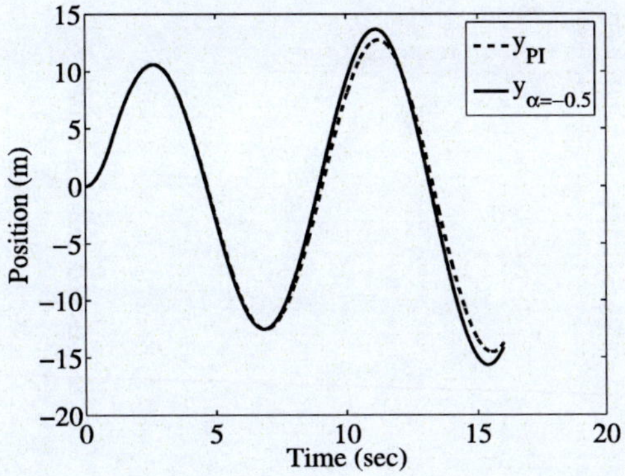

Fig. 5.26 Position responses for $\alpha = -1/2$

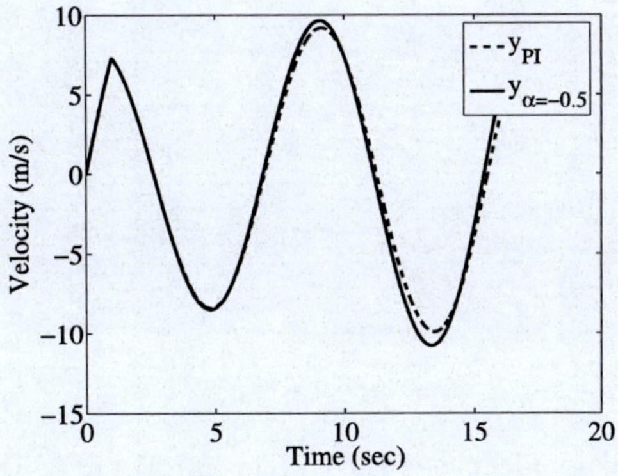

Fig. 5.27 Velocity responses for $\alpha = -1/2$

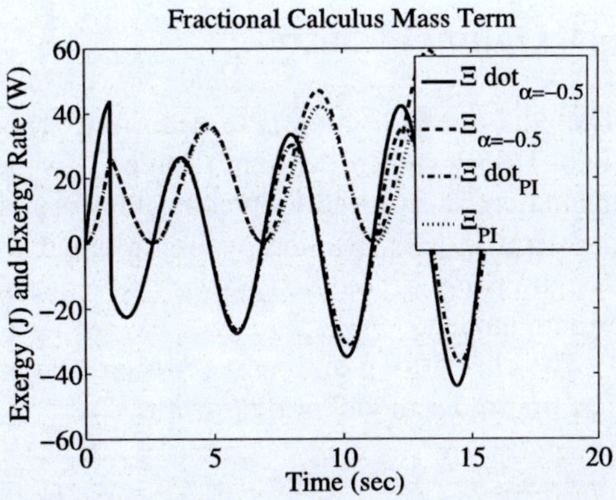

Fig. 5.28 Fractional calculus mass term responses for $\alpha = -1/2$

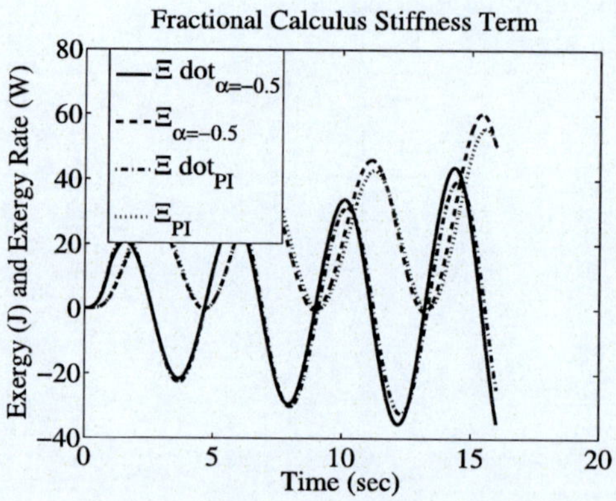

Fig. 5.29 Fractional calculus stiffness term responses for $\alpha = -1/2$

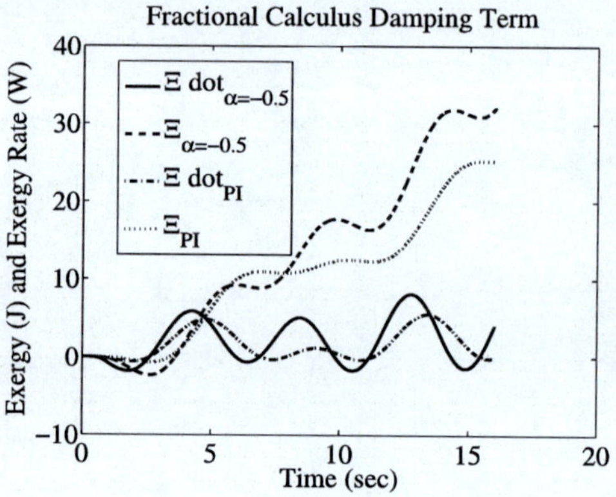

Fig. 5.30 Fractional calculus damping term responses for $\alpha = -1/2$

controller can create a PID controller in two terms instead of three as

$$u = -K_P \frac{d^{1/2}x}{dt^{1/2}} - K_I \frac{d^{-1/2}x}{dt^{-1/2}}.$$

5.4 Optimal Control

The goal of this section is to determine when HSSPF controllers are optimal. Optimal HSSPF controllers are designed for both open-loop and closed-loop implementation. The closed-loop controllers are inspired by a discussion and homework problems in [32]. Homotopy is combined with dynamic programming [61, 62] to design open-loop optimal power flow controllers that are nonlinear even for linear system models.

The closed-loop optimal controller design follows the discussion in Ogata [32]. The first-order model of the system is

$$\dot{\mathbf{x}} = \mathbf{f}(\mathbf{x}, \mathbf{u}), \quad \mathbf{u} = \mathbf{g}(\mathbf{x}) \quad \Rightarrow \quad \dot{\mathbf{x}} = \mathbf{f}\big(\mathbf{x}, \mathbf{g}(\mathbf{x})\big),$$

which is assumed to be asymptotically stable to $\mathbf{x} = \mathbf{0}$.

The first step is to choose a cost functional and assume a Lyapunov function to create the Control Hamiltonian Function for the optimal control formulation. The cost functional has an infinite final time which leads to a feedback regulator solution

$$\begin{aligned} J &= \int_0^\infty L(\mathbf{x}, \mathbf{u})\, dt, \\ \tilde{\mathcal{H}}(\mathbf{x}, \mathbf{u}) &= \dot{V} + L(\mathbf{x}, \mathbf{u}) \geq 0, \\ \min_{\mathbf{u}} \tilde{\mathcal{H}}(\mathbf{x}, \mathbf{u}) &= \min_{\mathbf{u}} \big[\dot{V} + L(\mathbf{x}, \mathbf{u})\big] = \dot{V}\big|_{\mathbf{u}=\mathbf{u}_1} + L(\mathbf{x}, \mathbf{u}_1) = 0. \end{aligned}$$

These optimal conditions and constraints lead to

$$\dot{V}\big|_{\mathbf{u}=\mathbf{u}_1} = -L(\mathbf{x}, \mathbf{u}_1).$$

Note that this relationship between the cost functional and the Lyapunov function leads to

$$\begin{aligned} V\big(\mathbf{x}(\infty)\big) &= 0 = V\big(\mathbf{x}(0)\big) - \int_0^\infty L\big(\mathbf{x}(t), \mathbf{u}_1(t)\big)\, dt, \\ V\big(\mathbf{x}(0)\big) &= \int_0^\infty L\big(\mathbf{x}(t), \mathbf{u}_1(t)\big)\, dt. \end{aligned}$$

Next, a couple of examples are examined to determine when a closed-loop HSSPFC controller is optimal.

Example #1 is a single DOF mass–spring system with optimal power flow control. The equation of motion becomes

$$m\ddot{x} + kx = u.$$

The cost functional is selected as

$$J = \int_0^\infty L(\dot{x}, x, u, t)\, dt = \int_0^\infty \big[k_1 P^2 + k_2 u^2\big]\, dt = \int_0^\infty \big[k_1 u^2 \dot{x}^2 + k_2 u^2\big]\, dt,$$

where P is the power defined as $P = u\dot{x}$, and k_1, k_2 are optimization weights and/or controller gains. The following candidate Lyapunov function uses the Hamiltonian:

$$V = \mathcal{H} = \frac{1}{2} m\dot{x}^2 + \frac{1}{2} kx^2$$

with the corresponding derivative determined as

$$\begin{aligned} \dot{V} &= [m\ddot{x} + kx]\dot{x} = [u\dot{x}] \\ &= -L = -\big[k_1 u^2 \dot{x}^2 + k_2 u^2\big] = u\dot{x}. \end{aligned}$$

Setting the bracketed term to zero

$$u\big[u\big(k_1 \dot{x}^2 + k_2\big) + \dot{x}\big] = 0$$

and solving for the nonzero optimal controller yields

$$u = -\dot{x}/\left[k_1\dot{x}^2 + k_2\right],$$

which gives

$$m\ddot{x} + \dot{x}/\left[k_1\dot{x}^2 + k_2\right] + kx = 0.$$

Example #2 is a mass–spring system with optimal feedback control from a quadratic cost functional. The equation of motion is

$$m\ddot{x} + kx = u.$$

The cost functional is

$$J = \int_0^\infty \left[k_1\dot{x}^2 + k_2x^2 + k_3u^2\right] dt,$$

where the gains are arbitrary. This cost functional leads to a constraint on the Lyapunov function as

$$\dot{V} = -L(x, u) = -k_1\dot{x}^2 - k_2x^2 - k_3u^2.$$

The Lyapunov function is chosen to be

$$V = \frac{1}{2}m\dot{x}^2 + \frac{1}{2}kx^2,$$

which produces the first time derivative as

$$\dot{V} = [m\ddot{x} + kx]\dot{x} = u\dot{x},$$

and equating leads to

$$-u\dot{x} = k_1\dot{x}^2 + k_2x^2 + k_3u^2.$$

Next, pick values for the arbitrary gains k_1, k_2, and k_3, to obtain an optimal closed-loop controller as

$$k_1 = k_2 = 0, \qquad u(k_3u + \dot{x}) = 0,$$

then the optimal controller becomes

$$u = -\frac{1}{k_3}\dot{x}.$$

If instead, all the gains are selected to be nonzero, then the optimal controller becomes

$$u_{1,2} = \frac{1}{2k_3}\left[-\dot{x} \pm \left(\dot{x}^2 - 4k_3\left(k_1\dot{x}^2 + k_2x^2\right)\right)^{1/2}\right].$$

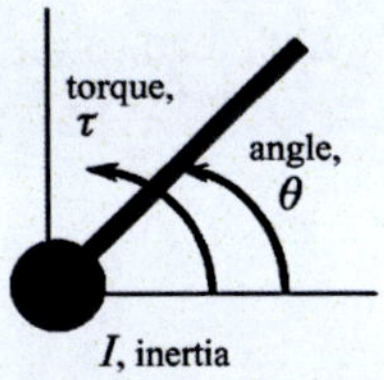

Fig. 5.31 Horizontal slewing link

The open-loop optimal controller design is for a fixed final time and a cost functional that includes squared power flow input (minimum power input) and squared control input (minimum control effort). A combination of homotopy and dynamic programming [61, 62] are used to design open-loop optimal power flow controllers that are nonlinear even for linear system models due to the fourth-order cost functional. The system is a single horizontal slewing link shown in Fig. 5.31 with an equation of motion as

$$I\ddot{\theta} = \tau = u$$

with a first-order model as

$$\dot{\mathbf{x}} = \mathbf{f}(\mathbf{x}, \mathbf{u}), \quad \mathbf{x} = \left[\theta \; \dot{\theta}\right]^T.$$

The fixed final time cost functional is

$$J = \int_0^{t_f} \frac{1}{2}\left[(ux_2)^2 + u^2\right] dt,$$

which produces an optimal control Hamiltonian as

$$\tilde{\mathcal{H}} = L + \boldsymbol{\lambda}^T \mathbf{f},$$

where λ here is a vector. The detailed first-order differential equations are

$$\dot{\mathbf{x}} = \begin{Bmatrix} \dot{x}_1 \\ \dot{x}_2 \end{Bmatrix} = \begin{Bmatrix} \dot{\theta} \\ \ddot{\theta} \end{Bmatrix} = \begin{Bmatrix} x_2 \\ u/I \end{Bmatrix} = \begin{bmatrix} 0 & 1 \\ 0 & 0 \end{bmatrix} \begin{Bmatrix} x_1 \\ x_2 \end{Bmatrix} + \begin{Bmatrix} 0 \\ 1/I \end{Bmatrix} u = \mathbf{A}\mathbf{x} + \mathbf{B}u,$$

which lead to a detailed optimal control Hamiltonian as

$$\tilde{\mathcal{H}} = \frac{1}{2}(ux_2)^2 + \frac{1}{2}u^2 + \lambda_1 x_2 + \lambda_2 \frac{u}{I}.$$

The first variation conditions are

$$0 = \frac{\partial \tilde{\mathcal{H}}}{\partial u} = ux_2^2 + u + \frac{1}{I}\lambda_2 \quad \Rightarrow \quad u = -\lambda_2 / I(x_2^2 + 1)$$

and

$$\begin{Bmatrix} \dot{\lambda}_1 \\ \dot{\lambda}_2 \end{Bmatrix} = \begin{Bmatrix} -\partial \tilde{\mathcal{H}}/\partial x_1 \\ -\partial \tilde{\mathcal{H}}/\partial x_2 \end{Bmatrix} = \begin{Bmatrix} 0 \\ -u^2 x_2 - \lambda_1 \end{Bmatrix},$$

which lead to the following nonlinear two-point boundary-value problem:

$$
\begin{aligned}
u &= -\lambda_2/I\left(x_2^2+1\right),\\
\lambda_1 &= \lambda_{1_0} = \text{constant},\\
\dot{\lambda}_2 &= -u^2 x_2 - \lambda_{1_0},\\
\dot{x}_1 &= x_2,\\
\dot{x}_2 &= \frac{1}{I}u = -\lambda_2/I^2\left(x_2^2+1\right),
\end{aligned}
$$

with

$$
\begin{aligned}
x_1(0) &= x_2(0) = 0,\\
x_1(t_f) &= x_{\text{desired}},\\
x_2(t_f) &= 0.
\end{aligned}
$$

This nonlinear two-point boundary-value problem is formulated for the combination of homotopy and dynamic programming as

$$
\begin{aligned}
&J = \phi(\xi) = \frac{1}{2}\int_0^{t_f} \left[W_1\alpha\left(u(\xi)x_2\right)^2 + W_2(1-\alpha)u^2(\xi)\right] dt,\\
&\dot{\mathbf{x}} = \mathbf{Ax} + \mathbf{Bu},\\
&\Psi_1(\xi) = \theta(t_f) - \theta_{\text{desired}} = 0,\\
&\Psi_2(\xi) = \dot{\theta}(t_f) = 0,\\
&\theta(t_0) = \dot{\theta}(t_0) = 0,
\end{aligned}
$$

with α as the homotopy parameter, and W_1 and W_2 as the weight values to show the effects of minimum power flow and minimum control effort. The results presented in Figs. 5.32 and 5.33 show the effects of minimum control effort, of hybrid minimum control effort and minimum power flow, and of minimum power flow. Note the singularity in the minimum power flow control profile which is apparent in the two-point boundary value equations as the minimum control effort is eliminated in the cost functional. In particular, the minimum control effort is the "straight-line" solution for the input torque that is deformed into the "singular" solution as the homotopy parameter is varied from zero to one.

5.5 Robust/Tracking Control

In this section, the basic concepts of robust HSSPFC tracking controllers are developed by determining the uncertainty bounds on the system parameters and assessing how these constraints affect the stability boundaries and performance of the system. Application of these basic concepts will be given in the next section on adaptive HSSPFC tracking controllers. If the system performance including power consumption is being limited by the robust HSSPFC tracking controller, an adaptive

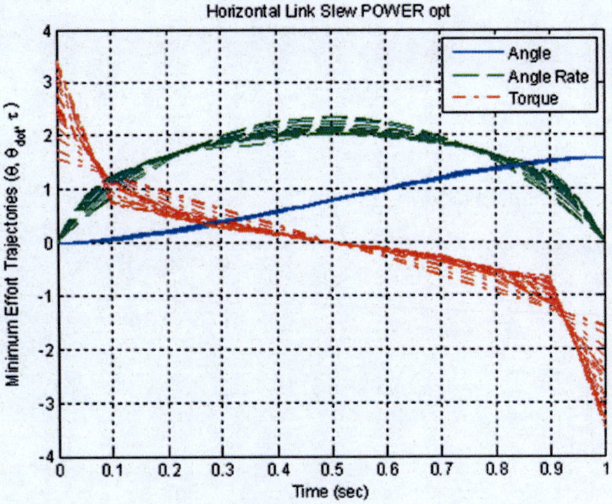

Fig. 5.32 Minimum control effort/power flow responses

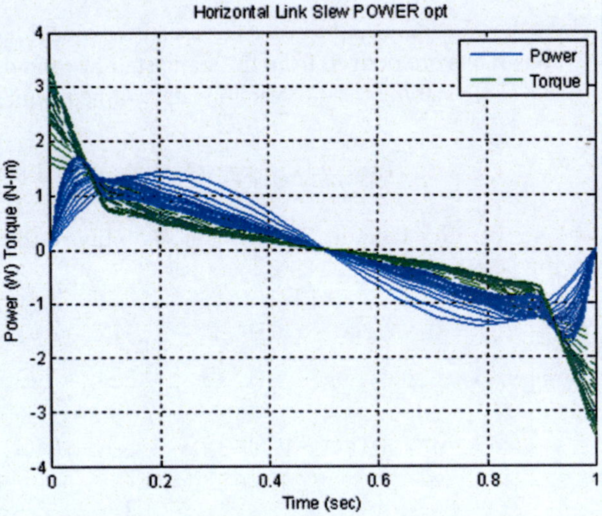

Fig. 5.33 Minimum control effort/power flow torque and power flow responses

HSSPFC tracking controller is designed by adding the appropriate information potential functions to the Hamiltonian. Both designs are compared and contrasted for a power engineering problem (i.e., nonlinear RLC in Chap. 9).

The robust HSSPFC tracking controller design begins with a nonlinear model (see Fig. 4.4)

$$m\ddot{x} + g(x) = -f(\dot{x}) + u, \quad g(x) = \frac{\partial V(x)}{\partial x},$$

which has the Hamiltonian

$$\mathcal{H} = \frac{1}{2}m(\dot{x} - \dot{x}_R)^2 + V(x - x_R).$$

The time derivative of the Hamiltonian is

$$\begin{aligned}\dot{\mathcal{H}} &= [m(\ddot{x} - \ddot{x}_R) + g(x - x_R)](\dot{x} - \dot{x}_R) \\ &= [-m\ddot{x}_R + g(x - x_R) - g(x) - f(\dot{x}) + u](\dot{x} - \dot{x}_R).\end{aligned}$$

The robust HSSPFC controller is defined as

$$u = u_{\text{ref}} + \Delta u$$

with

$$u_{\text{ref}} = \hat{m}\ddot{x}_R - \hat{g}(x - x_R) + \hat{g}(x) + \hat{f}(\dot{x}), \quad \hat{g}(x) = \frac{\partial \hat{V}(x)}{\partial x},$$

and

$$\Delta u = -\hat{g}(x - x_R) - K_I \int_0^t (x - x_R)\, d\tau - K_D(\dot{x} - \dot{x}_R),$$

where the hats indicate the best estimate of the model parameters. The robust stability constraints are derived from the static stability and dynamic stability conditions.

The static stability of the system is determined from the Hamiltonian as

$$\hat{\mathcal{H}} = \frac{1}{2}m(\dot{x} - \dot{x}_R)^2 + \hat{V}(x - x_R) + V(x - x_R),$$

which is positive definite. The dynamic stability of the system is determined from the time derivative of the Hamiltonian as

$$\begin{aligned}\dot{\hat{\mathcal{H}}} &= \big[m(\ddot{x} - \ddot{x}_R) + g(x - x_R) + \hat{g}(x - x_R)\big](\dot{x} - \dot{x}_R) \\ &= \big[-m\ddot{x}_R + g(x - x_R) + \hat{g}(x - x_R) - g(x) - f(\dot{x}) + u\big](\dot{x} - \dot{x}_R) \\ &= \Bigg[(\hat{m} - m)\ddot{x}_R + \big(g(x - x_R) - \hat{g}(x - x_R)\big) + \big(\hat{g}(x) - g(x)\big) + \big(\hat{f}(\dot{x}) - f(x)\big) \\ &\quad - K_I \int_0^t (x - x_R)\, d\tau - K_D(\dot{x} - \dot{x}_R)\Bigg](\dot{x} - \dot{x}_R).\end{aligned}$$

The robustness constraint takes the form of

$$\begin{aligned}\big\{K_D(\dot{x} - \dot{x}_R)^2\big\}_{\text{ave}} > \Bigg\{\Bigg[&-K_I \int_0^t (x - x_R) d\tau + (\hat{m} - m)\ddot{x}_R \\ &+ \big(\hat{g}(x - x_R) - g(x - x_R)\big) + \big(\hat{g}(x) - g(x)\big) + \big(\hat{f}(\dot{x}) - f(\dot{x})\big)\Bigg](\dot{x} - \dot{x}_R), \Bigg\}_{\text{ave}}\end{aligned}$$

which ensures dynamic stability. The next section on adaptive HSSPFC controller design will demonstrate the stability and performance of robust HSSPFC controllers by comparing and contrasting the two designs.

5.6 Adaptive/Tracking Control

The next natural extension is to investigate when the parameters are adapted or adaptive control as physical and information exergies. The same example is linearized and used to compare and contrast robust and adaptive HSSPFC controllers. A nonlinear example is demonstrated in more detail in Chap. 9. Once again, we start with the mass–spring–damper (with a PID tracking controller) problem (also see Fig. 4.4)

$$m\ddot{x} + kx = -c\dot{x} + u$$

with the Hamiltonian given as

$$\begin{aligned}\mathcal{H} &= \mathcal{T} + \mathcal{V} + \mathcal{V}_C + \mathcal{V}_I \\ &= \frac{1}{2}m(\dot{x} - \dot{x}_r)^2 + \frac{1}{2}k(x - x_r)^2 + \frac{1}{2}K_P(x - x_r)^2 + \frac{1}{2}\tilde{\Phi}^T \Gamma^{-1}\tilde{\Phi},\end{aligned}$$

where the last two terms are associated with the exergy potential, $\mathcal{V}_c$, feedback control term and the adaptive parameter estimation terms or the information exergy potential, $\mathcal{V}_I$,

$$\begin{aligned}\mathcal{V}_c &= \frac{1}{2}K_p(x - x_r)^2, \\ \mathcal{V}_I &= \frac{1}{2}\tilde{\Phi}^T \Gamma^{-1}\tilde{\Phi},\end{aligned}$$

which are also discussed in Chap. 7. In this case, the difference between the estimated and "true" parameters makes up the information flow. The goal is to drive the estimated parameters to the "true" parameters for which specific performance criteria can then be established. Next, the Hamiltonian rate becomes

$$\begin{aligned}\dot{\mathcal{H}} &= [m(\ddot{x} - \ddot{x}_r) + (k + K_P)(x - x_r)](\dot{x} - \dot{x}_r) + \tilde{\Phi}^T \Gamma^{-1}\dot{\tilde{\Phi}} \\ &= [-m\ddot{x}_r - kx_r + K_P(x - x_r) - c\dot{x} + u](\dot{x} - \dot{x}_r) + \tilde{\Phi}^T \Gamma^{-1}\dot{\tilde{\Phi}}. \qquad (5.9)\end{aligned}$$

The controller is selected as

$$u = u_{\text{ref}} + \Delta u = u_{\text{ref}} + u_P + u_G + u_D, \qquad (5.10)$$

where

$$\begin{aligned}u_{\text{ref}} &= \hat{m}\ddot{x}_r + \hat{k}x_r + \hat{c}\dot{x}, \\ \Delta u &= u_P + u_G + u_D, \\ u_P &= -K_P(x - x_r), \\ u_G &= -K_I\int_0^t (x - x_r)\,d\tau, \\ u_D &= -K_D(\dot{x} - \dot{x}_r).\end{aligned}$$

Note that parameters with a hat (ˆ) represent the estimate of the true parameter. Substituting (5.10) into (5.9) yields

$$\dot{\mathcal{H}} = \left[(\hat{m} - m)\ddot{x}_r + (\hat{k} - k)x_r + (\hat{c} - c)\dot{x} - K_I \int_0^t (x - x_r)\,d\tau - K_D(\dot{x} - \dot{x}_r) \right](\dot{x} - \dot{x}_r) + \tilde{\Phi}^T \Gamma^{-1} \dot{\tilde{\Phi}}. \tag{5.11}$$

Next identify and set

$$(\hat{m} - m)\ddot{x}_r + (\hat{k} - k)x_r + (\hat{c} - c)\dot{x} = Y\tilde{\Phi}. \tag{5.12}$$

Then substituting (5.12) into (5.11) and simplifying gives

$$\dot{\mathcal{H}} = \left[-K_I \int_0^t (x - x_r)\,d\tau - K_D(\dot{x} - \dot{x}_r) \right](\dot{x} - \dot{x}_r) + \tilde{\Phi}^T \left[Y^T(\dot{x} - \dot{x}_r) + \Gamma^{-1}\dot{\tilde{\Phi}} \right], \tag{5.13}$$

where

$$\begin{aligned} \tilde{\Phi}^T &= \left[(\hat{m} - m)\ (\hat{k} - k)\ (\hat{c} - c) \right], \\ \dot{\tilde{\Phi}}^T &= \dot{\hat{\Phi}}^T = \left[\dot{\hat{m}}\ \dot{\hat{k}}\ \dot{\hat{c}} \right], \\ Y &= [\ddot{x}_r\ x_r\ \dot{x}]. \end{aligned}$$

To ensure the information exergy potential is a conservative term, the last term in (5.13) vanishes with

$$\dot{\hat{\Phi}} = -\Gamma Y^T(\dot{x} - \dot{x}_r).$$

This also results in the following adaptive parameter update equations:

$$\begin{aligned} \dot{\hat{m}} &= -\gamma_1 \ddot{x}_r(\dot{x} - \dot{x}_r), \\ \dot{\hat{k}} &= -\gamma_2 x_r(\dot{x} - \dot{x}_r), \\ \dot{\hat{c}} &= -\gamma_3 \dot{x}(\dot{x} - \dot{x}_r), \end{aligned}$$

which concludes in the following passively stable condition:

$$\oint_\tau \left[-K_I \int_0^t (x - x_r)\,d\tau' \right](\dot{x} - \dot{x}_r)\,dt < \oint_\tau \left[K_D(\dot{x} - \dot{x}_r)^2 \right] dt. \tag{5.14}$$

This derivation demonstrates that when the adaptive parameter estimates or information exergies are driven to the "true" parameters, then the system reduces to the actual physical exergy or available storage in the system for which the final constraint (5.14) determines the system performance and passive stability. The inherent trade-off is the additional cost of the adaptive controller hardware and software versus a robust control design.

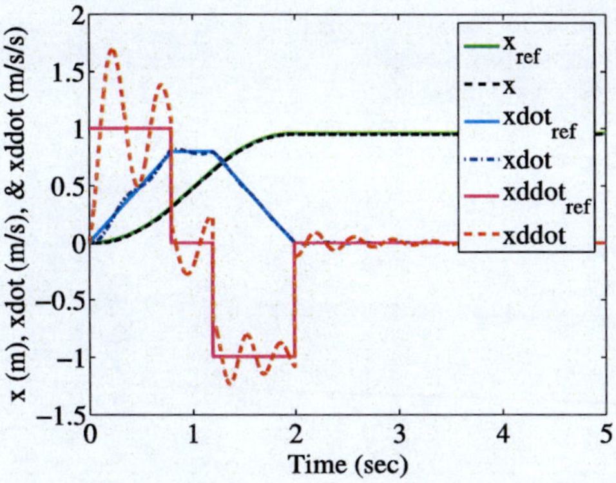

Fig. 5.34 Adaptive control mass–spring–damper responses: position, velocity, and acceleration

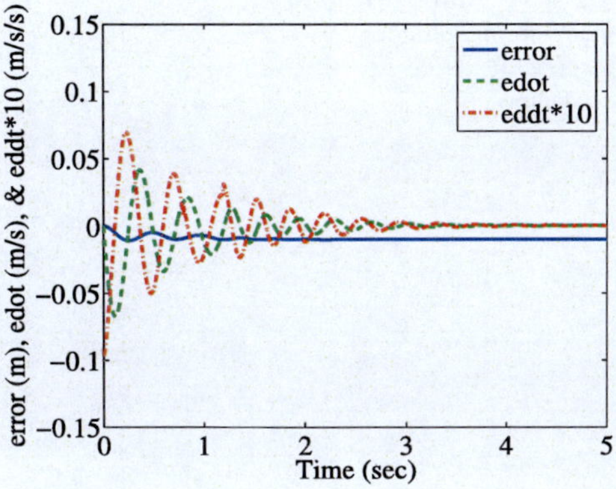

Fig. 5.35 Adaptive control mass–spring–damper responses: position, velocity, and acceleration errors

To summarize the analysis a numerical simulation was performed. The following numerical values were selected for the model: $m = 10$ kg, $k = 10$ N/m, and $c = 1$ N s/m. The control gains were selected as: $K_P = 550$, $K_I = 20$, $K_D = 31$, $\gamma_1 = 1000$, $\gamma_2 = 2500$, and $\gamma_3 = 1000$. The system was initially at rest ($x_0 = 0$, $\dot{x}_0 = 0$). The control system gains were selected to provide a critically damped response and the adaptive gains used to provide quick convergence to the "true" parameters. As with most standard adaptive control, this provides guaranteed stability, but not necessarily converging to the exact "true" parameters. A standard bang-coast-bang acceleration profile was used to generate the reference inputs and provide rich signal content. The position, velocity and acceleration responses are illustrated in Fig. 5.34. The corresponding errors for position, velocity, and acceleration, are also shown in Fig. 5.35. The total input force (u), reference force (u_{ref}), and control force (Δu) are shown in Fig. 5.36. The adaptive estimated parameter responses for $\hat{m}$, $\hat{k}$, and $\hat{c}$ are shown in Fig. 5.37. For the adaptive control mass–spring–damper system, the exergy-rate responses are illustrated in Fig. 5.38 with the corresponding exergy responses in Fig. 5.39. For both responses, the passivity terms are greater (in magnitude) than the generative terms which satisfy the inequality in (5.14).

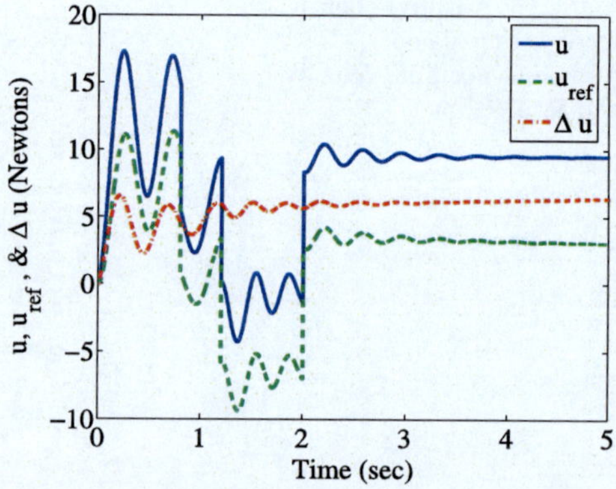

Fig. 5.36 Adaptive control mass–spring–damper responses: input force, reference force, and control force

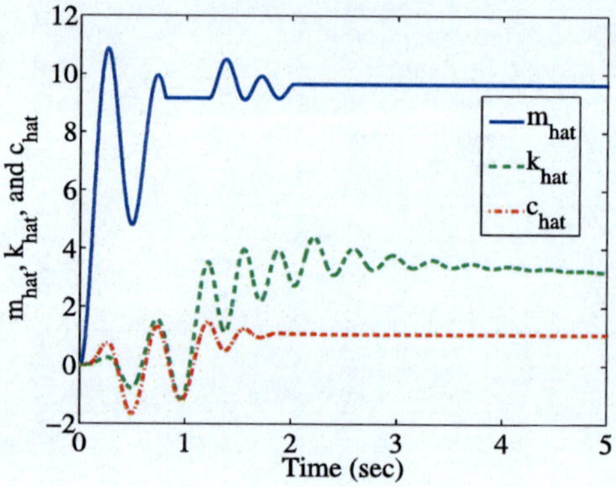

Fig. 5.37 Adaptive control mass–spring–damper responses: adaptive parameter estimates

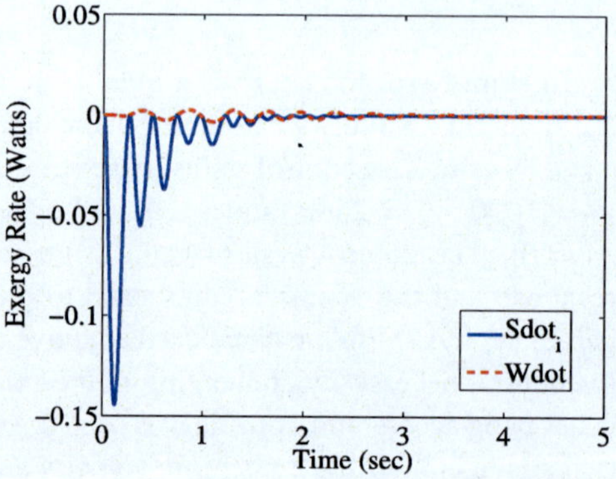

Fig. 5.38 Adaptive control mass–spring–damper responses: exergy-rate

5.7 Chapter Summary

HSSPFC was used to design advanced control laws for distributed parameter systems (PDEs), fractional calculus control, optimal power flow control, robust tracking control, and adaptive tracking control. The design of HSSPFC controllers for distributed parameter systems was shown to be a straightforward extension of the

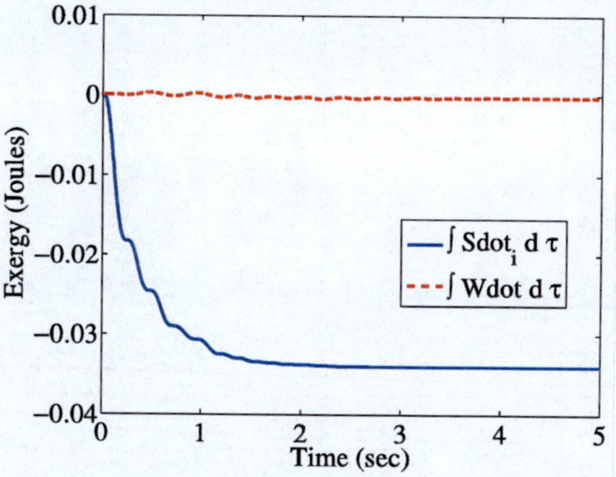

Fig. 5.39 Adaptive control mass–spring–damper responses: exergy

results of the previous chapter by applying calculus of variations to the Extended Hamilton's Principle. Fractional calculus control was presented as an excellent mathematical tool for the sorting of power flows into generators, dissipators, and storage based on the order of the derivative. Optimal HSSPFC controllers were designed for both open-loop and closed loop implementation. Robust HSSPFC tracking controllers were developed by determining the uncertainty bounds on the system parameters and assessing how these constraints affect the stability boundaries and performance of the system. If the system performance including power consumption was limited, an adaptive HSSPFC tracking controller was designed by adding the appropriate information potential functions to the Hamiltonian.

Chapter 6
Case Study #1: Control Design Issues

6.1 Introduction

Many control design issues exist when designing controllers for nonlinear systems. Chapter 6 focuses on a few major issues of interest including single nonlinear feedback terms, MIMO systems, and noncollocated control. A sinusoidal damping/nonlinear feedback term is investigated to show the effect of nonlinear feedback terms that change sign as a function of velocity. Multiple coupled mass spring dampers are used to demonstrate the HSSPFC design process for linear and nonlinear MIMO systems. These coupled mass spring damper systems are used to present the effects of noncollocated control on stability and performance based on HSSPFC. Coupled mass-spring-damper systems were selected for the example MIMO cases because they provide a clear presentation for control volume analysis and *do not* naturally decouple into SISO control designs like gyroscopic systems (such as those investigated in Chaps. 11, 12). Chapter 6 is based on several papers by Robinett and Wilson [7, 12, 34, 37, 46, 50, 66, 67].

Chapter 6 is subdivided into six sections. Section 6.2 introduces a second-order nonlinear system for which multiple limit cycles are identified. Section 6.3 extends the eigenanalysis principle to MIMO nonlinear systems. Section 6.4 investigates controller designs for several two-mass numerical examples for both linear and nonlinear MIMO systems. Section 6.5 investigates noncollocated control designs with numerical simulations. Finally, in Sect. 6.6 the chapter results are summarized with concluding remarks.

6.2 Nonlinear Second-Order System with Sinusoidal Damping

Consider the following nonlinear differential equation with unity mass and stiffness and a nonlinear damping term $\sin(\dot{x})$ which can be formulated as a nonlinear feedback controller $u = \sin(\dot{x})$, to be defined as

$$\ddot{x} - \sin(\dot{x}) + x = 0.$$

R.D. Robinett III, D.G. Wilson, *Nonlinear Power Flow Control Design*, Understanding Complex Systems,
DOI 10.1007/978-0-85729-823-2_6,

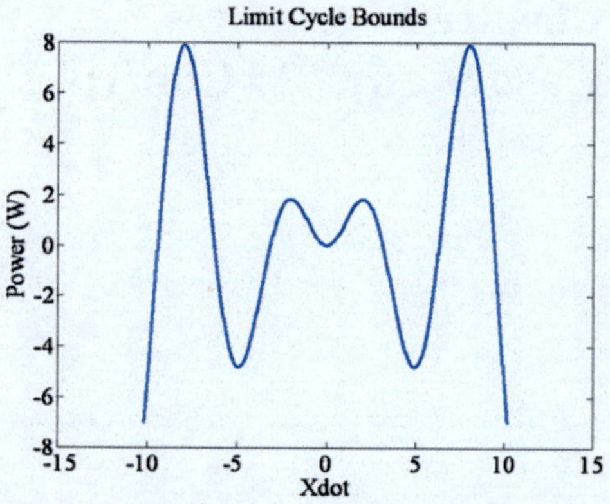

Fig. 6.1 Power flow identifies both limit cycle bounds at $\pm\pi$ and $\pm 3\pi$. Robinett III, R.D. and Wilson, D.G. [67], reprinted with the permission of the American Astronautical Society

The appropriate Hamiltonian function is positive definite about the origin and defined as

$$\mathcal{H} = V = \frac{1}{2}\dot{x}^2 + \frac{1}{2}x^2,$$

which is statically stable. Then the corresponding time derivative of the Hamiltonian function becomes

$$\dot{\mathcal{H}} = \dot{V} = [\ddot{x} + x]\dot{x} = [\sin(\dot{x})]\dot{x}.$$

A limit cycle is given by

$$\mathcal{H}_{cyclic} = \oint_{\tau} \sin(\dot{x})\dot{x}\, dt = 0,$$

where the power flow balances at $\dot{x} = \pm n\pi$ for $n = 1, 3, \ldots$ and is shown in Fig. 6.1.

Three separate regions (for Cases 1–3) can be observed by investigating several initial conditions inside, on, and outside the first limit cycle. Figure 6.2 presents all three of these regions with the corresponding numerical values given in Table 6.1.

The responses are plotted on the Hamiltonian 3D surface (top) with the projection onto the phase plane illustrated on the 2D plot (bottom). The system trajectories are constrained to move along the Hamiltonian surface as a function of power flow into the system versus the energy dissipation rate within the system. In Fig. 6.1, for power versus velocity, the zero crossings identify the limit cycles at $\pm\pi$ and $\pm 3\pi$. For the case inside the first limit cycle, or Case 1, the term $\sin(\dot{x})$ becomes a generator term and drives power into the dynamically unstable system as given in Fig. 6.3. For Case 2, the dynamically neutral stable system is on the stability boundary and the power flow is balanced as presented in Fig. 6.4. For Case 3, the system is initially outside the limit cycle, and the term $\sin(\dot{x})$ becomes a dissipator term, and power is dissipated out of the dynamically stable system as illustrated in Fig. 6.5.

The next set of initial conditions (see Table 6.1, Cases 4–6) demonstrate that the system response bifurcates as a result of the power flow to the next higher energy level limit cycle with the same characteristics as the previous investigation (see

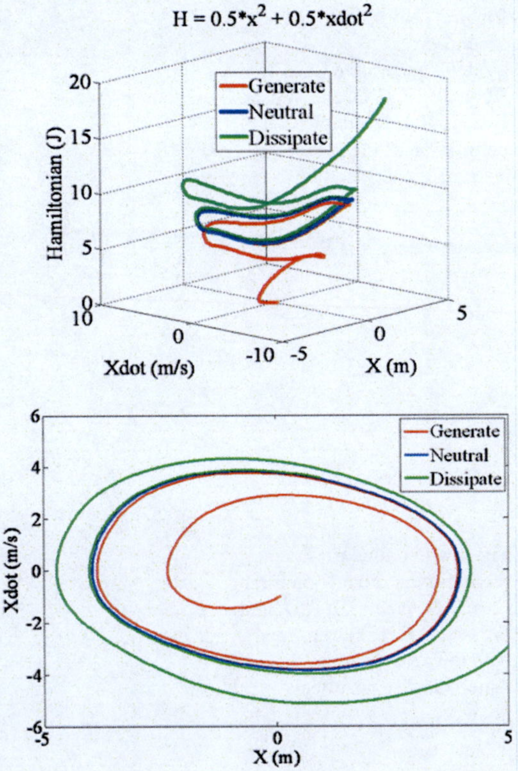

Fig. 6.2 Nonlinear second-order model: 3D Hamiltonian (*top*) and 2D phase-plane (*bottom*) responses. Robinett III, R.D. and Wilson, D.G. [67], reprinted with the permission of the American Astronautical Society

Table 6.1 Nonlinear second-order model numerical values. Robinett III, R.D. and Wilson, D.G. [67], reprinted with the permission of the American Astronautical Society

Case	Description	x_0 (m)	$\dot{x}_0$ (m/s)
1	Generate	0.1	−0.1
2	Neutral	3.98	0.0
3	Dissipate	5.0	−3.0
4	Generate	5.0	−5.0
5	Neutral	10.25	0.0
6	Dissipate	10.0	−6.0

Fig. 6.6 for the combined 3D Hamiltonian and phase-plane responses for the second limit cycle). For Case 4, the generator term is exhibited in Fig. 6.7. For Case 5, the system is again on the next stability boundary as demonstrated in Fig. 6.8. Case 6 gives the dissipation behavior as the response returns back to the limit cycle boundary (see Fig. 6.9).

Once again, Fig. 6.1 presents the power versus velocity responses which capture both limit cycles, the first one again at $\pm\pi$ and the next one at $\pm 3\pi$, respectively. The 3D Hamiltonian and the 2D phase plane responses that illustrate both the first and second limit cycles are given in Fig. 6.10, respectively.

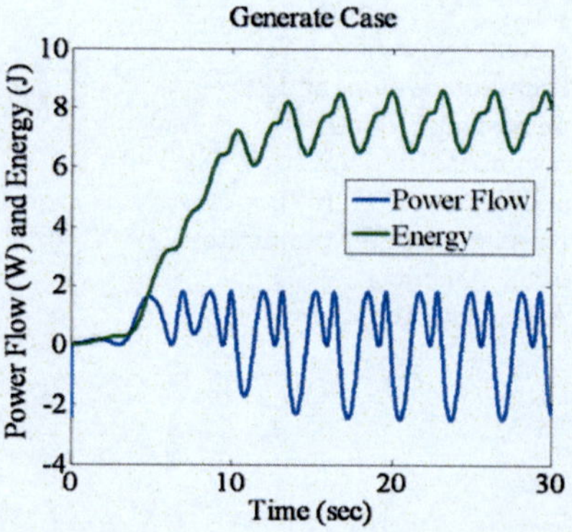

Fig. 6.3 Nonlinear second-order model—generative Case 1. Robinett III, R.D. and Wilson, D.G. [67], reprinted with the permission of the American Astronautical Society

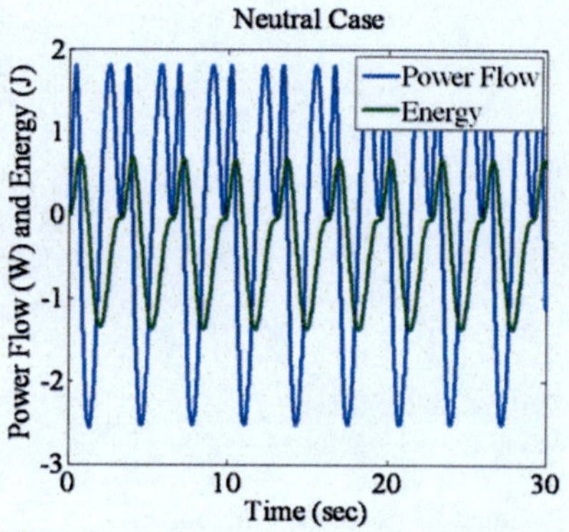

Fig. 6.4 Nonlinear second-order model—neutral Case 2. Robinett III, R.D. and Wilson, D.G. [67], reprinted with the permission of the American Astronautical Society

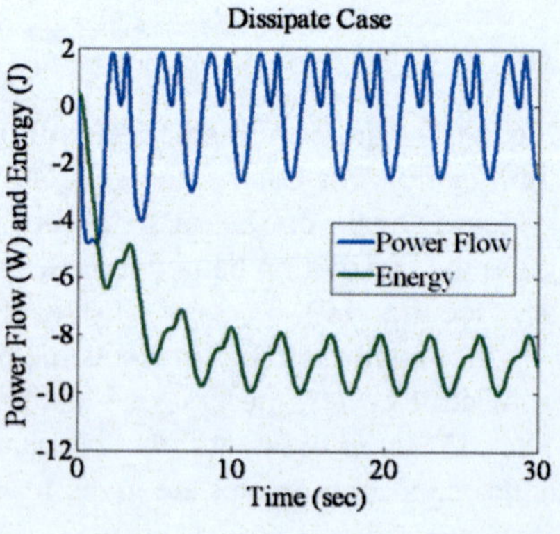

Fig. 6.5 Nonlinear second-order model—dissipative Case 3. Robinett III, R.D. and Wilson, D.G. [67], reprinted with the permission of the American Astronautical Society

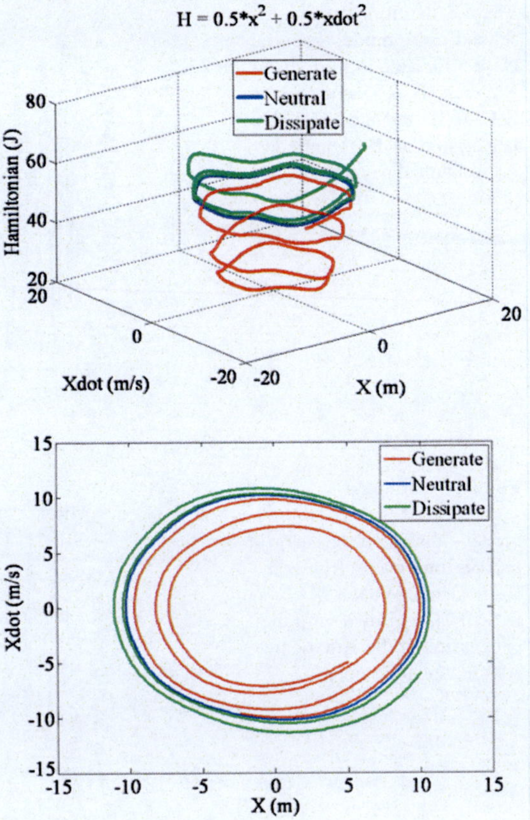

Fig. 6.6 Nonlinear second-order model: 3D Hamiltonian (*top*) and 2D phase-plane (*bottom*) responses for second limit cycle. Robinett III, R.D. and Wilson, D.G. [67], reprinted with the permission of the American Astronautical Society

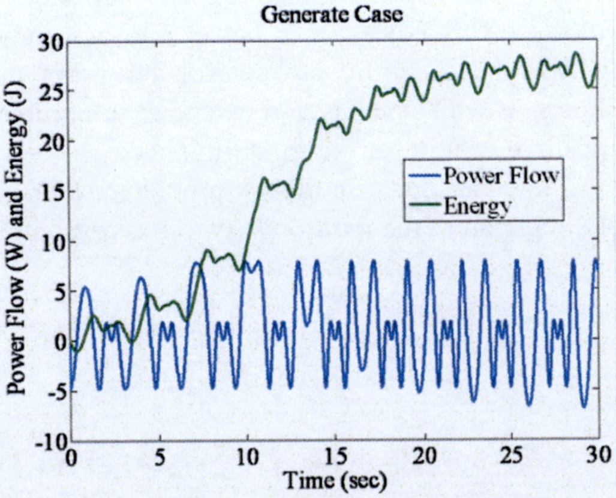

Fig. 6.7 Nonlinear second-order model—generative Case 4 for second limit cycle. Robinett III, R.D. and Wilson, D.G. [67], reprinted with the permission of the American Astronautical Society

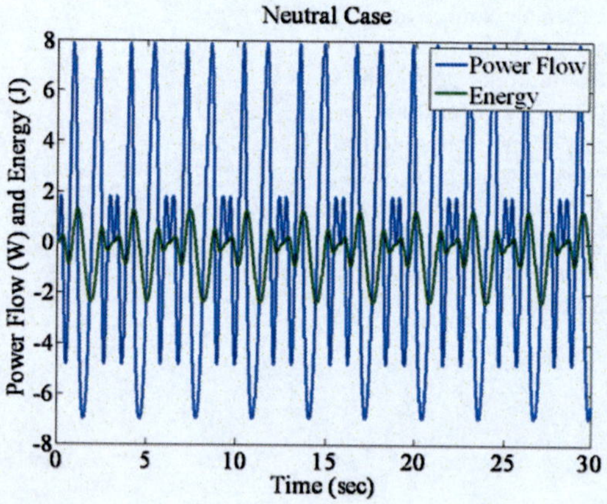

Fig. 6.8 Nonlinear second-order model—neutral Case 5 for second limit cycle. Robinett III, R.D. and Wilson, D.G. [67], reprinted with the permission of the American Astronautical Society

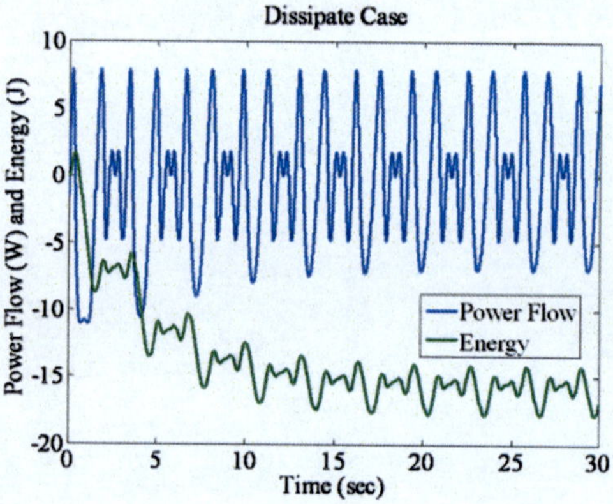

Fig. 6.9 Nonlinear second-order model—dissipative Case 6 for second limit cycle. Robinett III, R.D. and Wilson, D.G. [67], reprinted with the permission of the American Astronautical Society

6.3 An Extension of Eigenanalysis to MIMO Nonlinear Systems

This section expands on the concepts in Chap. 4, Sect. 4.3.5 of relating eigenanalysis of linear SISO systems to power flow and limit cycles in order to assess nonlinear SISO system stability and control. The problem is cast as an eigenvalue problem for linear MIMO systems and extended to nonlinear MIMO systems by employing a positive definite surface and limit cycles.

First consider a multibody problem presented in Fig. 6.11. The Hamiltonian and Lagrangian of the multibody two-mass system are given as

$$\mathcal{H}=\mathcal{T}+\mathcal{V}=\frac{1}{2}m_1\dot{x}_1^2+\frac{1}{2}m_2\dot{x}_2^2+V_1(x_1)+V_2(x_2)+V_{12}(x_2-x_1), \tag{6.1}$$

$$\mathcal{L}=\mathcal{T}-\mathcal{V}=\frac{1}{2}m_1\dot{x}_1^2+\frac{1}{2}m_2\dot{x}_2^2-V_1(x_1)-V_2(x_2)-V_{12}(x_2-x_1), \tag{6.2}$$

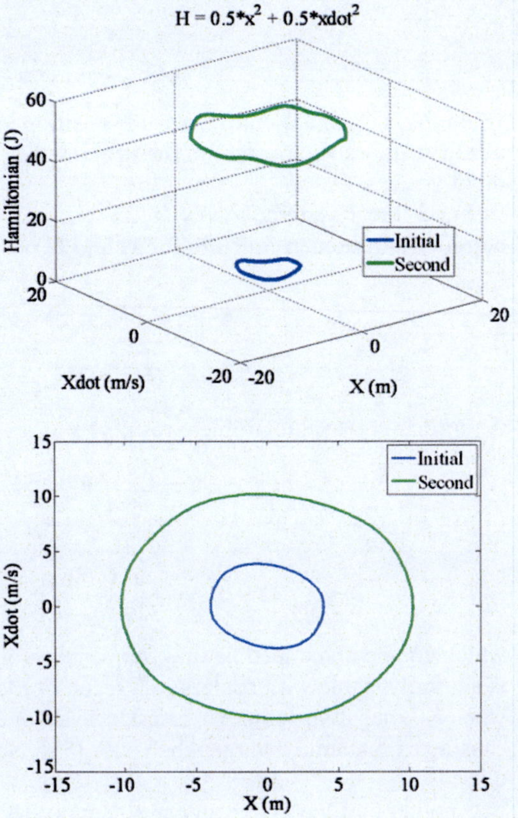

Fig. 6.10 Nonlinear second-order model: 3D Hamiltonian (*top*) and 2D phase-plane (*bottom*) responses for both limit cycles. Robinett III, R.D. and Wilson, D.G. [67], reprinted with the permission of the American Astronautical Society

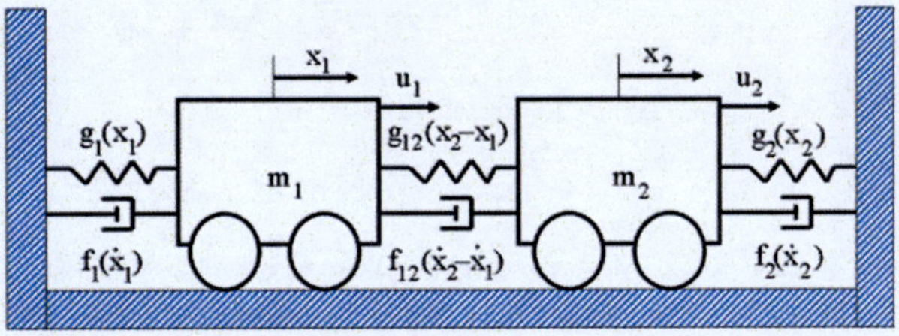

Fig. 6.11 Simple two-body system

where

$$\mathcal{T} = \text{kinetic energy} = \frac{1}{2}m_1\dot{x}_1^2 + \frac{1}{2}m_2\dot{x}_2^2,$$

$$\mathcal{V} = \text{potential energy} = V_1(x_1) + V_2(x_2) + V_{12}(x_2 - x_1),$$

and the equations of motion are determined from Lagrange's equation

$$\frac{d}{dt}\left(\frac{\partial\mathcal{L}}{\partial\dot{x}_i}\right) - \frac{\partial\mathcal{L}}{\partial x_i} = Q_i \tag{6.3}$$

as

$$\mathbf{M}\ddot{\mathbf{x}} + \mathbf{g}(\mathbf{x}) = \mathbf{u} - \mathbf{f}(\dot{\mathbf{x}}), \tag{6.4}$$

where $\mathbf{M}$ is the mass matrix, $\mathbf{x}$ is the position vector, $\mathbf{g}(\mathbf{x}) = \partial\mathcal{V}(\mathbf{x})/\partial\mathbf{x}$ is the generalized stiffness vector, $\mathbf{f}(\dot{\mathbf{x}})$ is the generalized damping vector, and $\mathbf{u}$ is the control input vector.

For a linear system, $\mathcal{V}(\mathbf{x}) = 1/2\mathbf{x}^T\mathbf{K}\mathbf{x}$ and $\mathbf{f}(\dot{\mathbf{x}}) = \mathbf{C}\dot{\mathbf{x}}$, which leads to a positive definite Hamiltonian function for $\mathbf{M}$ and $\mathbf{K}$ positive definite:

$$\begin{aligned} \mathcal{H} &= \frac{1}{2}\dot{\mathbf{x}}^T\mathbf{M}\dot{\mathbf{x}} + \frac{1}{2}\mathbf{x}^T\mathbf{K}\mathbf{x}, \\ \dot{\mathcal{H}} &= \dot{\mathbf{x}}^T[\mathbf{M}\ddot{\mathbf{x}} + \mathbf{K}\mathbf{x}] = \dot{\mathbf{x}}^T[\mathbf{u} - \mathbf{C}\dot{\mathbf{x}}]. \end{aligned} \tag{6.5}$$

A familiar eigenvalue problem results from

$$\mathbf{u} - \mathbf{C}\dot{\mathbf{x}} = \mathbf{0} \quad \text{and} \quad \dot{\mathcal{H}} = 0$$

as

$$\begin{aligned} \mathbf{M}\ddot{\mathbf{x}} + \mathbf{K}\mathbf{x} &= \mathbf{0}, \\ \left[\omega^2\right] &= \mathbf{M}^{-1}\mathbf{K}, \end{aligned} \tag{6.6}$$

which are the undamped natural frequencies (eigenvalues) of $\mathbf{M}$ and $\mathbf{K}$. The system is neutrally stable with centers of a second-order system [46]. An instability occurs when $\mathbf{K}$ is negative definite because, at least, one of the roots has a positive real value causing exponential divergent behavior [53]. Notice that $\mathcal{V}(\mathbf{x})$ is negative definite in this case.

Another familiar eigenvalue problem occurs when H is positive definite:

$$\mathcal{H} = \frac{1}{2}\dot{\mathbf{x}}^T\mathbf{M}\dot{\mathbf{x}} + \frac{1}{2}\mathbf{x}^T\mathbf{K}\mathbf{x} \tag{6.7}$$

for $\mathbf{M}$, $\mathbf{K}$ positive definite, and

$$\dot{\mathcal{H}} = \dot{\mathbf{x}}^T[\mathbf{M}\ddot{\mathbf{x}} + \mathbf{K}\mathbf{x}] = -\dot{\mathbf{x}}^T[\mathbf{K_D} + \mathbf{C}]\dot{\mathbf{x}} \tag{6.8}$$

for $\mathbf{u} = -\mathbf{K_D}\dot{\mathbf{x}}$. The system is (asymptotically) stable if the real parts of the eigenvalues meet the positive definite condition

$$\mathbf{K_D} + \mathbf{C} > \mathbf{0}$$

and unstable (negative definite condition) for

$$\mathbf{K_D} + \mathbf{C} < \mathbf{0}.$$

An extension of these eigenvalue problems to nonlinear systems is accomplished by replacing the neutral stability condition of the linear system, a center, with a positive definite surface [7, 46]. The Hamiltonian

$$\mathcal{H} = \frac{1}{2}\dot{\mathbf{x}}^T\mathbf{M}\dot{\mathbf{x}} + \mathcal{V}(\mathbf{x}) \tag{6.9}$$

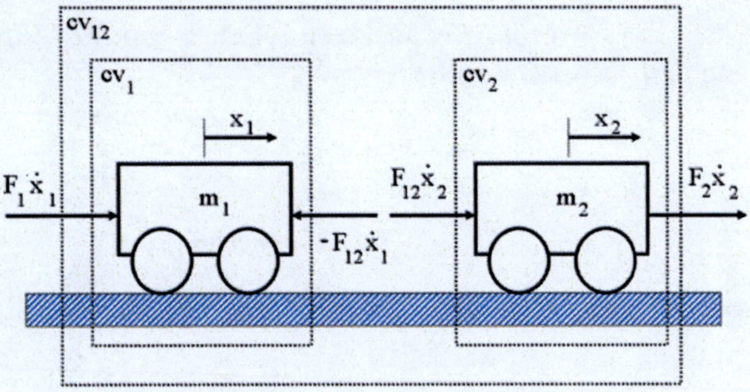

Fig. 6.12 Simple two-body system with indicated control volumes

is positive definite for positive definite $\mathbf{M}$ and $\mathcal{V}(\mathbf{x})$, and a limit cycle exists for

$$\oint_\tau \dot{\mathcal{H}}\,dt = \oint_\tau \dot{\mathbf{x}}^T[\mathbf{M}\ddot{\mathbf{x}} + \mathbf{g}(\mathbf{x})]\,dt = \oint_\tau \dot{\mathbf{x}}^T[\mathbf{u} - \mathbf{f}(\dot{\mathbf{x}})]\,dt = 0. \tag{6.10}$$

The first term describes the nonlinear frequency content [32] of the undamped/undriven system

$$\mathbf{M}\ddot{\mathbf{x}} + \mathbf{g}(\mathbf{x}) = \mathbf{0}, \tag{6.11}$$

which effectively extends the undamped natural frequency to nonlinear systems. Notice that if $\mathcal{V}(\mathbf{x})$ is negative definite, the nonlinear system is unstable. The second term describes the extension of the real parts of the eigenvalues of (6.8) to nonlinear systems

$$\frac{1}{\tau_c}\int_0^{\tau_c} \dot{\mathbf{x}}^T[\mathbf{u} - \mathbf{f}(\dot{\mathbf{x}})]\,dt \gtrless 0, \tag{6.12}$$

which determines stability of the nonlinear system for positive definite $\mathcal{V}(\mathbf{x})$. Now, it is time for a MIMO example.

6.4 Two-Mass Numerical Example

In this section several numerical simulations for the MIMO two-body system (see Fig. 6.12) with both decoupled/coupled controls are investigated. The equation of motion based on the control volume for the first mass (cv_1) is

$$m_1\ddot{x}_1 = F_1 - F_{12},$$

and for the second mass (cv_2), it is

$$m_2\ddot{x}_2 = F_2 + F_{12}.$$

Next, the power flow is analyzed for each control volume where the Hamiltonians and derivatives for each are:

$$\begin{aligned}\mathcal{H}_1 &= \frac{1}{2}m_1\dot{x}_1^2,\\ \dot{\mathcal{H}}_1 &= m_1\ddot{x}_1\dot{x}_1\\ &= [F_1 - F_{12}]\dot{x}_1;\end{aligned} \tag{cv$_1$}$$

$$\begin{aligned}\mathcal{H}_2 &= \frac{1}{2}m_2\dot{x}_2^2,\\ \dot{\mathcal{H}}_2 &= m_2\ddot{x}_2\dot{x}_2\\ &= [F_2 + F_{12}]\dot{x}_2;\end{aligned} \tag{cv$_2$}$$

and

$$\begin{aligned}\mathcal{H}_{12} &= \frac{1}{2}m_1\dot{x}_1^2 + \frac{1}{2}m_2\dot{x}_2^2,\\ \dot{\mathcal{H}}_{12} &= m_1\ddot{x}_1\dot{x}_1 + m_2\ddot{x}_2\dot{x}_2\\ &= [F_1 - F_{12}]\dot{x}_1 + [F_2 + F_{12}]\dot{x}_2\\ &= F_1\dot{x}_1 + F_2\dot{x}_2 + F_{12}[\dot{x}_2 - \dot{x}_1].\end{aligned} \tag{cv$_{12}$}$$

These results can be extended to an N-body system as

$$\mathcal{H}_{1N} = \sum_{i=1}^{N}\frac{1}{2}m_i\dot{x}_i^2$$

and

$$\dot{H}_{1N} = \sum_{i=1}^{N}F_i\dot{x}_i + \sum_{j=1}^{N-1}F_{j(j+1)}[\dot{x}_{j+1} - \dot{x}_j].$$

An interesting control design occurs when comparing a coupled system to a decoupled system.

6.4.1 Linear MIMO System Controller Design

The linear MIMO system is defined as

$$\mathbf{M}\ddot{\mathbf{x}} = \mathbf{F} = \begin{Bmatrix} F_1 - F_{12} \\ F_2 + F_{12} \end{Bmatrix} = -\mathbf{K}\mathbf{x} - \mathbf{C}\dot{\mathbf{x}} + \mathbf{u},$$

where

$$\mathbf{M}=\begin{bmatrix} m_1 & 0 \\ 0 & m_2 \end{bmatrix}, \qquad \mathbf{K}=\begin{bmatrix} k_{11} & -k_{12} \\ -k_{12} & k_{22} \end{bmatrix}, \qquad \mathbf{C}=\begin{bmatrix} c_{11} & -c_{12} \\ -c_{12} & c_{22} \end{bmatrix},$$

and $k_{ii}=(k_i+k_{12})$, $c_{ii}=(c_i+c_{12})$ for $i=1,2$. A straightforward way to decouple this system is to cancel F_{12}. A decoupling PID controller is defined as

$$\mathbf{u}=-\mathbf{K_P}\mathbf{x}-\mathbf{K_I}\int_0^t \mathbf{x}\,d\tau-\mathbf{K_D}\dot{\mathbf{x}}-\begin{Bmatrix} -\hat{F}_{12} \\ \hat{F}_{12} \end{Bmatrix}$$

that for ideal parameter matching $\hat{F}_{12}=F_{12}$ leads to

$$\begin{aligned}
&\begin{bmatrix} m_1 & 0 \\ 0 & m_2 \end{bmatrix}\begin{Bmatrix} \ddot{x}_1 \\ \ddot{x}_2 \end{Bmatrix}+\begin{bmatrix} k_1 & 0 \\ 0 & k_2 \end{bmatrix}\begin{Bmatrix} x_1 \\ x_2 \end{Bmatrix} \\
&\quad=-\begin{bmatrix} c_1 & 0 \\ 0 & c_2 \end{bmatrix}\begin{Bmatrix} \dot{x}_1 \\ \dot{x}_2 \end{Bmatrix}-\begin{bmatrix} K_{P_1} & 0 \\ 0 & K_{P_2} \end{bmatrix}\begin{Bmatrix} x_1 \\ x_2 \end{Bmatrix}-\begin{bmatrix} K_{I_1} & 0 \\ 0 & K_{I_2} \end{bmatrix}\begin{Bmatrix} \int_0^t x_1\,d\tau \\ \int_0^t x_2\,d\tau \end{Bmatrix} \\
&\qquad-\begin{bmatrix} K_{D_1} & 0 \\ 0 & K_{D_2} \end{bmatrix}\begin{Bmatrix} \dot{x}_1 \\ \dot{x}_2 \end{Bmatrix},
\end{aligned}$$

which produces two independent SISO controller designs. The proportional gains are chosen such that $K_{P_1}, K_{P_2}>0$ to ensure static stability, while $K_{D_i}, K_{I_i}>0$ for $i=1,2$ are chosen to ensure dynamic stability.

Another way to decouple this system is to pick the PID controller

$$\mathbf{u}=-\mathbf{K_P}\mathbf{x}-\mathbf{K_I}\int_0^t \mathbf{x}\,d\tau-\mathbf{K_D}\dot{\mathbf{x}}$$

with the following controller gain matrices:

$$\mathbf{K_P}=\begin{bmatrix} K_{P_1} & k_{12} \\ k_{12} & K_{P_2} \end{bmatrix}, \quad \mathbf{K_I}=\begin{bmatrix} K_{I_1} & 0 \\ 0 & K_{I_2} \end{bmatrix}, \quad \text{and} \quad \mathbf{K_D}=\begin{bmatrix} K_{D_1} & c_{12} \\ c_{12} & K_{D_2} \end{bmatrix}.$$

This controller only cancels out the actual cross-coupling terms as opposed to canceling out the whole F_{12}. The neutral stability condition for each SISO controller is given by [13, 30]

$$\omega_i^2=\frac{k_{ii}+K_{P_i}}{m_i}=\frac{K_{I_i}}{c_{ii}+K_{D_i}} \quad \text{for } i=1,2. \tag{6.13}$$

Numerical simulations are performed with the following numerical values: $m_1=10$ kg, $m_2=20$ kg, $c_1=2$ kg/s, $c_2=1$ kg/s, $k_1=3.94$ N/m, $k_2=7.8957$ N/m, $c_{12}=1$ kg/s, and $k_{12}=5$ N/m. The following SISO controller gains are: $K_{P_1}=K_{P_2}=1$ N/m, $K_{D_1}=K_{D_2}=1$ kg/s, and $K_{I_1}=3.976$ kg/s^3 and $K_{I_2}=$

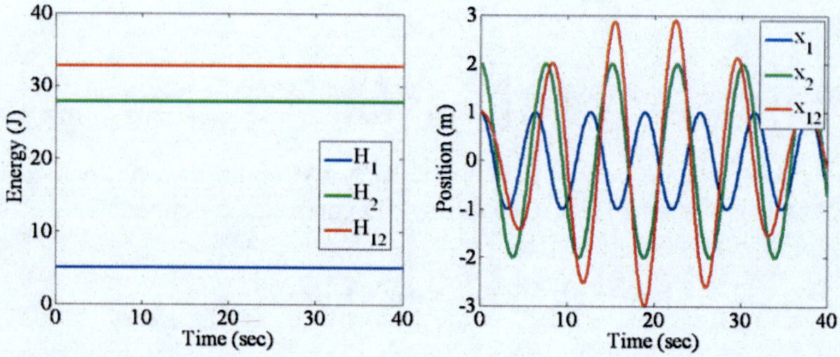

Fig. 6.13 Two-body dynamically neutral stability decoupled control—energy (*left*) and position (*right*) responses

2.0844 kg/s^3. The initial conditions are $x_{01} = 1$ m, $x_{02} = 2$ m, and all velocities are set to zero.

These neutral stability results are shown in Fig. 6.13 (left) for the energy responses which for the positions in Fig. 6.13 (right) include x_1, x_2, and x_{12} (note that the energy is constant and the corresponding power flow is zero). The dissipation and generation terms are balanced and offset for all three groups $\mathcal{H}_1$, $\mathcal{H}_2$, and $\mathcal{H}_{12}$ as shown in Fig. 6.14.

Next, for both asymptotically stable and unstable SISO cases, the K_{I_i} gains were modified. For asymptotically stable cases, the integral gains were reduced to $K_{I_1} = 0.994$ kg/s^3 and $K_{I_2} = 1.0422$ kg/s^3. For the unstable run, the integral gains were increased to $K_{I_1} = 5.964$ kg/s^3 and $K_{I_2} = 3.1266$ kg/s^3. Note that all other model and controller gains remained unchanged. The position responses for both stable (left) and unstable (right) are shown in Fig. 6.15. For asymptotically stable runs, the dissipation terms are greater than the generation terms as shown for all three groups $\mathcal{H}_1$, $\mathcal{H}_2$, and $\mathcal{H}_{12}$ in Fig. 6.16 (left column). For the unstable runs, just the opposite is demonstrated: the generation terms are greater than the dissipation terms as shown in Fig. 6.16 (right column). The total energy and power flow for the combined system are shown for both the asymptotically stable (left column) and unstable (right column) runs in Fig. 6.17.

In comparison, a coupled controller is derived from the combined Hamiltonian

$$\mathcal{H}_{12} = \frac{1}{2}\dot{\mathbf{x}}^T\mathbf{M}\dot{\mathbf{x}} + \frac{1}{2}\mathbf{x}^T[\mathbf{K} + \mathbf{K_P}]\mathbf{x}$$

and the time derivative

$$\begin{aligned}\dot{\mathcal{H}}_{12} &= \dot{\mathbf{x}}^T[\mathbf{M}\ddot{\mathbf{x}} + [\mathbf{K} + \mathbf{K_P}]\mathbf{x}] \\ &= \dot{\mathbf{x}}^T\left[-[\mathbf{C} + \mathbf{K_D}]\dot{\mathbf{x}} - \mathbf{K_I}\int_0^t \mathbf{x}\,d\tau\right].\end{aligned}$$

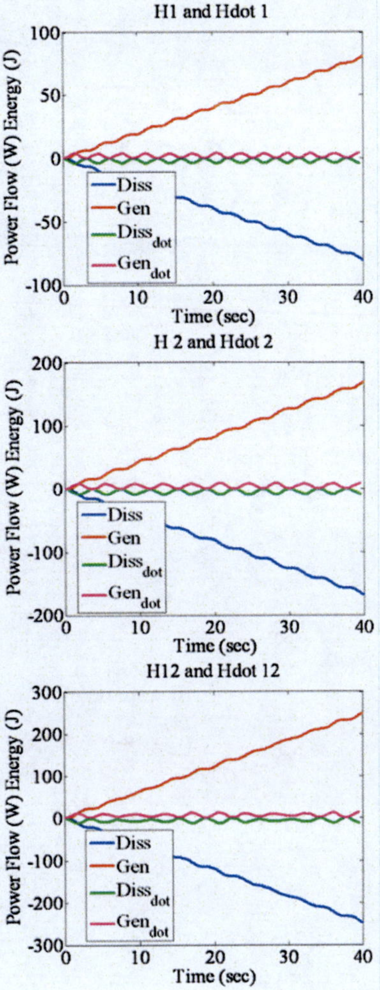

Fig. 6.14 Two-body dynamically neutral stability decoupled control dissipation versus generation

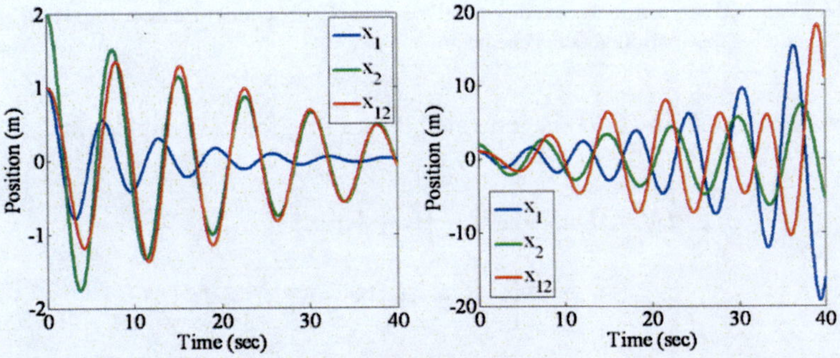

Fig. 6.15 Two-body dynamically asymptotically stable (*left*) and unstable (*right*) decoupled control—position responses

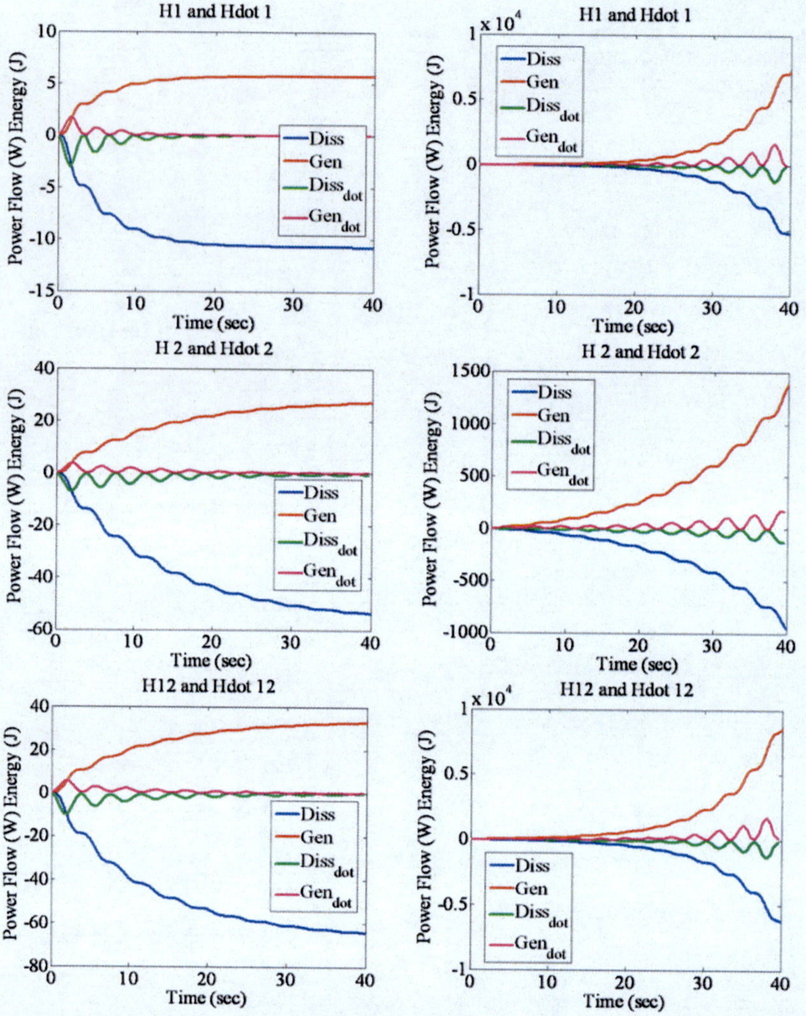

Fig. 6.16 Two-body dynamically asymptotically stable (*left column*) and unstable (*right column*) decoupled control dissipation versus generation

A matrix equivalent of (6.13) is generated by solving for the coupled integral gains from

$$\dot{\mathcal{H}}_{12} = 0 = \dot{\mathbf{x}}^T \left[\mathbf{M}\ddot{\mathbf{x}} + [\mathbf{K} + \mathbf{K_P}]\mathbf{x}\right]$$
$$= \dot{\mathbf{x}}^T \left[-[\mathbf{C} + \mathbf{K_D}]\dot{\mathbf{x}} - \mathbf{K_I}\int_0^t \mathbf{x}\, d\tau\right],$$

which matches up the eigenvalues of the system when

$$\mathbf{K_I} = [\mathbf{C} + \mathbf{K_D}]\mathbf{M}^{-1}[\mathbf{K} + \mathbf{K_P}].$$

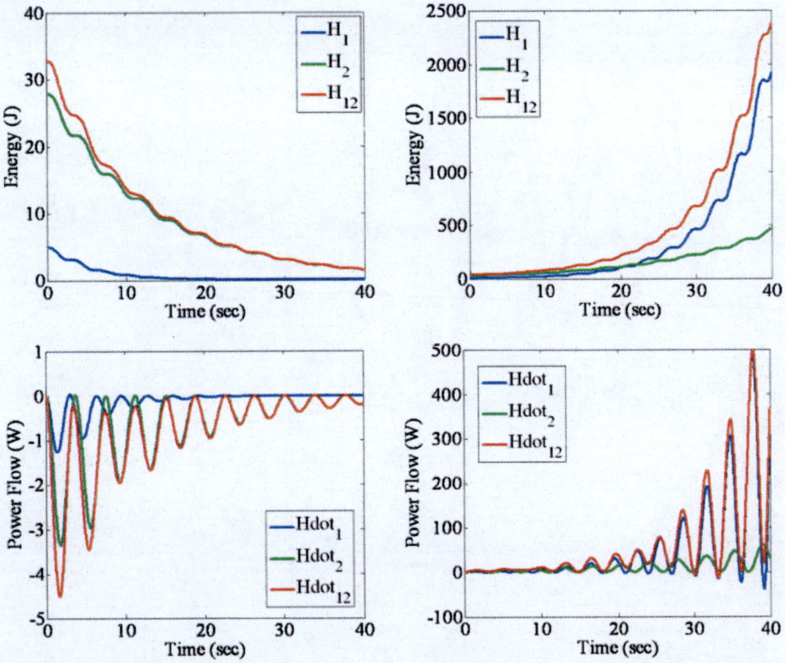

Fig. 6.17 Two-body dynamically asymptotically stable (*left column*) and unstable (*right column*) decoupled control

The resultant $\mathbf{K_I}$ matrix is fully populated with differing off-diagonal terms determined as

$$\mathbf{K_I} = \begin{bmatrix} 4.226 & -2.6948 \\ -1.744 & 2.5844 \end{bmatrix}. \tag{6.14}$$

This system is effectively decoupled into the eigenspace defined by the undamped/undriven natural frequencies. The other numerical values and controller gains $\mathbf{K_P}$ and $\mathbf{K_D}$ are given from the previous example.

These results are shown in Fig. 6.18 for the positions (upper left), power flow and energy responses for both the dissipation and generation terms (upper right), and the total energy $\mathcal{H}_{12}$ (bottom left) and power flow $\dot{\mathcal{H}}_{12}$ (bottom right), respectively. Note that the energy remains constant (see energy plot for $\mathcal{H}_{12}$, bottom left). The coupled integral control gain matrix (6.14) decouples the system in the eigenspace [68]

$$\mathbf{q} = \boldsymbol{\Phi}^{-1}\mathbf{x} \quad \text{and} \quad \dot{\mathbf{q}} = \boldsymbol{\Phi}^{-1}\dot{\mathbf{x}},$$

where $\mathbf{q}$ and $\dot{\mathbf{q}}$ are the modal positions and velocities, $\boldsymbol{\omega}$ the eigenvalues, and $\boldsymbol{\Phi}$ the corresponding eigenvectors from

$$\left[[\mathbf{K} + \mathbf{K_P}] - \omega^2 \mathbf{M}\right]\boldsymbol{\Phi} = \left[\mathbf{K_I} - \omega^2[\mathbf{C} + \mathbf{K_D}]\right]\boldsymbol{\Phi} = \mathbf{0}.$$

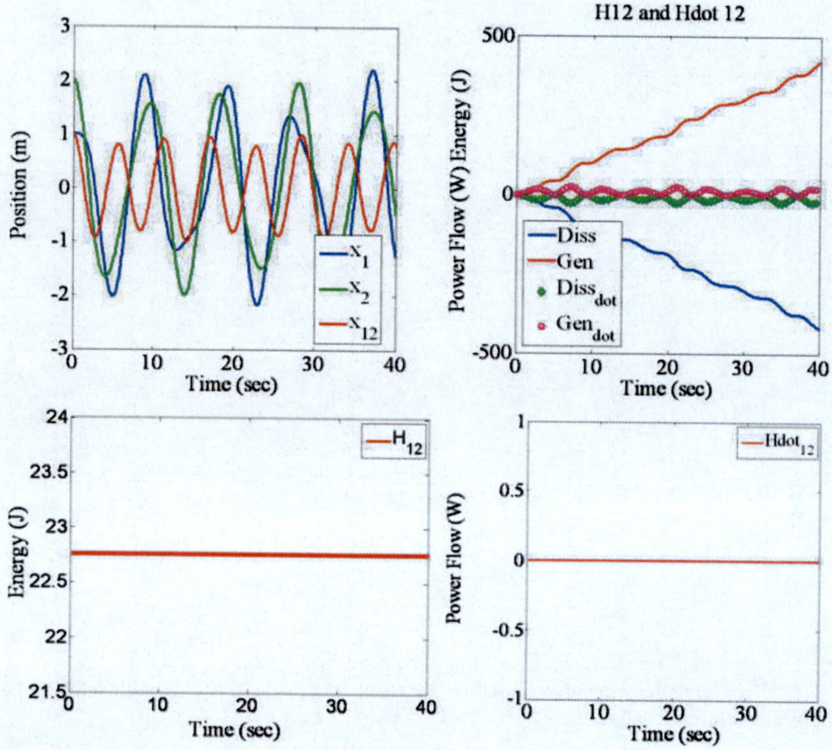

Fig. 6.18 Two-body dynamically neutral stability

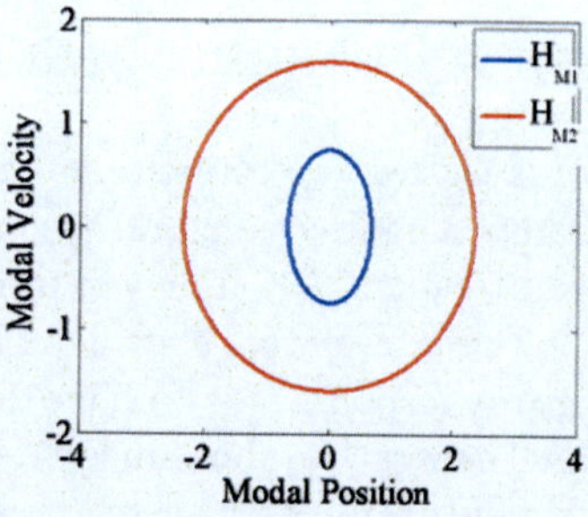

Fig. 6.19 Two-body dynamically neutral eigenspace phase plane response

The decoupled phase plane plots are shown in Fig. 6.19 for both $\mathcal{H}_{M_1}$ and $\mathcal{H}_{M_2}$ in the modal domain.

Similar results are shown in Fig. 6.20 for the dissipative (decaying responses) case (with $K_{I_{11}} = 1.0565$, $K_{I_{12}} = K_{I_{21}} = 0.0$, $K_{I_{22}} = 1.033776$, and $K_{D_1} = 10.0$) and Fig. 6.21 for the generative (growing responses) case (with $K_{I_{11}} = 5.0712$, $K_{I_{12}} = K_{I_{21}} = 0.0$, $K_{I_{22}} = 3.10128$, and $K_{D_1} = 10.0$).

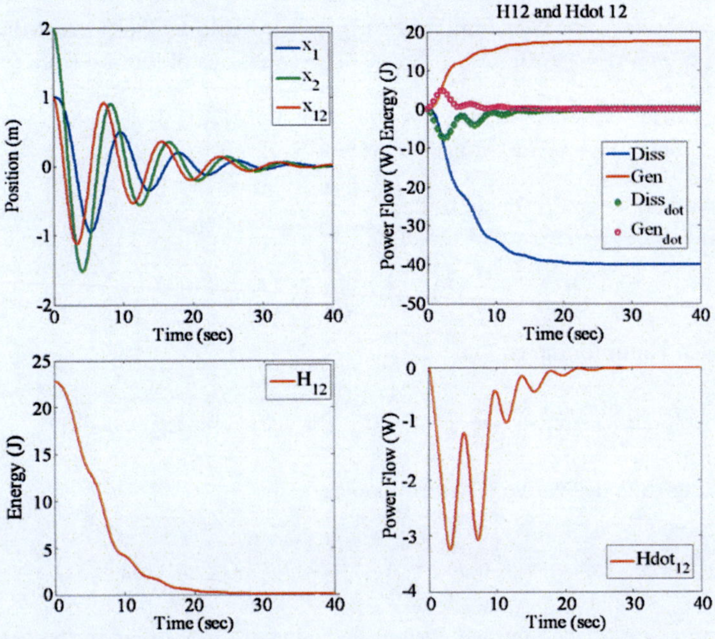

Fig. 6.20 Two-body dynamically stable

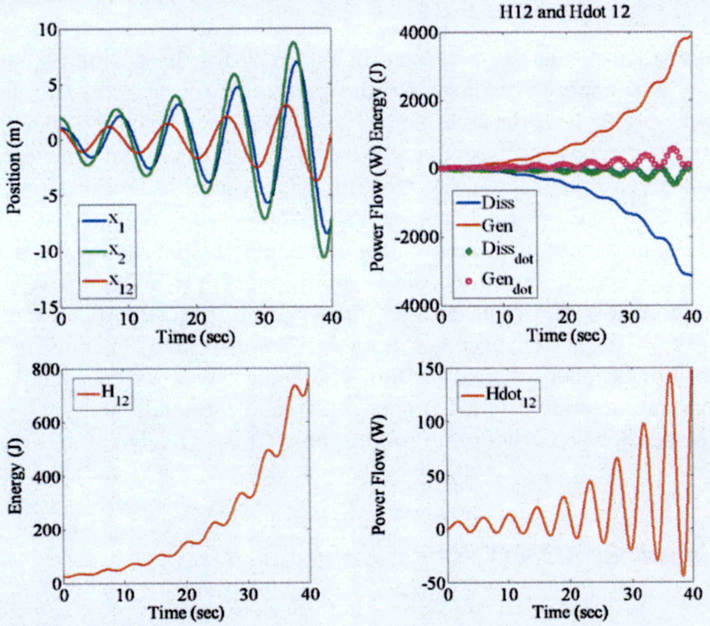

Fig. 6.21 Two-body dynamically unstable

Finally, the last linear example (see Fig. 6.12) explores the linear MIMO case where $m_1 > m_2$, $\mathbf{u} = [u_1\ 0]^T$, and $F_2 = 0$. The equations of motion from (6.3) are

$$\begin{aligned} m_1\ddot{x}_1 &= -k_1x_1 - c_1\dot{x}_1 + k_{12}(x_2 - x_1) + c_{12}(\dot{x}_2 - \dot{x}_1) + u_1, \\ m_2\ddot{x}_2 &= -k_{12}(x_2 - x_1) - c_{12}(\dot{x}_2 - \dot{x}_1), \end{aligned}$$

with a single collocated PID controller

$$u_1 = -K_{P_1}x_1 - K_{I_1}\int_0^t x_1\,d\tau - K_{D_1}\dot{x}_1. \tag{6.15}$$

The coupled Hamiltonian is

$$\mathcal{H}_{12} = \frac{1}{2}m_1\dot{x}_1^2 + \frac{1}{2}m_2\dot{x}_2^2 + \frac{1}{2}(k_1 + K_{P_1})x_1^2 + \frac{1}{2}k_{12}(x_2 - x_1)^2,$$

and resulting time derivative is determined as

$$\dot{\mathcal{H}}_{12} = \left[-(c_1 + K_{D_1})\dot{x}_1 - K_{I_1}\int_0^t x_1\,d\tau\right]\dot{x}_1 - c_{12}(\dot{x}_2 - \dot{x}_1)^2.$$

The performance of the coupled system is evaluated. In particular, the limit cycle is identified for the coupled system $\mathcal{H}_{12}$ with $\oint_\tau \dot{\mathcal{H}}_{12}\,dt = 0$. Numerical simulations are performed with the following numerical values: $m_1 = 20$ kg, $m_2 = 10$ kg, $c_1 = 2$ kg/s, $k_1 = 3.94$ N/m, $c_2 = k_2 = 0$, $c_{12} = 1$ kg/s, and $k_{12} = 5$ N/m. The following u_1 controller gains are: $K_{P_1} = 1$ N/m, $K_{I_1} = 0.605$ kg/s^3, and $K_{D_1} = 2$ kg/s. The initial conditions are $x_{01} = 1$ m, $x_{02} = 2$ m, and all velocities are set to zero.

The numerical results are presented in Fig. 6.22 for the positions (upper left), power flow and energy responses for the partitioned dissipation and generation terms (upper right), and the total energy $\mathcal{H}_{12}$ (bottom left) and power flow $\dot{\mathcal{H}}_{12}$ (bottom right) responses, respectively. Note that after the initial transient, the energy remains constant (see energy plot for $\mathcal{H}_{12}$, bottom left). For an increase to $K_{D_1} = 4.0$ kg/s, the system becomes stable. The results are given in Fig. 6.23 for the positions (upper left), power flow and energy responses for the partitioned dissipation and generation terms (upper right), and the total energy $\mathcal{H}_{12}$ (bottom left) and power flow $\dot{\mathcal{H}}_{12}$ (bottom right) responses, respectively. For a reduction to $K_{D_1} = 1.0$ kg/s, the system becomes unstable. The results are illustrated in Fig. 6.24 for the positions (upper left), power flow and energy responses for the partitioned dissipation and generation terms (upper right), and the total energy $\mathcal{H}_{12}$ (bottom left) and power flow $\dot{\mathcal{H}}_{12}$ (bottom right) responses, respectively.

6.4.2 Nonlinear MIMO System Controller Design

As a first nonlinear example, the last linear case is extended by adding a Duffing oscillator and Coulomb friction effects to the system. The equations of motion

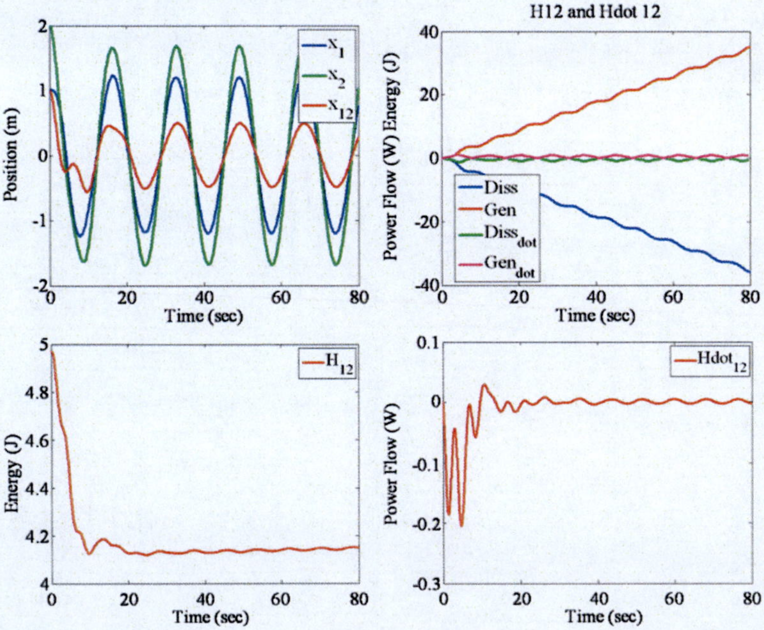

Fig. 6.22 Two-body dynamically neutral stability for ($m_1 > m_2$, $u_2 = F_2 = 0$)—linear MIMO system

Fig. 6.23 Two-body dynamically asymptotically stable for ($m_1 > m_2$, $u_2 = F_2 = 0$)—linear MIMO system

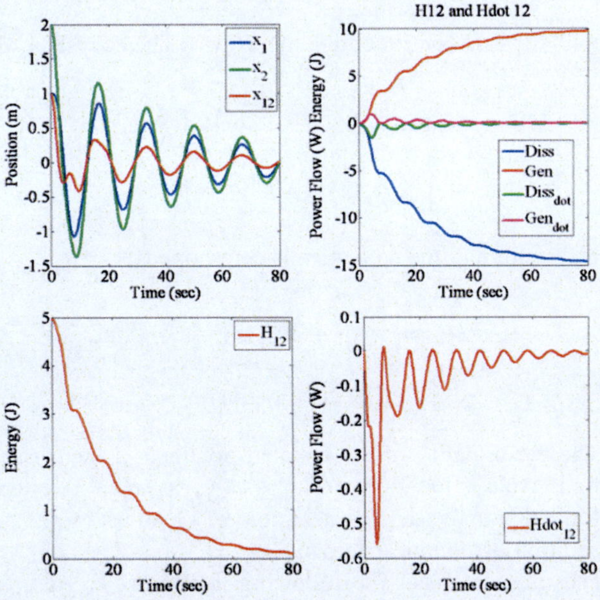

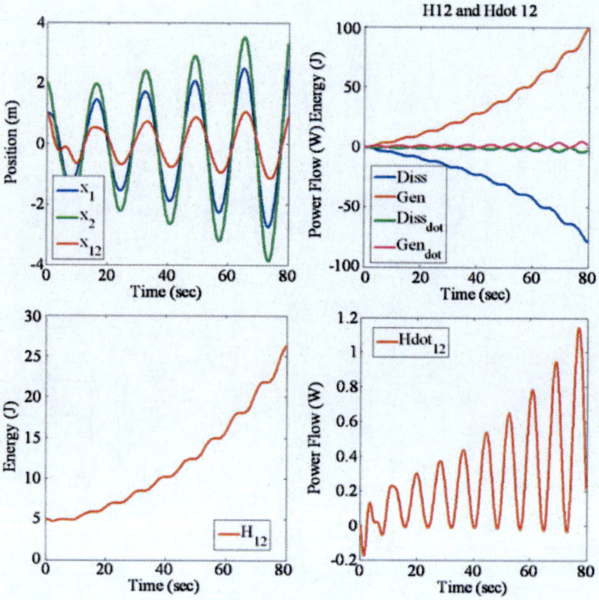

Fig. 6.24 Two-body dynamically unstable for ($m_1 > m_2$, $u_2 = F_2 = 0$)—linear MIMO system

from (6.3) are

$$\begin{aligned} m_1\ddot{x}_1 &= -k_1x_1 - k_{NL_1}x_1^3 - c_1\dot{x}_1 - c_{NL_1}\,\mathrm{sign}(\dot{x}_1) + k_{12}(x_2 - x_1) \\ &\quad + k_{NL_{12}}(x_2 - x_1)^3 + c_{12}(\dot{x}_2 - \dot{x}_1) + c_{NL_{12}}\,\mathrm{sign}(\dot{x}_2 - \dot{x}_1) + u_1, \\ m_2\ddot{x}_2 &= -k_{12}(x_2 - x_1) - k_{NL_{12}}(x_2 - x_1)^3 - c_{12}(\dot{x}_2 - \dot{x}_1) - c_{NL_{12}}\,\mathrm{sign}(\dot{x}_2 - \dot{x}_1), \end{aligned}$$

with the PID controller given in (6.15). The coupled Hamiltonian is

$$\begin{aligned} \mathcal{H}_{12} &= \frac{1}{2}m_1\dot{x}_1^2 + \frac{1}{2}m_2\dot{x}_2^2 + \frac{1}{2}(k_1 + K_{P_1})x_1^2 + \frac{1}{4}k_{NL_1}x_1^4 + \frac{1}{2}k_{12}(x_2 - x_1)^2 \\ &\quad + \frac{1}{4}k_{NL_{12}}(x_2 - x_1)^4, \end{aligned}$$

and resulting time derivative is determined as

$$\begin{aligned} \dot{\mathcal{H}}_{12} &= \left[-(c_1 + K_{D_1})\dot{x}_1 - c_{NL_1}\,\mathrm{sign}(\dot{x}_1) - K_{I_1}\int_0^t x_1\,d\tau\right]\dot{x}_1 \\ &\quad + [-c_{12}(\dot{x}_2 - \dot{x}_1) - c_{NL_{12}}\,\mathrm{sign}(\dot{x}_2 - \dot{x}_1)](\dot{x}_2 - \dot{x}_1). \end{aligned}$$

The performance of the coupled nonlinear system is evaluated. In particular, the limit cycle is identified for $\mathcal{H}_{12}$ as $\oint_\tau \dot{\mathcal{H}}_{12}\,dt = 0$. Numerical simulations are performed with the same system numerical values for the previous linear case with the addition of nonlinear terms: $k_{NL_1} = 100\ \mathrm{N/m^3}$, $c_{NL_1} = 2$ N, $k_{NL_{12}} = 10\ \mathrm{N/m^3}$, and $c_{NL_{12}} = 1$ N. The following u_1 controller gains are: $K_{P_1} = 1$ N/m, $K_{I_1} =$

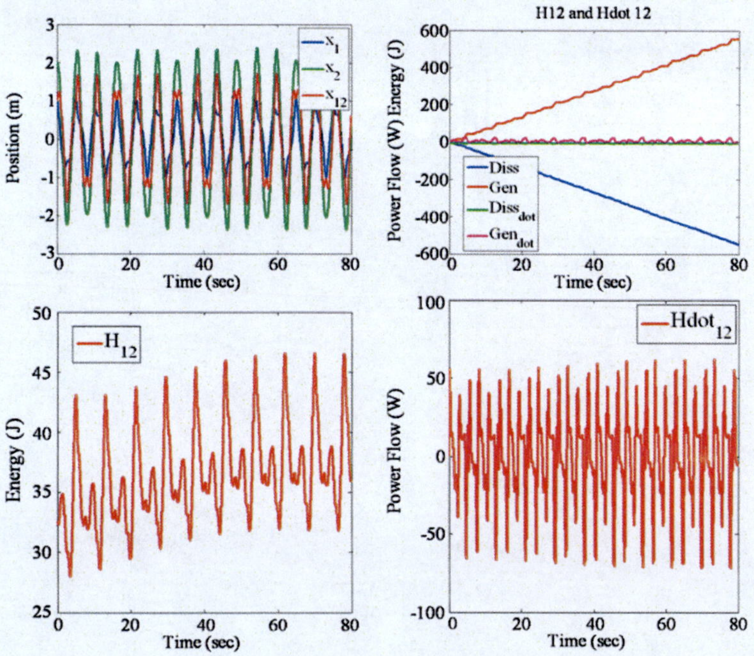

Fig. 6.25 Two-body dynamically neutral stability for ($m_1 > m_2, u_2 = F_2 = 0$)—nonlinear MIMO system

19.5 kg/s^3, and $K_{D_1} = 2$ kg/s. The initial conditions are $x_{01} = 1$ m, $x_{02} = 2$ m, and all velocities are set to zero.

The numerical results for the nonlinear system are exhibited in Fig. 6.25 for the positions (upper left), power flow and energy responses for the partitioned dissipation and generation terms (upper right), and the total energy $\mathcal{H}_{12}$ (bottom left) and power flow $\dot{\mathcal{H}}_{12}$ (bottom right) responses, respectively. Note that after the initial transient, the energy oscillates about a constant (see energy plot for $\mathcal{H}_{12}$, bottom left). For a decrease of $K_{I_1} = 2.0$, the system becomes stable. The results are given in Fig. 6.26 for the positions (upper left), power flow and energy responses for the partitioned dissipation and generation terms (upper right), and the total energy $\mathcal{H}_{12}$ (bottom left) and power flow $\dot{\mathcal{H}}_{12}$ (bottom right) responses, respectively. For an increase of $K_{I_1} = 100.0$ kg/s^3, the system becomes unstable and rises to another limit cycle [46]. The results are presented in Fig. 6.27 for the positions (upper left), power flow and energy responses for the partitioned dissipation and generation terms (upper right), and the total energy $\mathcal{H}_{12}$ (bottom left) and power flow $\dot{\mathcal{H}}_{12}$ (bottom right) responses, respectively. A van der Pol MIMO system is reviewed next.

The nonlinear MIMO system is defined as

$$\mathbf{M}\ddot{\mathbf{x}} + \mathbf{K}\mathbf{x} = -\mathbf{f}(\mathbf{x}, \dot{\mathbf{x}}) + \Delta\mathbf{u},$$

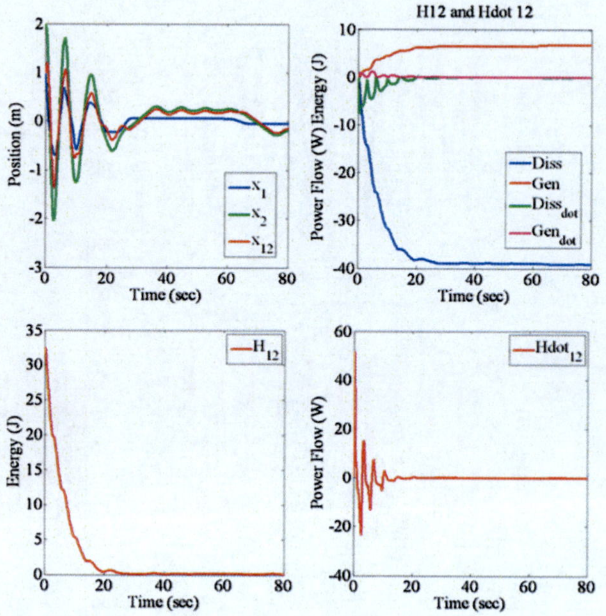

Fig. 6.26 Two-body dynamically asymptotically stable for ($m_1 > m_2$, $u_2 = F_2 = 0$)—nonlinear MIMO system

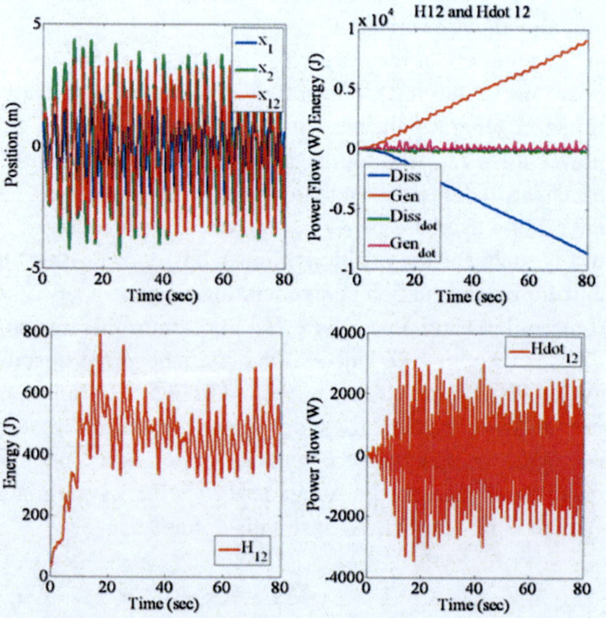

Fig. 6.27 Two-body dynamically unstable for ($m_1 > m_2$, $u_2 = F_2 = 0$)—nonlinear MIMO system

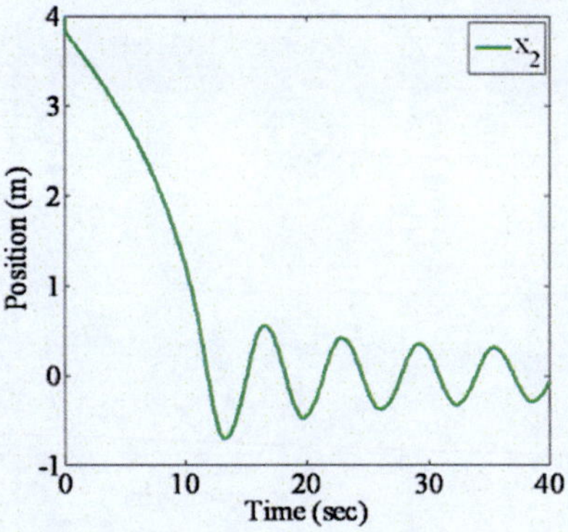

Fig. 6.28 Decoupled SISO control van der Pol systems: stable position response

where the van der Pol [46, 69, 70] damping is treated as the controller

$$\mathbf{f}(\mathbf{x},\dot{\mathbf{x}}) = \begin{Bmatrix} (K_{G_1} - K_{D_1}x_1^2)\dot{x}_1 - (K_{G_{12}} - K_{D_{12}}\Delta x^2)\Delta\dot{x} \\ (K_{G_2} - K_{D_2}x_2^2)\dot{x}_2 + (K_{G_{12}} - K_{D_{12}}\Delta x^2)\Delta\dot{x} \end{Bmatrix},$$

and $\Delta x = x_2 - x_1$. A straightforward way to decouple this system is to cancel F_{12}. A decoupling controller is defined as

$$\Delta\mathbf{u} = -\begin{Bmatrix} -\hat{F}_{12} \\ \hat{F}_{12} \end{Bmatrix}$$

that for ideal parameter matching $\hat{F}_{12} = F_{12}$ leads to

$$\begin{bmatrix} m_1 & 0 \\ 0 & m_2 \end{bmatrix}\begin{Bmatrix} \ddot{x}_1 \\ \ddot{x}_2 \end{Bmatrix} + \begin{bmatrix} k_1 & 0 \\ 0 & k_2 \end{bmatrix}\begin{Bmatrix} x_1 \\ x_2 \end{Bmatrix} = \begin{Bmatrix} (K_{G_1} - K_{D_1}x_1^2)\dot{x}_1 \\ (K_{G_2} - K_{D_2}x_2^2)\dot{x}_2 \end{Bmatrix},$$

which produces two independent SISO controller designs. Numerical simulations are performed with the following numerical values $m_1 = m_2 = 1.0$ kg, $k_1 = k_2 = 1.0$ N/m, and $k_{12} = 0$ N/m. To define the SISO controller, the second DOF is investigated for both stable and unstable solutions. First set the following values: $K_{D_2} = 1.5$ and $K_{G_2} = K_{D_1} = K_{G_1} = 0.0$. The initial conditions are $x_{01} = 0.1$ m, $\dot{x}_{01} = 0.1$ m/s, $x_{02} = 4.0$ m, and $\dot{x}_{02} = 4.0$ m/s. These parameters are used for the following numerical simulations, except where otherwise specified. The stable solution is demonstrated through numerical simulation with the decaying position response presented in Fig. 6.28. The corresponding phase plane projection (left) and power flow and energy responses (right) are shown in Fig. 6.29, respectively. For the unstable case, $K_{D_2} = 0.0$ and $K_{G_2} = 0.1$. This results in an unstable response given in Fig. 6.30 with the corresponding phase plane projection (left) and power flow and energy responses (right) are shown in Fig. 6.31.

The neutral stability condition for each SISO system is defined by

$$\oint_\tau \dot{\mathcal{H}}_i\,dt = 0 = \oint_\tau \dot{x}_i\left[(K_{G_i} - K_{D_i}x_i^2)\dot{x}_i\right]dt.$$

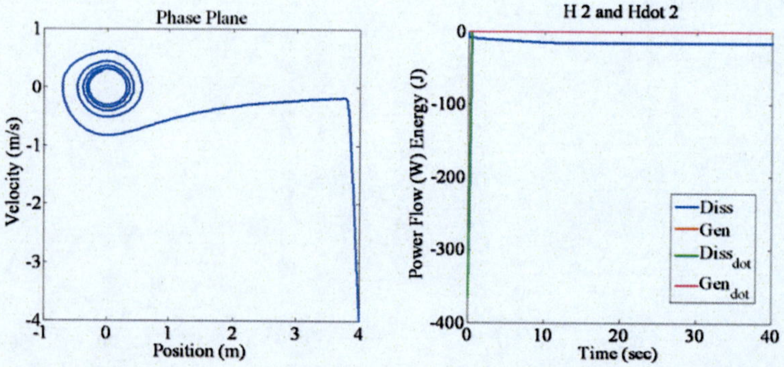

Fig. 6.29 Decoupled SISO control van der Pol systems: stable phase plane, power flow and energy responses

Fig. 6.30 Decoupled SISO control van der Pol systems: unstable position response

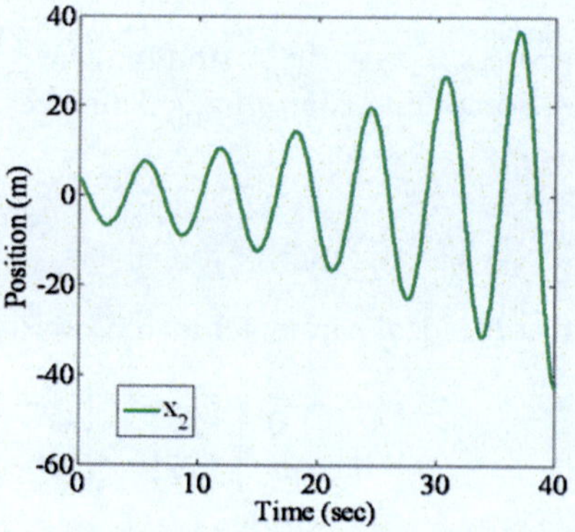

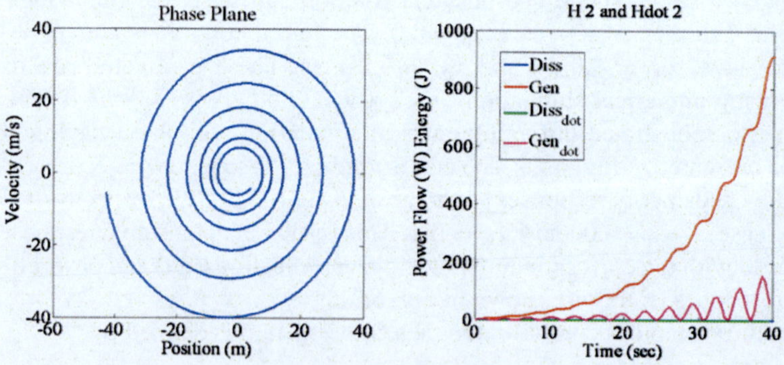

Fig. 6.31 Decoupled SISO control van der Pol systems: unstable phase plane, power flow, and energy responses

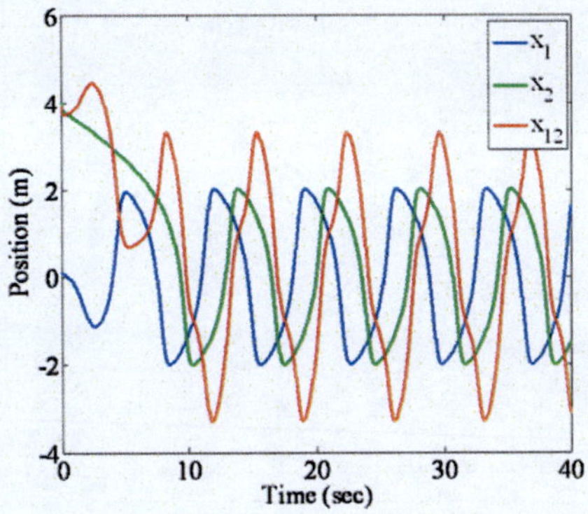

Fig. 6.32 Two-body dynamically neutral stability decoupled control van der Pol system

This is demonstrated through numerical simulation with the following controller gain settings: $K_{G_1} = K_{G_2} = K_{D_1} = K_{D_2} = 1.5$, respectively. These results are shown in Fig. 6.32 for the positions. Limit cycles and power flow partitioned into dissipation and generation are shown for x_1, x_2, and x_{12}. Once again, the dissipation and generation are balanced after the transients settle out. x_1 was started inside the limit cycle and generates power to move onto the limit cycle, and x_2 was started outside the limit cycle and dissipates power to move onto the limit cycle (see Fig. 6.33). The total energy (left) and power flow (right) for x_{12} are presented in Fig. 6.34.

In comparison, a coupled controller is derived from the combined Hamiltonian

$$\mathcal{H}_{12} = \frac{1}{2}\dot{\mathbf{x}}^T \mathbf{M}\dot{\mathbf{x}} + \frac{1}{2}\mathbf{x}^T \mathbf{K}\mathbf{x}$$

and the time derivative

$$\dot{\mathcal{H}}_{12} = \dot{\mathbf{x}}^T [\mathbf{M}\ddot{\mathbf{x}} + \mathbf{K}\mathbf{x}] = \dot{\mathbf{x}}^T [-\mathbf{f}(\mathbf{x}, \dot{\mathbf{x}})]$$

with $\Delta\mathbf{u} = \mathbf{0}$. The neutral stability condition for the coupled system is defined by

$$\oint_\tau \dot{\mathcal{H}}_{12}\, dt = 0 = \oint_\tau \dot{\mathbf{x}}^T [-\mathbf{f}(\mathbf{x}, \dot{\mathbf{x}})]\, dt.$$

Numerical simulations are performed with the previous example numerical values. The coupled system also includes the off-diagonal stiffness terms $k_{12} = 5.0$ N/m and controller gains $K_{G_{12}} = K_{D_{12}} = 1.5$, respectively.

These results are given in Fig. 6.35 for the positions. Phase plane projections and power flow partitioned into dissipation and generation are presented for the coupled 12 system. The dissipation and generation are balanced after the transients settle out. The individual phase plane projections and combined phase plane projections are illustrated in Fig. 6.36. In addition, the power flow balance for the combined system is exhibited in the bottom right of Fig. 6.36. The total energy (left) and power flow (right) for the combined system are shown in Fig. 6.37. The neutral boundary condition or limit cycle for the nonlinear MIMO system was demonstrated.

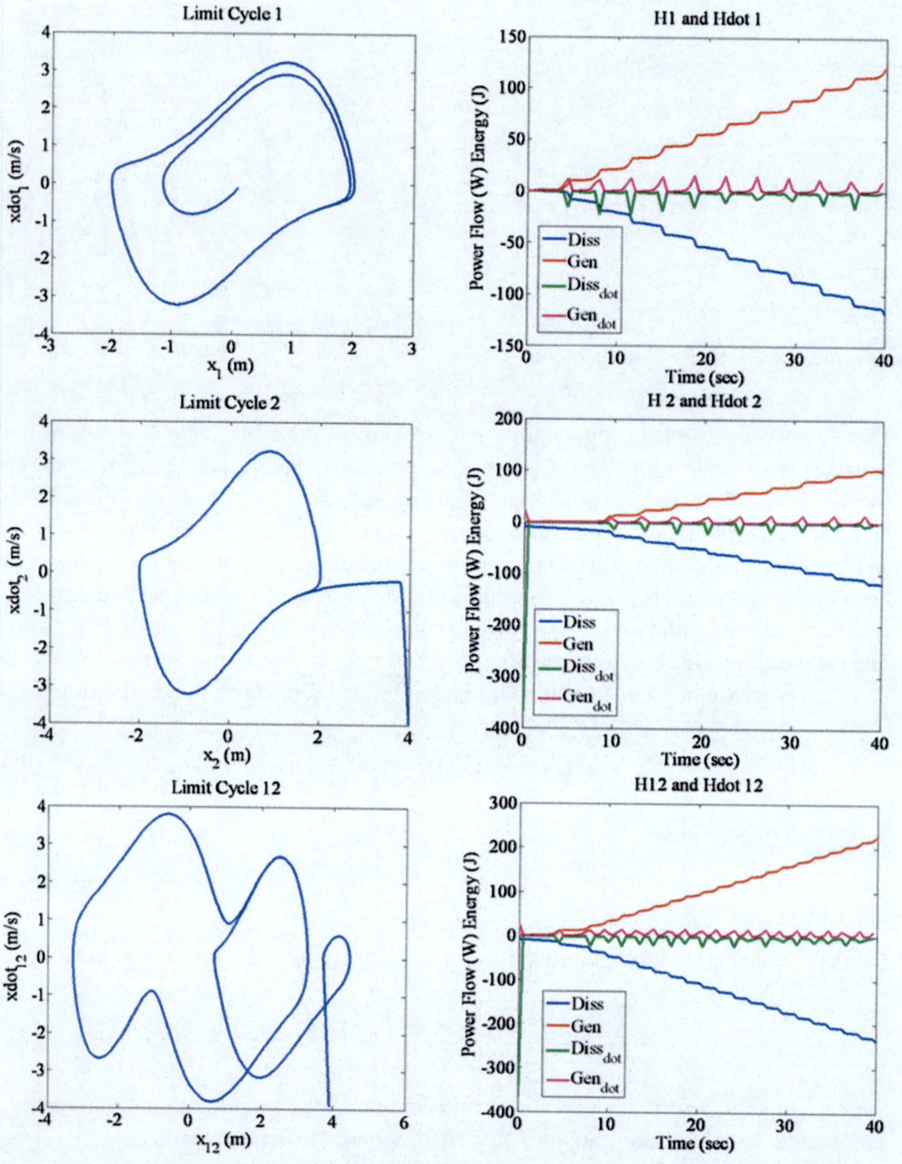

Fig. 6.33 Two-body dynamically neutral stability decoupled van der Pol system

6.5 Noncollocated Control

The next control design issue to be addressed is noncollocated control. Typically, this happens when the actuator is separated from the sensor by a flexible connection. For a linear system, the transfer function is nonminimum phase with zeros

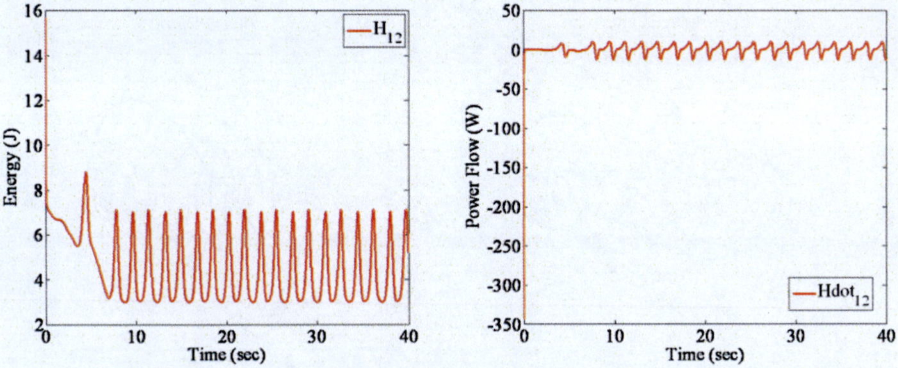

Fig. 6.34 Two-body dynamically neutral stability decoupled control van der Pol system

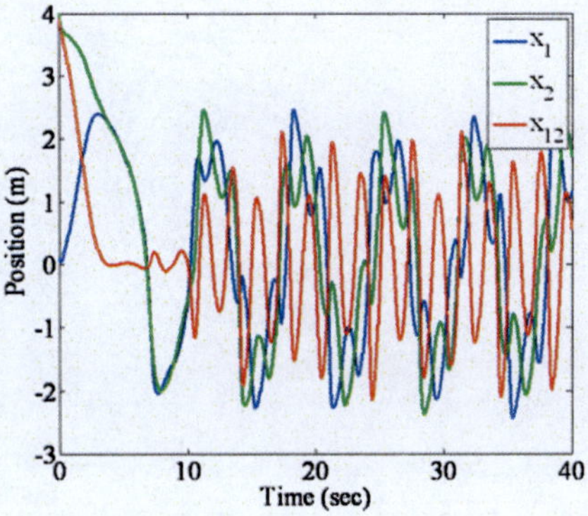

Fig. 6.35 Two-body dynamically neutral stability coupled control van der Pol system

in the right-half plane. This section extends this concept to nonlinear systems with HSSPFC.

The previous two-mass system (see Figs. 6.11 and 6.12) is used to demonstrate the effects of noncollocated control on nonlinear systems. The dynamic model is:

$$\begin{aligned}
m_1\ddot{x}_1 &= -g_1(x_1) - f_1(\dot{x}_1) - g_{12}(x_1 - x_2) - f_{12}(\dot{x}_1 - \dot{x}_2) + u_1,\\
m_2\ddot{x}_2 &= -g_{12}(x_2 - x_1) - f_{12}(\dot{x}_2 - \dot{x}_1) + u_2,\\
g_2(x_2) &= f_2(\dot{x}_2) = 0.
\end{aligned}$$

The individual Hamiltonian and time derivatives are:

$$\begin{aligned}
\mathcal{H}_1 &= \frac{1}{2}m_1\dot{x}_1^2 + \mathcal{V}_1(x_1),\\
\dot{\mathcal{H}}_1 &= [m_1\ddot{x}_1 + g_1(x_1)]\dot{x}_1 = [-f_1(\dot{x}_1) - f_{12}(\dot{x}_1 - \dot{x}_2) - g_{12}(x_1 - x_2) + u_1]\dot{x}_1,\\
\mathcal{H}_2 &= \frac{1}{2}m_2\dot{x}_2^2,\\
\dot{\mathcal{H}}_2 &= [m_2\ddot{x}_2]\dot{x}_2 = [-g_{12}(x_2 - x_1) - f_{12}(\dot{x}_2 - \dot{x}_1) + u_2]\dot{x}_2.
\end{aligned}$$

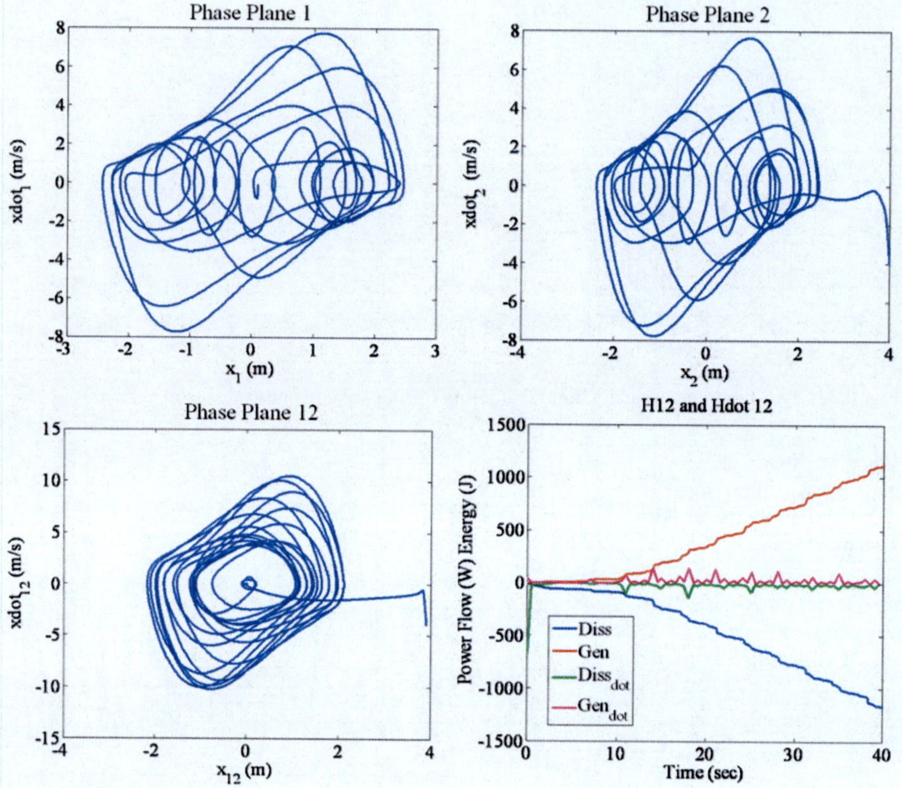

Fig. 6.36 Two-body dynamically neutral stability coupled control van der Pol system

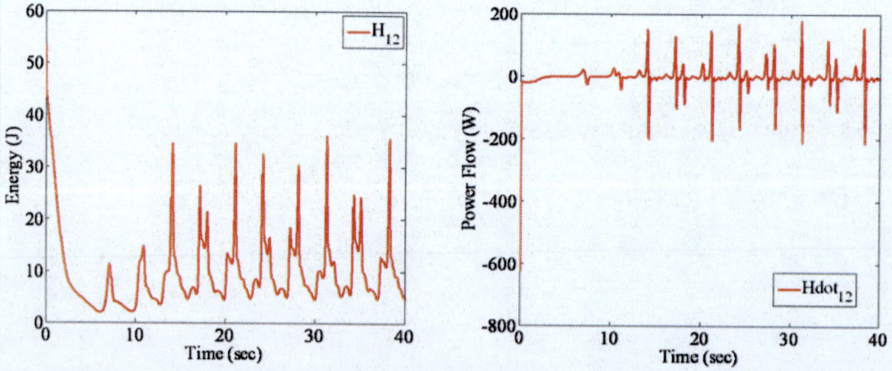

Fig. 6.37 Two-body dynamically neutral stability coupled control van der Pol system

The coupled Hamiltonian and time derivative are:

$$\begin{aligned}
\mathcal{H}_{12} &= \frac{1}{2}m_1\dot{x}_1^2 + \frac{1}{2}m_2\dot{x}_2^2 + \mathcal{V}_1(x_1) + \mathcal{V}_{12}(x_1 - x_2),\\
\dot{\mathcal{H}}_{12} &= [m_1\ddot{x}_1 + g_1(x_1)]\dot{x}_1 + [m_2\ddot{x}_2]\dot{x}_2 + g_{12}(x_1 - x_2)(\dot{x}_1 - \dot{x}_2)\\
&= [m_1\ddot{x}_1 + g_1(x_1) + g_{12}(x_1 - x_2)]\dot{x}_1 + [m_2\ddot{x}_2 + g_{12}(x_2 - x_1)]\dot{x}_2\\
&= [-f_1(\dot{x}_1) - f_{12}(\dot{x}_1 - \dot{x}_2) + u_1]\dot{x}_1 + [-f_{12}(\dot{x}_2 - \dot{x}_1) + u_2]\dot{x}_2\\
&= [-f_1(\dot{x}_1) + u_1]\dot{x}_1 + [u_2]\dot{x}_2 + [-f_{12}(\dot{x}_1 - \dot{x}_2)](\dot{x}_1 - \dot{x}_2).
\end{aligned}$$

Now, one can design a controller based on picking $\mathcal{V}_1(x_1)$ and $\mathcal{V}_{12}(x_1 - x_2)$ as positive definite which implies driving the system to (0,0) while maintaining static stability. To enable a comparison between collocated and noncollocated, the first controller is collocated with

$$\begin{aligned}
u_1 &= -K_{P_1}x_1 - K_{I_1}\int_0^t x_1\,d\tau - K_{D_1}\dot{x}_1,\\
u_2 &= -K_{P_2}x_2 - K_{I_2}\int_0^t x_2\,d\tau - K_{D_2}\dot{x}_2 \quad \text{set to} \quad u_2 = 0,
\end{aligned}$$

which results in the following coupled Hamiltonian and time derivative:

$$\begin{aligned}
\mathcal{H}_{12} &= \frac{1}{2}m_1x_1^2 + \frac{1}{2}m_2x_2^2 + \frac{1}{2}K_{P_1}x_1^2 + \mathcal{V}_1(x_1) + \mathcal{V}_{12}(x_1 - x_2)\\
\dot{\mathcal{H}}_{12} &= [m_1\ddot{x}_1 + g_1(x_1) + K_{P_1}x_1]\dot{x}_1 + [m_2\ddot{x}]\dot{x}_2 + g_{12}(x_1 - x_2)(\dot{x}_1 - \dot{x}_2)\\
&= [m_1\ddot{x}_1 + g_1(x_1) + g_{12}(x_1 - x_2) + K_{P_1}x_1]\dot{x}_1 + [m_2\ddot{x}_2 + g_{12}(x_2 - x_1)]\dot{x}_2\\
&= \left[-f_1(\dot{x}_1) - K_{D_1}\dot{x}_1 - K_{I_1}\int_0^t x_1 d\tau\right]\dot{x}_1 + [-f_{12}(\dot{x}_1 - \dot{x}_2)](\dot{x}_1 - \dot{x}_2).
\end{aligned}$$

The collocated controller is both statically stable and dynamically stable for the correct choices of controller gains based on previous examples.

The second controller is noncollocated with

$$\begin{aligned}
u_1 &= -K_{P_1}x_2 - K_{I_1}\int_0^t x_2\,d\tau - K_{D_1}\dot{x}_2,\\
u_2 &= 0,
\end{aligned}$$

which results in the following coupled Hamiltonian and time derivative:

$$\begin{aligned}
\mathcal{H}_{12} &= \frac{1}{2}m_1\dot{x}_1^2 + \frac{1}{2}m_2\dot{x}_2^2 + \mathcal{V}_1(x_1) + \mathcal{V}_{12}(x_1 - x_2),\\
\dot{\mathcal{H}}_{12} &= [m_1\ddot{x}_1 + g_1(x_1) + g_{12}(x_1 - x_2)]\dot{x}_1 + [m_2\ddot{x}_2 + g_{12}(x_2 - x_1)]\dot{x}_2\\
&= [-f_1(\dot{x}_1) - f_{12}(\dot{x}_1 - \dot{x}_2) + u_1]\dot{x}_1 + [-f_{12}(\dot{x}_2 - \dot{x}_1)]\dot{x}_2
\end{aligned}$$

$$= [-f_1(\dot{x}_1) + u_1]\dot{x}_1 + [-f_{12}(\dot{x}_1 - \dot{x}_2)](\dot{x}_1 - \dot{x}_2)$$

$$= \left[-f_1(\dot{x}_1) - K_{P_1}x_2 - K_{I_1}\int_0^t x_2 d\tau - K_{D_1}\dot{x}_2\right]\dot{x}_1$$

$$+ [-f_{12}(\dot{x}_1 - \dot{x}_2)](\dot{x}_1 - \dot{x}_2).$$

To provide better insight into the problem, let $\Delta x = x_2 - x_1$, which implies that $x_2 = x_1 + \Delta x$; then the power flow becomes

$$\dot{\mathcal{H}}_{12} = \left[-f_1(\dot{x}_1) - K_{P_1}(x_1 + \Delta x) - K_{I_1}\int_0^t (x_1 + \Delta x)\, d\tau - K_{D_1}(\dot{x}_1 + \Delta\dot{x})\right]\dot{x}_1$$

$$+ [-f_{12}(\Delta\dot{x})]\Delta\dot{x}$$

$$= \left[-f_1(\dot{x}_1) - K_{P_1}x_1 - K_{I_1}\int_0^t x_1\, d\tau - K_{D_1}\dot{x}_1\right]\dot{x}_1 + [-f_{12}(\Delta\dot{x})]\Delta\dot{x}$$

$$+ \left\{-K_{P_1}\Delta x - K_{I_1}\int_0^t \Delta x\, d\tau - K_{D_1}\Delta\dot{x}\right\}\dot{x}_1,$$

where the term in the brackets from the previous equation is defined as

$$\Delta\dot{\mathcal{H}}_{12} = -\left\{K_{P_1}\Delta x + K_{I_1}\int_0^t \Delta x d\tau + K_{D_1}\Delta\dot{x}\right\}\dot{x}_1 = \Delta u_{\Delta x},$$

which is the destabilizing power flow that limits the feedback gains. The question of interest is: "How large is this term?" The following discussion and plots show the effects of noncollocated control relative to collocated control.

The numerical examples presented in the following figures are based on the Duffing oscillator combined with Coulomb friction models with parameter values given as: $m_1 = m_2 = 1.0$ kg, $k_1 = 1.0$ N/m, $k_2 = 0.0$ N/m, $c_1 = 0.1$ kg/s, $c_2 = 0.0$ kg/s, $k_{NL_1} = 100.0$ N/m^3, $k_{NL_2} = 0.0$ N/m^3, $c_{NL_1} = 1.0$ N, $c_{NL_2} = 0.0$ N, $k_{12} = 5.0$ N/m, and $c_{12} = 1.0$ kg/s. The PID controller gain values are given in Table 6.2 for each case. Specifically, the model functions are

$$\begin{aligned} g_1(x_1) &= k_1x_1 + k_{NL}x_1^3, \\ g_{12}(x_1 - x_2) &= k_{12}(x_1 - x_2), \\ f_1(\dot{x}_1) &= c_1\dot{x}_1 + c_{NL}\text{sign}(\dot{x}_1), \\ f_{12}(\dot{x}_1 - \dot{x}_2) &= c_{12}(\dot{x}_1 - \dot{x}_2). \end{aligned}$$

The effects of noncollocated control are determined relative to a reference collocated control case with feedback on both masses. The parameter values are previously given, and the PID control gains for the stable collocated control are given in Table 6.2, Case 1.

Three noncollocated control cases are simulated to determine the stability boundary for the integral gain. Specifically, the u_1 control is noncollocated, and $u_2 = 0$, which produce the unstable Case 2, the neutral Case 3, and the stable Case 4 in Table 6.2, respectively.

Table 6.2 Collocated/noncollocated PID gain values

Case	Description	K_{P_1} N/m	K_{I_1} kg/s^3	K_{D_1} kg/s	K_{P_2} N/m	K_{I_2} kg/s^3	K_{D_2} kg/s
1	Collocated (stable)	1.0	1.2	1.0	1.0	1.2	1.0
2	Noncollocated (unstable)	1.0	1.2	1.0	0.0	0.0	0.0
3	Noncollocated (neutral)	1.0	1.038	1.0	0.0	0.0	0.0
4	Noncollocated (stable)	1.0	0.12	1.0	0.0	0.0	0.0

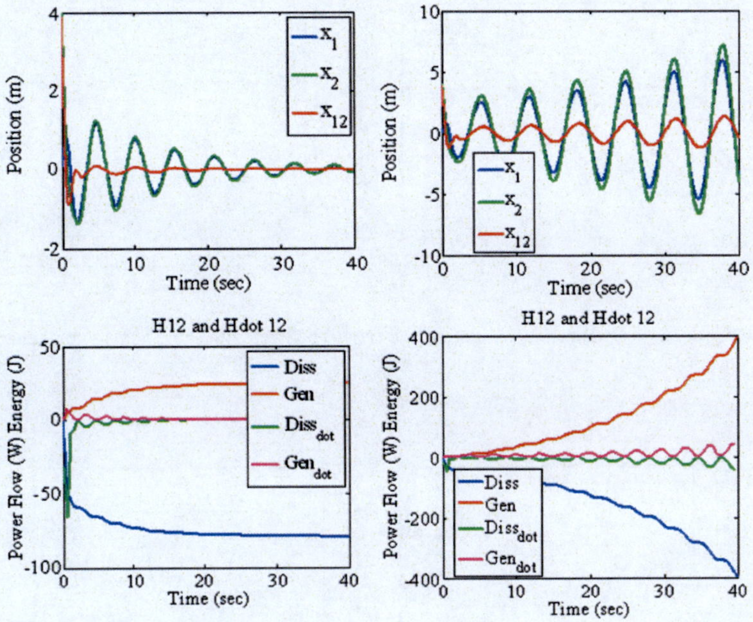

Fig. 6.38 Collocated (*left*) versus noncollocated (*right*) control with same gains—positions and power flows

Figures 6.38, 6.39, and 6.40 compare the performances of the reference collocated case and the unstable noncollocated case. Note that the parameter values and controller gains are the same except for $u_2 = 0$. One can observe the typical behavior of stable and unstable systems which include damped and amplified motion, decreasing and increasing energy, and decreasing and increasing power flows. Lastly, the destabilizing effects of noncollocated control are presented in Fig. 6.39 (bottom right).

Figures 6.41, 6.42, and 6.43 compare the performances of neutral and stable noncollocated control cases. From Table 6.2 note that the integral gain is reduced to 1.038 for neutral stability and 0.12 for the stable case. This set of figures presents the dynamic stability boundary for the noncollocated control of the two-mass system. Also, the integral gain for the stable case was reduced such that the noncollocated

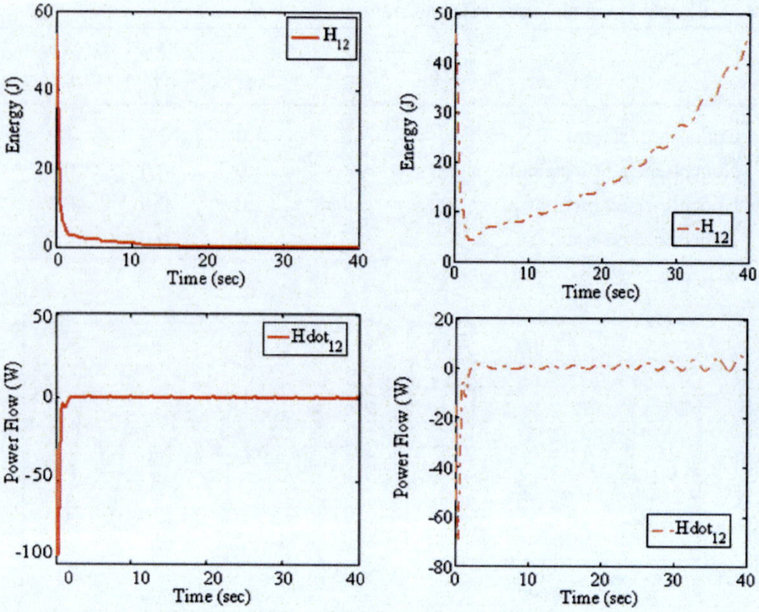

Fig. 6.39 Collocated (*left*) versus noncollocated (*right*) control with same gains—total energy and power

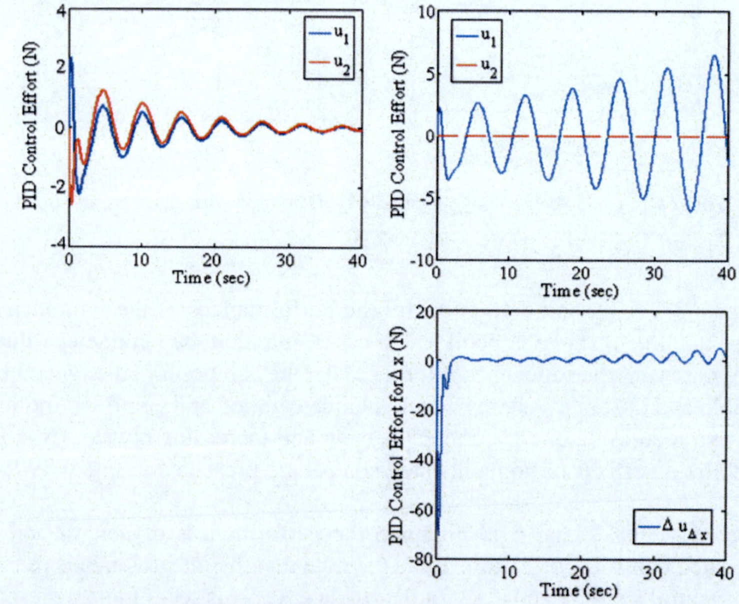

Fig. 6.40 Collocated (*left*) versus noncollocated (*right*) control with same gains—control effort and destabilizing control

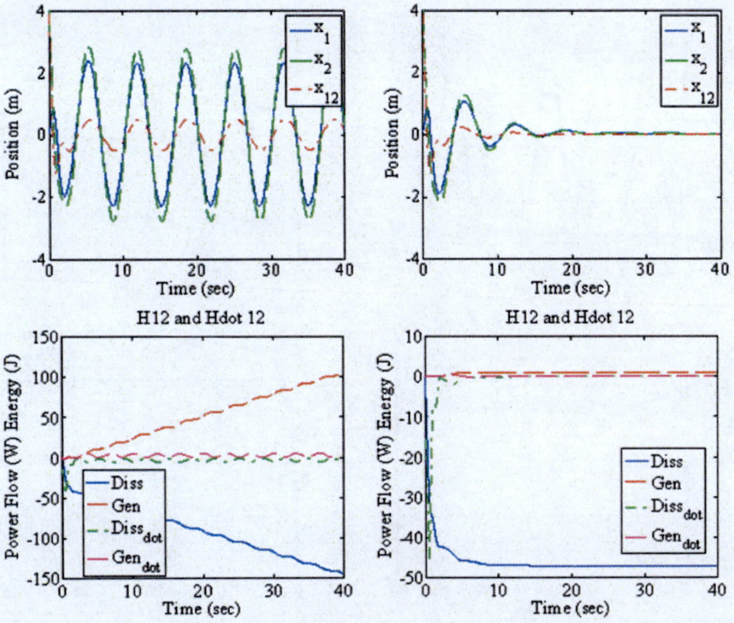

Fig. 6.41 Noncollocated control: neutral (*left*) versus stable (*right*)—positions and power flows

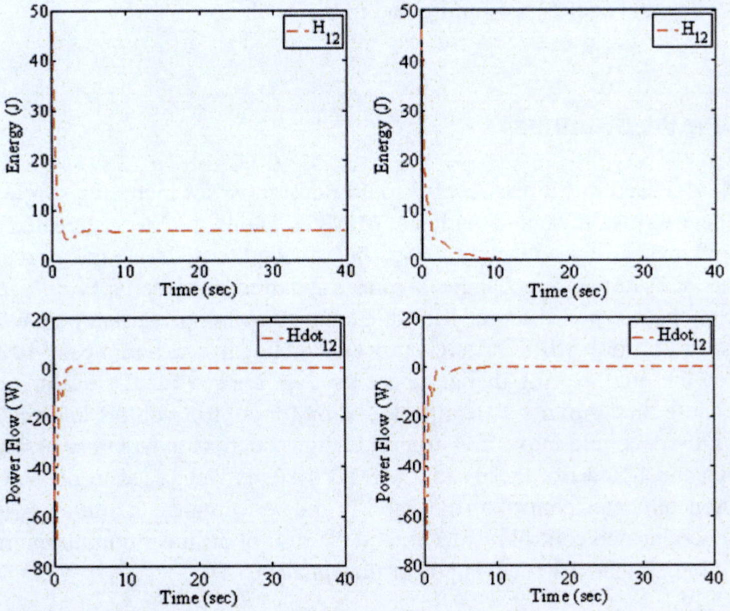

Fig. 6.42 Noncollocated control: neutral (*left*) versus stable (*right*)—total energy and power

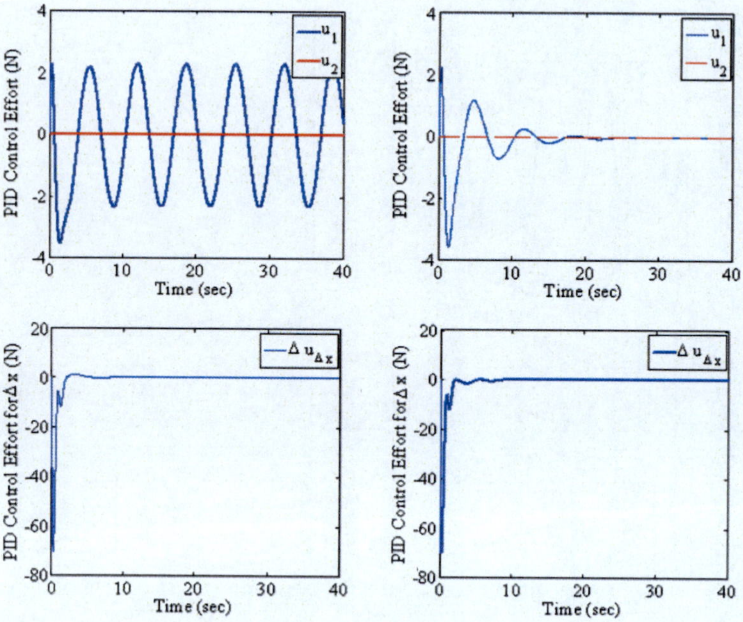

Fig. 6.43 Noncollocated control: neutral (*left*) versus stable (*right*)—control effort and destabilizing control

control damps faster than the collocated control. Finally, the destabilizing effects of the noncollocated control are suppressed in Fig. 6.43.

6.6 Chapter Summary

HSSPFC was used to address several control design issues including single nonlinear feedback terms, linear and nonlinear MIMO systems, and noncollocated control. A sinusoid nonlinear feedback term was investigated to illustrate the effect of nonlinear feedback terms that changed sign as a function of velocity. Limit cycles can result from these types of terms. Multiple coupled mass-spring-dampers were used to demonstrate the HSSPFC design process for linear and nonlinear MIMO systems. Coupled mass-spring-damper systems were chosen for the example MIMO cases because they provide a clear presentation for control volume analysis and do not naturally decouple into SISO control designs such as gyroscopic systems (see Chaps. 11 and 12). Also, these same MIMO systems were used to present the effects of noncollocated control on the stability and performance of nonlinear systems. These noncollocated control results extend the insight of linear nonminimum phase systems to nonlinear systems via power flow analysis.

Chapter 7
Case Study #2: Collective Systems and Controls

7.1 Introduction

Collective behaviors exist in many diversified fields from man-made systems to biological environments. Most of these collective behaviors are created by distributed decentralized control architectures on each agent. To make multiagent collective systems efficient new challenges need to be addressed in cooperative decentralized distributed control strategies. These strategies need to optimally leverage the large number of agents, be robust to single-points of failure, and use effectively the information that is communicated interactively among each individual in the collective system. Many times a complex system of many agents, such as a robotic or spacecraft collectives, are desired to act in a coordinated collective emerging behavior with autonomous decisions capabilities to achieve a specific mission objective. This mission space may include, for example: chemical plume tracing, multirobot planetary exploration, cooperative multi-UAV systems, and satellite collectives for cooperative/coordinated observations.

Distributed cooperative control of multiagent or robot collective systems is an active area of research. A unified view of control and communication is investigated by [71] that clarifies many of the issues underlying the distributed control problem. The formulation is cast as a dynamic programming problem that aides in understanding the fundamental limits to performance in distributed systems when there are channel constraints. Ilaya [72] suggested a theoretical framework for multivehicle control where the collective flock was modeled using geometric methods. The multivehicle control problem was constructed as a hierarchical control with a high-level supervisory control perspective and local vehicle guidance, navigation with control components. A decentralized control scheme based on decentralized model predictive control was introduced at the vehicle level to synthesize cooperative and self-organized behavior. The results demonstrated optimal closed-loop stable and self-organized behavior. Izzo and Pettazzi [73] investigated satellite path-planning to make a group of spacecraft acquire a given configuration. A behavior-based coordination mechanism approach achieved an autonomous and distributed control capability while using only limited sensor information. With a limited number of

R.D. Robinett III, D.G. Wilson, *Nonlinear Power Flow Control Design*,
Understanding Complex Systems,
DOI 10.1007/978-0-85729-823-2_7,

defined behaviors, it was possible to establish several varying numbers of flight formations.

Chapter 7 develops and extends the collective robotic research at Sandia [35, 38, 74–80] with the application of HSSPFC and information theory applied to distributed decentralized control of collective plume tracing. The collective robotics research at Sandia includes the work of Feddema, Robinett, and Byrne [74], where a three-step process is presented that creates locally optimal distributed controls for multiple robot vehicles. This approach was tested in simulation with hardware that included distribution of multiple robots [74–79], following lines/curves, unconstrained/constrained 2-D planes, elliptical curves, and convergence on the source of a plume in a plane and a 3-D volume. This approach is based on the optimization of a global performance index that includes only nearest-neighbor information. One of the hardware demonstrations utilized miniature robot platforms to demonstrate cooperative control of robot collectives [79].

In Feddema et al. [75], stability relationships are determined for cooperative robots with respect to control and communication. Berg [38] considered the problem of optimal decentralized estimation and developed practical aspects of nodal network implementation and identified important benchmarks for functionally decentralized systems designs. The results include an optimal, distributed algorithm that runs in parallel on systems of decentralized processors to accomplish robust, fault tolerant estimation, tracking, state update, and sensor fusion. Further improvements and developments by Robinett and Hurtado [35] include distributed feedback control laws for cooperative robotic systems whose task was to localize unknown sources. In addition, Hurtado, Robinett, Dohrmann, and Goldsmith [36] further developed distributed control for cooperative robotic systems that are tasked to localize a time-invariant, stationary source that emits a measurable scalar field.

The main contribution of Chap. 7 is to present the application of HSSPFC for the design and analysis of distributed decentralized control of collective systems in the context of equilibrium and nonequilibrium thermodynamics and information theory. The Hamiltonian function (see Fig. 7.1) is a basic ingredient in the design of collective systems and the concepts of physical and information exergies (see Chaps. 2, 3, 5, 14 and references [7, 12, 30, 33, 34, 37, 50]) intrinsic to collective systems. In particular, physical and information exergies are used for Hamiltonian surface shaping and are shown to be equivalent based on thermodynamics, Hamiltonian mechanics, and Fisher Information.

This chapter is divided into six sections. Section 7.2 applies equilibrium thermodynamics to the collective to determine a bounding solution. Section 7.3 designs a collective kinematic controller for a team of robot agents, describes the testing robot characteristics, and presents the information flow trade-off between processing, memory, and communications requirements. Section 7.4 designs a collective kinetic controller for a team of robots and presents the stability plus performance analysis with numerical simulations of the collective kinetic control law for plume tracing. Section 7.5 presents the connection between Fisher Information and physical and information exergies while defining the Fisher Information Equivalency. Finally, Sect. 7.6 summarizes the results with concluding remarks.

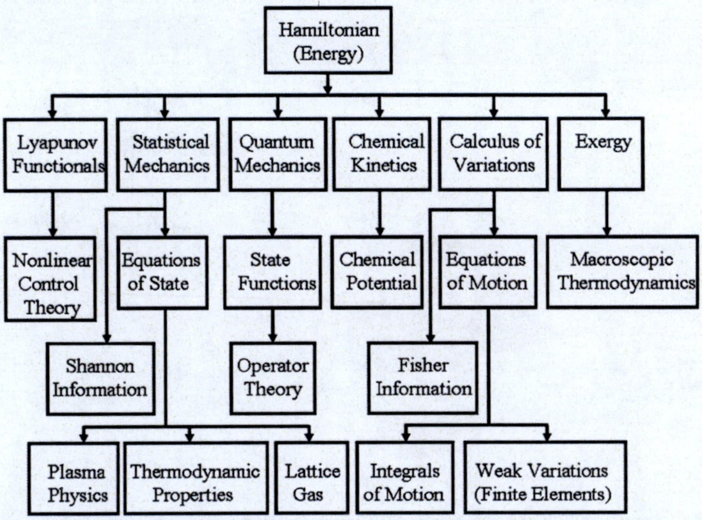

Fig. 7.1 Flowchart describing mechanics based approaches for collective systems. Robinett III, R.D. and Wilson, D.G. [37], reprinted by permission of the publisher (Interscience Publishers)

7.2 Equilibrium Collective Systems

One way to describe collective systems is by way of a mechanics-based analogy. For example, the aerodynamic coefficients (lift, drag, etc.) for a reentry vehicle, such as the Space Shuttle, moving from the exoatmosphere (which is modeled using rarified gas dynamics) to sea level (which is modeled using continuum mechanics) are analogous to the parameters that describe the dynamics of small numbers of robots to millions. Figure 7.2 depicts a fluid mechanics analogy of a swarm of robotic agents. The basic controlling parameter is the mean free path between collisions of molecules which relates to the size of particles (or robots) and density. To be specific, the mean free path is defined as [81]

$$\lambda = \frac{1}{\sqrt{2}\pi d^2 \bar{n}},$$

where

$$d = \text{diameter of the molecule},$$
$$\bar{n} = \text{number of molecules per unit volume}.$$

The mean free path determines the method used to calculate the aerodynamic coefficients from rarified gas dynamics which models many individual molecules in a Monte Carlo setting, through a transitional phase which mixes individual molecule models with continuum models, to continuum mechanics. This analogy, robot diameter and robot density, leads directly to the applications of statistical mechanics, continuum mechanics, calculus of variations, chemical kinetics, and quantum mechanics, which address the analysis of a single particle (agent) to millions of particles (agents).

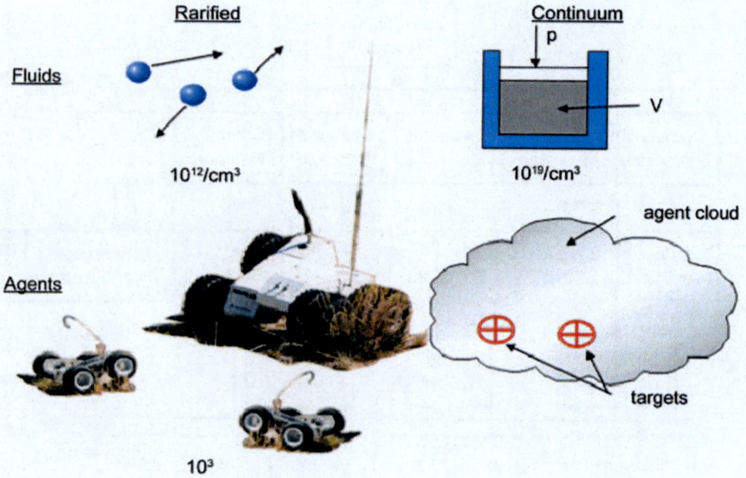

Fig. 7.2 Transition from discrete to continuum models (illustrative example) [7]

The basic mechanics-based strategy of analyzing and designing collective systems is presented in Fig. 7.1. The fundamental building block is the Hamiltonian function which is the total energy for conservative systems (i.e., external forces can be modeled as potential functions) [26]. In Sect. 3.5, the Hamiltonian was redefined as the stored exergy in a system based on an irreversible thermodynamics interpretation.

In general, these techniques can be divided into microscopic and macroscopic tools. Calculus of variations and quantum mechanics are typically used for microscopic analysis, whereas statistical mechanics and chemical kinetics are used for macroscopic analysis, even though there exists some overlap among these techniques. *Microscopic tools are necessary to complete the inverse problem: transform the desired or designed macroscopic/collective behavior into the microscopic/individual behavior of each agent.*

To be more specific, at the core of the mechanics-based strategy is Hamilton's principle. The Hamiltonian is a scalar function that is used to develop the evolution of dynamical systems; these dynamical systems can be either deterministic or statistical. Hamilton's principle assumes that the systems under consideration are characterized by two energy (stored exergy) functions, a kinetic energy and a potential energy. This chapter utilizes the extended Hamilton's principle [6] (also see Chap. 3) which accounts for nonconservative forces to connect Hamiltonian mechanics, irreversible and nonequilibrium thermodynamics, nonlinear control theory (from Lyapunov functionals), and self-organizing systems to collective systems by way of information theory. The rest of this chapter develops the statistical mechanics and calculus of variations pathways which culminate in equivalences between physical and information exergies for equilibrium and nonequilibrium collective systems.

Statistical and continuum mechanics involve the analysis of the collective behavior of large groups of objects, usually molecules, by relatively simple macroscopic

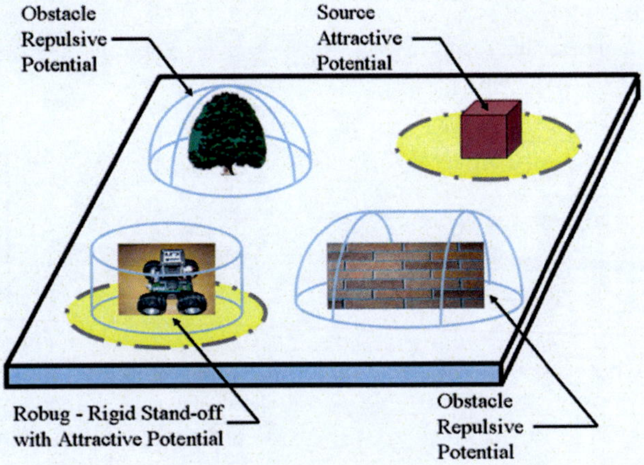

Fig. 7.3 Source/target detection and localization [7]

means. Many mathematical and intuitive rules have been developed that allow relatively easy handling of problems that would become quickly intractable if particles were to be analyzed individually.

One of the simplest examples of these relationships is the ideal gas law, which provides a simple algebraic relationship between the pressure, volume, and temperature of an ideal gas. In reality, an ideal gas consists of many individual particles, on the order of 10^{20} molecules per cubic centimeter, with the particles following much more complex relationships than the ideal gas law itself. However, the application of statistical mechanics to derive a continuum approximation allows this otherwise intractable problem to be handled easily. This is accomplished by generating a Hamiltonian, forming a partition function, and calculating the mean values. This framework is used to solve an inverse control problem. The inverse control problem involves envisioning a collective behavior of a swarm of robots that is desirable and from this determining the set of rules that individual robots must follow to produce the desired swarm behavior in the form of interaction potential fields.

To demonstrate these concepts with an example, it is given that a swarm of robots will be used to search a volume for a target (see Fig. 7.3). One can analyze the long-term (as $t \to \infty$) collective behavior of the robot swarm by analogy with equilibrium thermodynamics. The application of statistical mechanics begins with generating a Hamiltonian for an ideal gas of noninteracting (noncommunicating) robots within an empty volume (see Fig. 7.4),

$$\mathcal{H} = \mathcal{T} + \mathcal{V} = E = \sum_{i=1}^{N} \frac{p_i^2}{2m} + \mathcal{V}(\mathbf{r}), \tag{7.1}$$

followed by forming a partition function [82]

$$Z = \iint e^{-\beta E} \frac{d^f \mathbf{p}\, d^f \mathbf{r}}{h^f} = \frac{1}{N!} \left[\iint e^{-\beta(\mathbf{p}^2/2m + \mathcal{V})} \frac{d^3\mathbf{p}\, d^3\mathbf{r}}{h^3} \right]^N, \tag{7.2}$$

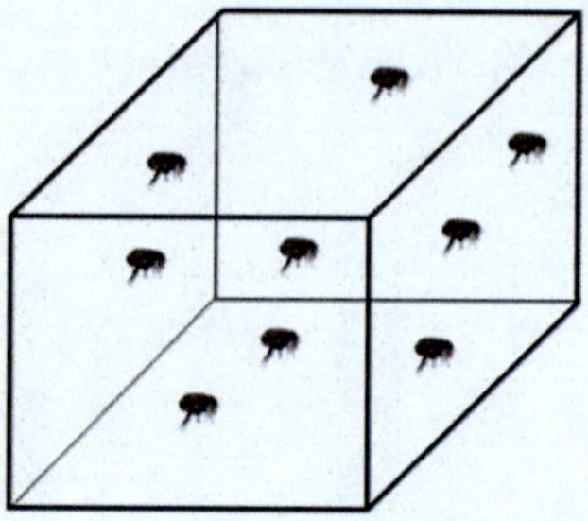

Fig. 7.4 Collection of noninteracting robots in limited volume box [7]

where $\mathcal{V} = 0$, $\beta = \frac{1}{kT}$, $T =$ temperature, which implies

$$Z = \frac{1}{N!}\left[\left(\frac{2\pi m}{h^2\beta}\right)^{3/2} V\right]^N$$

culminating in the calculation of the ideal gas law from the mean values

$$\begin{aligned}
\bar{p} &= \frac{1}{\beta}\frac{\partial \ln}{\partial Z} Z = \frac{1}{\beta}\frac{\partial}{\partial V}\left[\ln\frac{1}{N!} + \ln\left(\frac{2\pi m}{h^2\beta}\right)^{3N/2} + N\ln V\right], \\
&= \frac{1}{\beta}\frac{N}{V} \\
\bar{p} &= NkT/V,
\end{aligned} \tag{7.3}$$

where $\bar{p}$ is pressure, N is the number of robots, k is the Boltzmann constant (scaled for our robot problem), V is the volume, and T is the temperature. For this application, the search space is equivalent to the volume, and the temperature is equivalent to information flow (rate), which is derived from the sensors on the individual robots. The equivalent temperature can be derived in at least two ways.

The first way is to recognize that the kinetic energy is proportional to temperature at equilibrium which can be simplified to one-dimension (1D) as [81]

$$\mathcal{T}_{\text{ave}} = \frac{1}{2} m \dot{x}_{\text{ave}}^2 = \frac{1}{2} kT,$$

which leads to

$$T = \frac{m}{k}\dot{x}_{\text{ave}}^2$$

and

$$\dot{x}_{\text{ave}} = \sqrt{\frac{kT}{m}}.$$

The obstacle detection sensor systems on each robot must have a sampling rate and channel capacity (Shannon information/entropy) sufficient to detect the walls of the search volume and redirect each robot to produce an emulation of an elastic impact. Figure 7.3 presents the layout of obstacles, walls, and sources/targets that will be

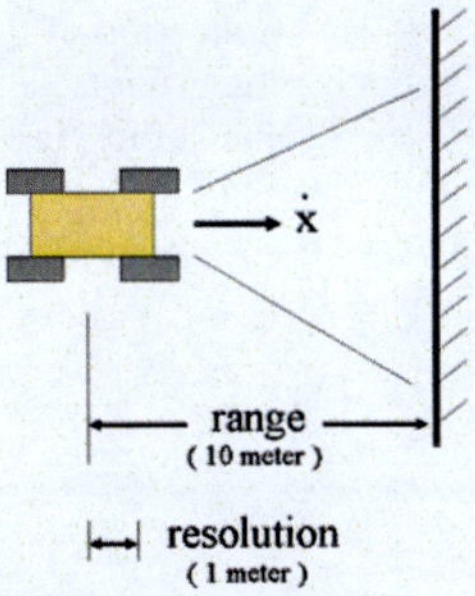

Fig. 7.5 Sensor system requirements [7]

used throughout this section. Figure 7.5 shows a simplified 1D model to support the determination of the bandwidth (Δf), number of bits (n), channel capacity (C_{ave}), and Shannon information/entropy rate ($\dot{H}_{\text{ave}}$) [39]. This results in

$$\begin{aligned} \Delta f &= \frac{1}{\tau} = \frac{\dot{x}}{\text{range}} = \frac{\dot{x}}{10\ \text{meter}} = 0.1\dot{x}\ \text{Hz}, \\ n &= \frac{\text{range}}{\text{resolution}} = \frac{10\ \text{meter}}{1\ \text{meter}} = 10\ \text{bits}, \\ C_{\text{ave}} &= \frac{1}{\tau}\log_2 n = \dot{H}_{\text{ave}}\ (\text{bits/s}). \end{aligned} \tag{7.4}$$

The second way is to utilize Fisher Information. From Frieden [41] it is observed that

$$I_{\text{ave}} = \frac{8}{m}\mathcal{T}_{\text{ave}} = 4\dot{x}^2_{\text{ave}} \tag{7.5}$$

and

$$I_{\text{ave}} = \frac{8}{m}\left[\frac{1}{2}kT\right] = 4\frac{kT}{m}, \tag{7.6}$$

which gives

$$T = \frac{m}{4k}I_{\text{ave}}, \tag{7.7}$$

where I_{ave} is the average Fisher information. Also, Frieden [41] provides another set of relationships for Fisher "temperature" as

$$\frac{1}{T_\theta} \equiv -k_\theta \frac{\partial I_{\text{ave}}}{\partial \theta}, \tag{7.8}$$

where θ is any parameter under measurement, and Fisher information per molecule for an ideal gas in a volume is

$$\frac{I_{\text{ave}}}{N} = \frac{kT}{k_E T_E} \tag{7.9}$$

with $\theta = E$ and $\frac{1}{T} \equiv \partial S/\partial E$.

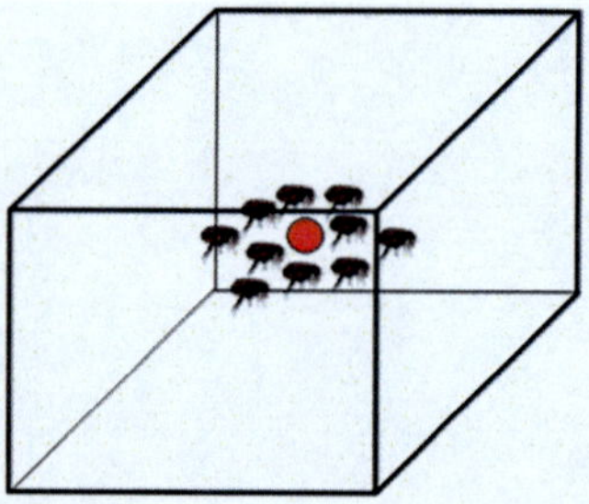

Fig. 7.6 Collection of noninteracting robots (noncommunicating) in limited volume box with a limited range attractive source [7]

For a volume, $\mathcal{T}_{\text{ave}} = \frac{3}{2}kT$ and $I_{\text{ave}} = \frac{12}{m}kT$, which produces

$$T = \frac{m}{12k} I_{\text{ave}}$$

and

$$\frac{I_{\text{ave}}}{N} = \frac{12}{mN}kT \quad \Rightarrow \quad \frac{1}{k_E T_E} = \frac{12}{mN} = -\frac{\partial I_{\text{ave}}}{\partial E}.$$

Returning to the ideal gas law, it is time to investigate the implications for the robot swarm. A necessary condition for all of the robots to swarm the target is for the pressure to be zero ($\bar{p} = 0$). This condition means that no robots hit the "walls" of the search volume. From (7.3) it is observed that the condition of $\bar{p} = 0$ is possible if $T = 0$, meaning that no information is being derived from the sensors since the robot is not moving. The pressure will also be zero if $V = \infty$, meaning an infinite search space exists.

The next step in the equilibrium analysis is to add a limited range attractive source/target (i.e., chemical plume) in the search volume (see Fig. 7.6). Following the same process above, the Hamiltonian is generated for an *ideal gas* of noninteracting (noncommunicating) robots within a volume containing a target

$$\mathcal{H} = \mathcal{T} + \mathcal{V} = E = \sum_{i=1}^{N} \frac{p_i^2}{2m} + \mathcal{V}(\mathbf{r}) \tag{7.10}$$

followed by forming a partition function

$$Z = \frac{1}{N!}\left[\left(\frac{2\pi m}{h^2 \beta}\right)^{3/2} \int e^{-\beta \mathcal{V}(\mathbf{r})}\, d^3\mathbf{r}\right]^N,$$

$$\mathcal{V}(\mathbf{r}) = \begin{cases} -\mathcal{V}_0 + \gamma \mathbf{r}, & 0 \le \mathbf{r} \le \mathbf{r}_s, \\ 0, & \mathbf{r}_s \le \mathbf{r} \le \mathbf{R}, \end{cases}$$

$$\gamma = \frac{\mathcal{V}_0}{\mathbf{r}_s}$$

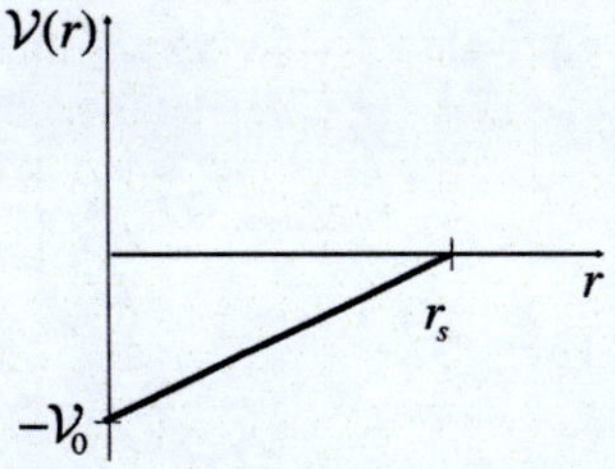

Fig. 7.7 Limited range attractive source/target [7]

(see Fig. 7.7 for plot of $\mathcal{V}(\mathbf{r})$ versus $\mathbf{r}$)

$$\begin{aligned}\int_0^R e^{-\beta\mathcal{V}(\mathbf{r})}\,d^3\mathbf{r} &= \int_0^{\mathbf{r}_s} e^{\beta(\mathcal{V}_0-\gamma\mathbf{r})}4\pi\mathbf{r}^2\,d\mathbf{r} + \int_{\mathbf{r}_s}^{\mathbf{R}} 4\pi\mathbf{r}^2\,d\mathbf{r}\\ &= 4\pi e^{\beta\mathcal{V}_0}\int_0^{\mathbf{r}_s}\mathbf{r}^2 e^{-\beta\gamma\mathbf{r}}\,d\mathbf{r} + V - V_s,\end{aligned}$$

where

$$\int \mathbf{r}^2 e^{\alpha\mathbf{r}}\,d\mathbf{r} = e^{-\beta\gamma\mathbf{r}}\left[-\frac{\mathbf{r}^2}{\beta\gamma} - \frac{2\mathbf{r}}{\beta^2\gamma^2} - \frac{2}{\beta^3\gamma^3}\right],$$

which implies

$$\int_0^{\mathbf{r}_s}\mathbf{r}^2 e^{\beta\gamma\mathbf{r}}\,d\mathbf{r} = -\frac{e^{-\beta\gamma\mathbf{r}_s}}{\beta\gamma}\left[\mathbf{r}_s^2 + \frac{2\mathbf{r}_s}{\beta\gamma} + \frac{2}{\beta^2\gamma^2}\right] + \frac{2}{\beta^3\gamma^3} = \nabla.$$

The resulting partition function is

$$\begin{aligned}Z &= \frac{1}{N!}\left[\left(\frac{2\pi m}{h^2\beta}\right)^{3/2}(\Delta + V - V_s)\right]^N,\\ \Delta &= 4\pi e^{\beta\mathcal{V}_0}\nabla,\end{aligned}$$

culminating in the calculation of the mean values

$$\begin{aligned}\bar{p} &= \frac{1}{\beta}\frac{\partial\ln}{\partial V}Z\\ &= \frac{1}{\beta}\frac{N}{(\Delta + V - V_s)}\\ &= \frac{NkT}{(\Delta + V - V_s)},\end{aligned}$$

where Δ is the source strength, and V_s is the volume of the attractive force potential. Once again, $\bar{p} = 0$ may be imposed, which implies that the robots do not hit the "walls." This means that $T = 0$, or $V = \infty$, as before, and that $\Delta = \infty$, which is equivalent to having an infinite source strength. For $\Delta \to \infty$:

$$\Delta = 4\pi \left[-\frac{1}{\beta\gamma}\left(\mathbf{r}_s^2 + \frac{2\mathbf{r}_s}{\beta\gamma} + \frac{2}{\beta^2\gamma^2} \right) + \frac{2e^{\beta\mathcal{V}_0}}{\beta^3\gamma^3} \right].$$

$\mathcal{V}_0 \to \infty$:

$$\lim_{\mathcal{V}_0\to\infty} \left(\frac{e^{\beta\mathcal{V}_0}}{\mathcal{V}_0^3} \right) = \lim_{\mathcal{V}_0\to\infty} \left(\frac{\beta^3 e^{\beta\mathcal{V}_0}}{6} \right) = \infty \quad \text{by L'Hôpital's Rule.}$$

This ideal gas formulation implies that the robots do not cooperate and perform target location independently as well as being able to take up the same position simultaneously (Bose–Einstein particles [82]).

The next step is to add cooperation (communication, sensing other robots, and taking up finite space) which can be accomplished by adding "real gas effects." Real gas effects can be modeled using van der Waal's equation and more generally, by the "virial expansion" of the equation of state [82]

$$\frac{\bar{p}}{kT} = n + B_2(T)n^2 + B_3(T)n^3 + \cdots,$$

where $n = N/V$ is the number of molecules (robots) per unit volume, $B_2 = B_3 = \cdots = 0$ is the ideal gas law, and $B_2 \neq 0$ provides for the van der Waal's equation which will be derived next. The higher-order terms enable the designer/analyst to evaluate and optimize different cooperation strategies.

The interaction potential fields presented in Fig. 7.3 are mathematical constructs of the microscopic/individual behaviors of each robot to produce the macroscopic/collective behaviors described by the viral expansion of the equation of state. This provides one step in the design/optimization of collective behaviors. A second step in the design/optimization process will be described in the rest of the chapter where nonequilibrium techniques are employed to ensure the time evolution of the collective behaviors meet the system constraints. Also, these potential fields can be treated as probability density functions in order to determine the likelihood of target detection and localization as well as inter-robot communication and cooperation by propagating them over time using Fisher Information [41] and the Fisher Lagrangian.

The Hamiltonian is generated for a nonideal gas of communicating/interacting robots within an empty search volume

$$\mathcal{H} = \mathcal{T} + \mathcal{V} = E = \sum_{i=1}^{N} \frac{p_i^2}{2m} + \mathcal{V}(\mathbf{r}) \tag{7.11}$$

followed by forming the partition function

$$Z = \frac{1}{N!}\left[\left(\frac{2\pi m}{h^2\beta} \right)^{3/2} \int e^{-\beta\mathcal{V}_B(\mathbf{r})}\, d^3\mathbf{r} \right]^N,$$

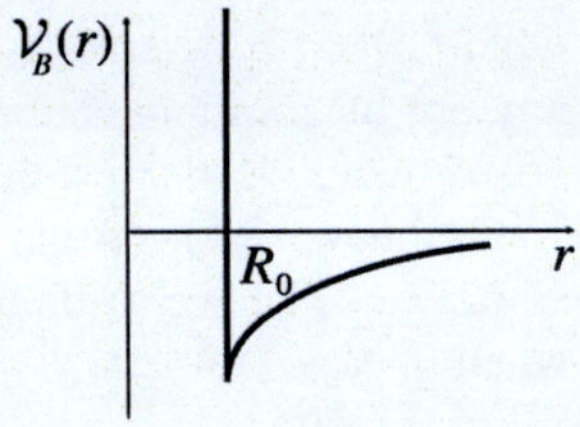

Fig. 7.8 Simplified Lennard-Jones potential [7]

$$\mathcal{V}_B(\mathbf{r}) = \begin{cases} \infty, & \mathbf{r} \le \mathbf{R}_0, \\ -\mathcal{V}_1(\frac{\mathbf{R}_0}{\mathbf{r}})^q, & \mathbf{r} > \mathbf{R}_0, \end{cases}$$

with a diagram of the interaction potential plotted in Fig. 7.8. This interaction potential which accounts for real gas effects requires an increase in sampling rate and channel capacity (Shannon entropy) due to the increase in closing speeds between robots. Basically, one can replace $\dot{x}$ with $2\dot{x}$ in (7.4) as a starting point.

Assume that $\mathcal{V}_B(\mathbf{r}) = \bar{\mathcal{V}}_B$ except within V_x, where $\mathcal{V}_B(\mathbf{r}) = \infty$, which implies

$$Z = \frac{1}{N!}\left[\left(\frac{2\pi m}{h^2\beta}\right)^{3/2}(V - V_x)e^{-\beta\bar{\mathcal{V}}_B}\right]^N,$$

where

$$\bar{\mathcal{V}}_B = \frac{1}{2}N\bar{U},$$

$$\bar{U} = \frac{1}{V}\int_{\mathbf{R}_0}^{\mathbf{R}} \mathcal{V}_B(\mathbf{r})4\pi\mathbf{r}^2\,d\mathbf{r}$$

$$= -\frac{4\pi}{3}\frac{\mathbf{R}_0^3}{V}\frac{3}{q-3}\mathcal{V}_1,$$

$$\bar{\mathcal{V}}_B = -\tilde{a}\frac{N}{V}, \quad \tilde{a} = \frac{2\pi}{3}\mathbf{R}_0^3\left(\frac{3}{q-3}\right)\mathcal{V}_1, \quad q > 3,$$

and

$$V_x = \tilde{b}N, \quad \tilde{b} = 4\left[\frac{4\pi}{3}\left(\frac{\mathbf{R}_0}{2}\right)^3\right],$$

culminating in the calculation of the mean values

$$\bar{p} = \frac{1}{\beta}\frac{\partial \ln}{\partial V}Z = \frac{1}{\beta}\frac{\partial}{\partial V}[N\ln(V - V_x) - N\beta\bar{\mathcal{V}}_B]$$

or

$$\bar{p} = \frac{NkT}{V - \tilde{b}N} - \tilde{a}\left(\frac{N^2}{V^2}\right), \tag{7.12}$$

where $\tilde{a}$ and $\tilde{b}$ determine the degree of interaction between the particles (i.e., the amount of communication and interaction between robots.) Again, $\bar{p} = 0$ is imposed, which leads to some new results and the old result of $V = \infty$.

One new result is that $T = 0$ does not satisfy $\bar{p} = 0$, which implies that this equation will have to be modified near $T = 0$. A second new result is that $\bar{p} = 0$ when

$$kT\frac{V^2}{N^2} - \tilde{a}\frac{V}{N} + \tilde{a}\tilde{b} = 0; \tag{7.13}$$

then

$$V = \frac{N}{2kT}\left[\tilde{a} \pm \sqrt{\tilde{a}(\tilde{a} - 4\tilde{b}kT)}\right],$$

which defines a phase transition or "emergent behavior." The robot swarm "condenses into a robot molecule."

This result is potentially problematic if the "robot molecule" inhibits the collective search process. However, the alpha–beta approach of reference [83] utilizes a robot molecule to conserve energy usage.

The final step in this example is to add a limited range attractive source in the search volume. A Hamiltonian is generated for a nonideal gas of communicating/interacting robots within a search volume containing the target

$$\mathcal{H} = \mathcal{T} + \mathcal{V} = E = \sum_{i=1}^{N} \frac{p_i^2}{2m} + \mathcal{V}(\mathbf{r}) \tag{7.14}$$

followed by forming the partition function

$$Z = \frac{1}{N!}\left[\left(\frac{2\pi m}{h^2\beta}\right)^{3/2} (\Delta + V - V_s - V_x)e^{-\beta\bar{\mathcal{V}}_B}\right]^N,$$

where

$$\bar{\mathcal{V}}_B = -\tilde{a}\frac{N}{V}, \quad \tilde{a} = \frac{2\pi}{3}\mathbf{R}_0^3\frac{3}{q-3}\mathcal{V}_1, \quad q > 3,$$

$$V_x = \tilde{b}N, \quad \tilde{b} = \frac{2\pi}{3}\mathbf{R}_0^3,$$

culminating in the calculation of the mean values

$$\begin{aligned}\bar{p} &= \frac{1}{\beta}\frac{\partial \ln Z}{\partial V} = \frac{1}{\beta}\frac{\partial}{\partial V}\left[N\ln(\Delta + V - V_s - V_x) - N\beta\bar{\mathcal{V}}_B\right] \\ &= NkT/(\Delta + V - V_s - V_x) - \tilde{a}N^2/V^2,\end{aligned}$$

which implies

$$\bar{p} = \frac{NkT}{(\Delta + V - V_s - \tilde{b}N)} - \tilde{a}\frac{N^2}{V^2}.$$

Once again, $\bar{p} = 0$ is imposed, which leads to new results and the old result of $V = \infty$. The result that $T = 0$ does not satisfy $\bar{p} = 0$ continues to be a problem.

A new result is that $\Delta = \infty$ does not satisfy $\bar{p} = 0$, which implies that the robots cannot take up the same position simultaneously (Fermi–Dirac particles [82]). A second new result is that $\bar{p} = 0$ when

$$NkTV^2 = \tilde{a}N^2(\Delta + V - V_s - \tilde{b}N),$$
$$kTV^2 - \tilde{a}NV + \tilde{a}N(\tilde{b}N + V_s - \Delta) = 0,$$

which implies

$$V = \frac{\tilde{a}N \pm [(\tilde{a}N)^2 - 4kT\tilde{a}N(\tilde{b}N + V_s - \Delta)]^{1/2}}{2kT}$$
$$= V_{1,2}.$$

This is a modified phase transition or emergent behavior that reference [83] exploits in alpha–beta variants.

7.3 Kinematic Collective Control

This section extends the results of collective plume tracing [35, 36] by employing HSSPFC and information theory. The goal is to provide unique insight into the design of distributed decentralized controllers for collective systems by applying HSSPFC combined with Shannon entropy. The basic concepts of [35, 36, 74, 78, 80] are reviewed, explained, and extended with HSSPFC and Shannon information theory.

7.3.1 Kinematic Control Design

The kinematic controller design begins with the realization that chemical plumes can be measured with a chemical sensor to determine a concentration level at a point in space and time, $c(\mathbf{x}, t)$ (where $c(\mathbf{x}, t)$ is a scalar, and $\mathbf{x}$ is a (2×1) vector), not range and bearing to the source (target). So, how does one get range and bearing to the source from $c(\mathbf{x}, t)$? One way is to create a "virtual potential field" (Hamiltonian surface shaping/information exergy) by flowing information through a distributed decentralized sensor and feedback control network in order to synthesize range and bearing to the source/target. In particular, a second-order approximation, $\hat{c}(\mathbf{x}, t)$, of the chemical plume field (see Fig. 7.9) is used to generate a quadratic potential in the Hamiltonian that takes the form of a second-order optimization method [36].

In this control design, it is assumed that the plume field is time-invariant and the robot dynamics are negligible (kinematics). The second-order approximation is

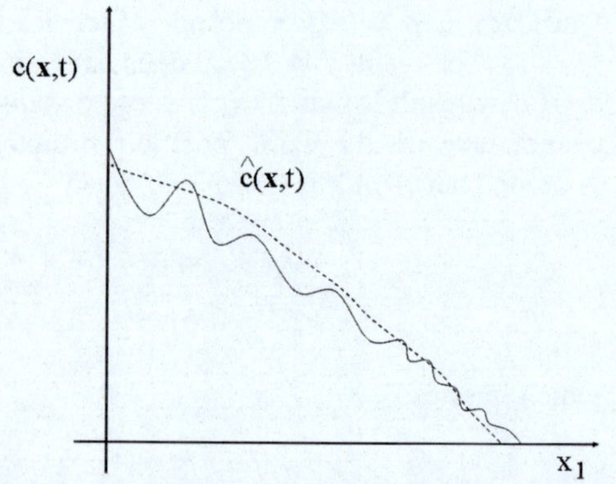

Fig. 7.9 Second-order approximation in positive x_1. Robinett III, R.D. and Wilson, D.G. [37], reprinted by permission of the publisher (Interscience Publishers)

[35, 36]

$$\hat{c}_i(\mathbf{x}_i) = \hat{c}_{0_i} + \hat{c}_{1_i}^T \mathbf{x}_i + \frac{1}{2}\mathbf{x}_i^T \hat{c}_{2_i} \mathbf{x}_i, \tag{7.15}$$

where $\hat{c}_0$ is a scalar, $\hat{c}_1$ is an $(N \times 1)$ vector, and $\hat{c}_2$ is an $(N \times N)$ symmetric positive definite matrix which is a result of fitting N robot sensor data points with

$$\frac{\partial \hat{c}_i(\mathbf{x}_i)}{\partial \mathbf{x}_i} = \mathbf{0} = \hat{c}_{1_i} + \hat{c}_{2_i}\mathbf{x}_i, \tag{7.16}$$

provides the source location (i.e., minimum/maximum)

$$\mathbf{x}_i^* = -\hat{c}_{2_i}^{-1}\hat{c}_{1_i}. \tag{7.17}$$

The virtual potential field for the ith robot which is positive definite becomes

$$\mathcal{V}_{c_i}(\mathbf{x}_i) = \frac{1}{2}\hat{c}_{1_i}^T \hat{c}_{2_i}^{-1}\hat{c}_{1_i} + \hat{c}_{1_i}^T \mathbf{x}_i + \frac{1}{2}\mathbf{x}_i^T \hat{c}_{2_i}\mathbf{x}_i, \tag{7.18}$$

which is the ith Hamiltonian, $\mathcal{H}_i = \mathcal{T}_i + \mathcal{V}_{c_i}$ with $\mathcal{T}_i = \frac{1}{2}m_i\dot{\mathbf{x}}_i^T\dot{\mathbf{x}}_i = 0$, for a kinematics solution, and is statically stable.

Static stability can be defined in terms of the energy storage surface which is defined by the Hamiltonian

$$\mathcal{H}_i = \mathcal{T}_i + \mathcal{V}_{c_i} = \frac{1}{2}m_i\dot{\mathbf{x}}_i^T\dot{\mathbf{x}}_i + \mathcal{V}_{c_i}(\mathbf{x}_i),$$

where m_i is the point mass of the ith robot. The system is statically stable if

$$\mathcal{V}_{c_i}(\mathbf{x}_i) > 0 \quad \forall \mathbf{x}_i \neq \mathbf{x}_i^* \quad \text{and} \quad \mathcal{V}_{c_i}(\mathbf{x}_\mathbf{i}^*) = 0,$$

statically unstable if

$$\mathcal{V}_{c_i}(\mathbf{x}_i) < 0 \quad \forall \mathbf{x}_i \neq \mathbf{x}_i^* \quad \text{and} \quad \mathcal{V}_{c_i}(\mathbf{x}_\mathbf{i}^*) = 0,$$

and neutrally stable if

$$\mathcal{V}_{c_i}(\mathbf{x}_i) = 0 \quad \forall \mathbf{x}_i .$$

The collective Hamiltonian (surface shaping) is

$$\mathcal{H} = \sum_{i=1}^{N} \mathcal{H}_i = \sum_{i=1}^{N} \mathcal{V}_{c_i}(\mathbf{x}_i),$$

which is positive definite and statically stable.

The time derivative of the Hamiltonian defines the power flow (see Chaps. 3, 4, and references [13, 37, 46]) for the ith robot

$$\dot{\mathcal{H}}_i = \dot{\mathbf{x}}_i^T \left[\hat{c}_{1_i} + \hat{c}_{2_i} \mathbf{x}_i\right],$$

which is dynamically stable for the kinematic controller

$$\dot{\mathbf{x}}_i = -\hat{c}_{2_i}^{-1}\left[\hat{c}_{1_i} + \hat{c}_{2_i} \mathbf{x}_i\right] = -\hat{c}_{2_i}^{-1} \hat{c}_{1_i} - \mathbf{x}_i = \mathbf{x}_i^* - \mathbf{x}_i ,$$

where

$$\dot{\mathcal{H}}_i = -\left[\hat{c}_{1_i} + \hat{c}_{2_i} \mathbf{x}_i\right]^T \hat{c}_{2_i}^{-1}\left[\hat{c}_{1_i} + \hat{c}_{2_i} \mathbf{x}_i\right] < 0.$$

The collective is dynamically stable for

$$\dot{\mathcal{H}} = \sum_{i=1}^{N} \dot{\mathcal{H}}_i < 0.$$

Further developments, analysis, and numerical validations for this controller are given in [35, 36, 79, 80].

7.3.2 Robot Description

This kinematic controller was implemented on each robot which performed a second-order fit to its neighboring robots information, position and sensor measurement. This information is obtained from a communications network where each robot takes a turn broadcasting its information to all robots within communications range [79]. To be more specific, the robots shared sensor information but individually decided a course of action based on their own estimate of the plume field. After a robot samples its environment, it broadcasts its information to others. The physical characteristics for a typical miniature mobile robot vehicle are $0.75 \times 0.71 \times 1.6$ inches. It includes an on-board temperature sensor for detecting temperature sources, an on-board 8-bit RISC processor with 4 Kbytes of RAM and 128 Kbytes of FLASH memory for program storage, four obstacle detection sensors, two brushless DC motors, a rechargeable lithium battery, a radio, and a step-up converter. The radio provided communication links to broadcast the sensor data to other robots and to receive periodic position updates from the base station. A CSMA (Carrier Sense Multiple Access) radio network was implemented which does not require global synchronization. The radio bit rate is 120 Kbits/sec, which gives an effective data rate of approximately 50 Kbits/sec, after the data is encoded. In the next section, these characteristics are used to analyze information flow.

7.3.3 Information Flow: A Trade-off Between Processing, Memory, and Communications

In order to effectively implement the kinematic controller, a fundamental design trade-off between processing, memory, and communications must be carried out on each robot of the team. In this case, all three were driven to their minimum values simultaneously which provides a lower limit. The minimum values for processing, memory, and communications are an 8-bit processor, no memory, and 3 words communicated. These values require a robot for every sensor measurement since the memory is zero. Further depth and development that summarizes the critical systems functions of communication, computation, and memory are given in [38].

The kinematic controller requires at least $n_c = (n^2 + 3n + 2)/2$ sensor measurements per control update to obtain an exact quadratic fit to the plume field, where $n =$ number of dimensions of the search space, and for a plane, $n = 2$, $n_c = 6$. If the number of robots is less than n_c, a minimum norm solution can be employed, but it will not converge due to the inherent noise in the system. This is an example of "emergent" behavior of collective systems where the designer continues adding robots to the collective until the collective behavior transitions from random motion to a converged solution: minimum norm to exact to least squares. Therefore, more than n_c robots should be used to enable a least squares solution to suppress noise effects.

During the testing phase, the robot team was slow to converge onto the source of the plume field. An information flow analysis was performed, and the temperature sensor was found to be the bottleneck in the system. The information flow is limited by the sensor because the sensor update rate is $1/\tau = 1/(60\ \text{s})$. To prove this, the channel or system capacity can be defined as the maximum rate of transmitting information under the assumption of an equiprobable source [39] or simply

$$C_{\text{ave}} = \frac{\text{information}}{T} = \frac{1}{\tau} \log_2 n \ \text{bits/s}.$$

Therefore, the system capacity is inversely proportional to the minimum interval τ over which signals can change and proportional to the logarithm of n. This is directly related to the Shannon information/entropy [40] flow rate, $\dot{H}_{\text{ave}}$, as

$$\dot{H}_{\text{ave}} = \frac{1}{\tau} \log_2 n = C_{\text{ave}}.$$

For the temperature sensor, the sensor bits are determined as $n = (T_{\text{high}} - T_{\text{low}})/\Delta T = 40/0.5 = 80$ bits, where T_{high} is the maximum temperature, T_{low} is the minimum temperature, and ΔT is the resolution of the sensor where the system capacity turns out to be less than a tenth of a bit/sec as compared to the communication rate of 50 Kbits/s. The performance of the robot collective is limited by the sensors. However, note that the collective kinematic control update law is stable despite the large delays due to the sensors as demonstrated in [35, 36, 80].

7.4 Kinetic Collective Control

The kinetic controller design begins by adding the point mass kinetic energy to the Hamiltonian of the kinematic controller. The Hamiltonian for the ith robot is

$$\mathcal{H}_i = \mathcal{T}_i + \mathcal{V}_{c_i} > 0 \quad \forall \mathbf{x}_i \neq \mathbf{x}_i^*,\ \dot{\mathbf{x}}_i \neq \mathbf{0}, \quad \mathcal{V}_{c_i}(\mathbf{x}_i^*) = 0, \quad \text{and} \quad \mathcal{T}_i(0) = 0,$$

which is statically stable. The Lagrangian becomes

$$\mathcal{L}_i = \mathcal{T}_i - \mathcal{V}_{c_i}.$$

The Hamiltonian of the robot collective is

$$\mathcal{H} = \sum_{i=1}^{N} \mathcal{H}_i = \sum_{i=1}^{N} [\mathcal{T}_i + \mathcal{V}_{c_i}],$$

which is positive definite and statically stable. The Lagrangian of the robot collective becomes

$$\mathcal{L} = \sum_{i=1}^{N} \mathcal{L}_i = \sum_{i=1}^{N} [\mathcal{T}_i - \mathcal{V}_{c_i}],$$

where

$$\begin{aligned}
\mathcal{T}_i &= \frac{1}{2} m_i \dot{\mathbf{x}}_i^T \dot{\mathbf{x}}_i, \\
\mathcal{V}_{c_i} &= \hat{c}_i(\mathbf{x}_i) - \hat{c}_i(\mathbf{x}_i^*), \\
&= \frac{1}{2} \hat{c}_{1_i}^T \hat{c}_{2_i}^{-1} \hat{c}_{1_i} + \hat{c}_{1_i}^T \mathbf{x}_i + \frac{1}{2} \mathbf{x}_i^T \hat{c}_{2_i} \mathbf{x}_i,
\end{aligned}$$

and $\mathcal{V}_{ci}$ is positive definite about the estimated minimum $\mathbf{x}_i^*$ for all $\mathbf{x}_i$ in the domain of $\mathbf{x}$ (see [35]). The equations of motion are derived from

$$\frac{d}{dt}\left(\frac{\partial \mathcal{L}_i}{\partial \dot{\mathbf{x}}_i}\right) - \frac{\partial \mathcal{L}_i}{\partial \mathbf{x}_i} = \mathbf{Q}_i,$$

which gives

$$m_i \ddot{\mathbf{x}}_i + \frac{\partial \mathcal{V}_{c_i}}{\partial \mathbf{x}_i} = \mathbf{u}_i. \tag{7.19}$$

The time derivative of the collective Hamiltonian becomes

$$\begin{aligned} \dot{\mathcal{H}} &= \sum_{i=1}^{N} \dot{\mathcal{H}}_i, \\ &= \sum_{i=1}^{N} \left[m_i \ddot{\mathbf{x}}_i + \frac{\partial \mathcal{V}_{c_i}}{\partial \mathbf{x}_i} \right]^T \dot{\mathbf{x}}_i, \\ &= \sum_{i=1}^{N} [\mathbf{u}_i]^T \dot{\mathbf{x}}_i, \end{aligned}$$

where the individual estimator/guidance algorithm for finding the source/target is

$$\hat{c}_i(\mathbf{x}_i) = \hat{c}_{0_i} + \mathbf{x}_i^T \hat{c}_{1_i} + \frac{1}{2} \mathbf{x}_i^T \hat{c}_{2_i} \mathbf{x}_i.$$

The feedback controller is

$$\mathbf{u}_i = -\mathbf{K}_{\mathbf{I}i} \int_0^t \mathbf{x}_i \, d\tau - \mathbf{K}_{\mathbf{D}i} \dot{\mathbf{x}}_i, \tag{7.20}$$

and the collective passivity controller design criteria becomes

$$\sum_{i=1}^{N} \left[[\mathbf{K}_{\mathbf{D}i} \dot{\mathbf{x}}_i]^T \dot{\mathbf{x}}_i \right]_{\text{ave}} > \sum_{i=1}^{N} \left[\left[-\mathbf{K}_{\mathbf{I}i} \int_0^t \mathbf{x}_i \, d\tau \right]^T \dot{\mathbf{x}}_i \right]_{\text{ave}}, \tag{7.21}$$

which is dynamically stable.

As a numerical example, a collective team of eight robots is used to implement the control law (7.20) and the virtual potential to find the source of the chemical plume. The actual controller implemented in the simulation is

$$\mathbf{U}_i = \mathbf{u}_i - \frac{\partial \mathcal{V}_{c_i}}{\partial \mathbf{x}_i}.$$

The dynamical entities are smart loitering autonomous robots. Initially, the eight robots are shown surrounding the chemical plume source (red target with center at $(0, 0)$) in Fig. 7.10. The overall collective performance results from utilizing one another's sensing, which improves the individual performance as well.

The collective system cooperatively localizes the chemical plume source, located at $(x_1, x_2) = (x, y) = (0, 0)$ (see Fig. 7.11). For this case, the collective dissipative term is greater than the collective generative term as shown by the decaying transient responses (see Fig. 7.12 X-positions and Fig. 7.13 Y-positions). In addition,

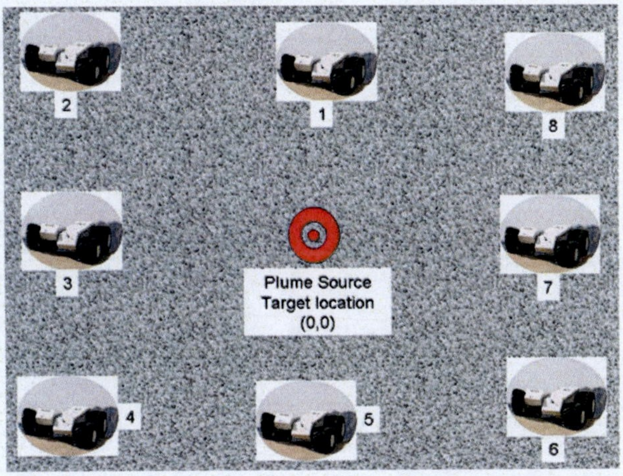

Fig. 7.10 The collective system of eight mobile RATLER^{TM} robots initially distributed away from the source. Robinett III, R.D. and Wilson, D.G. [37], reprinted by permission of the publisher (Interscience Publishers)

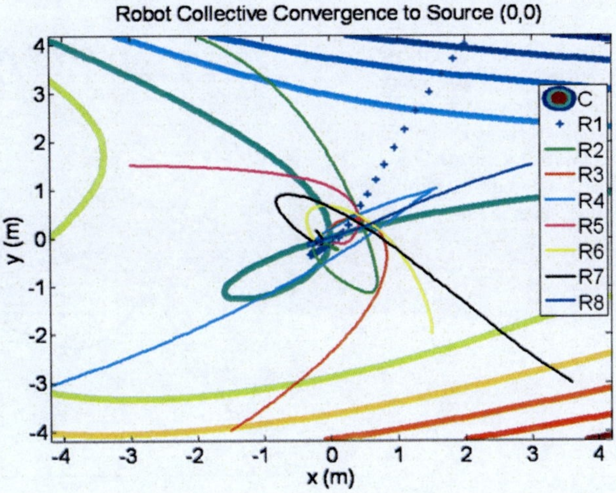

Fig. 7.11 The collective system of mobile robots cooperatively localized convergence to the chemical plume source located at $(x, y) = (0, 0)$. Note that in the legend: C represents the contour lines (*thick lines*) associated with the plume surface, and R1–R8 represent the traces of each robot (*thin lines*) which converges to *dark green contour* crossing at the origin. Robinett III, R.D. and Wilson, D.G. [37], reprinted by permission of the publisher (Interscience Publishers)

the mobile robot collectives Center of Mass (CM) transient responses are shown for both X- and Y-positions in Fig. 7.14 with the corresponding phase plane trajectory in Fig. 7.15. This demonstrates that the mobile robot collectives converge to the chemical plume source at approximately 14.0 seconds. Mobile robots $\{1, 3, 5, 8\}$ are modeled as unity point mass (1.0 kg), while robots $\{2, 4, 6, 7\}$ are modeled with increased (3.0 kg) mass. The governing dynamic equations of motion are given by (7.19).

7.5 Fisher Information and Equivalency

Following the same process as the kinematic controller design, it is time to analyze the information flow in the kinetic controller. This step requires using Fisher Infor-

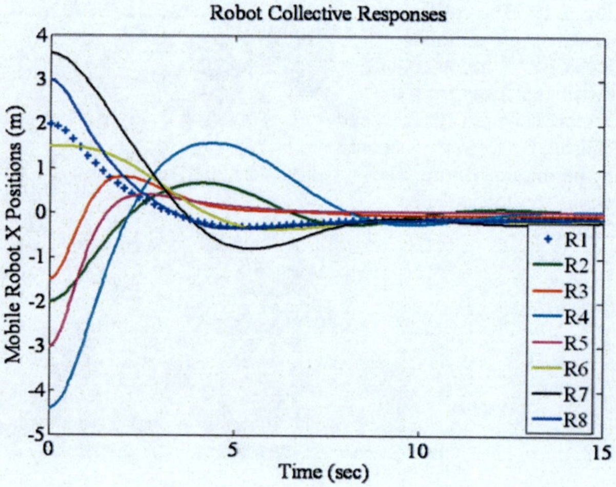

Fig. 7.12 Mobile robots transient responses for dissipative case X-positions. Robinett III, R.D. and Wilson, D.G. [37], reprinted by permission of the publisher (Interscience Publishers)

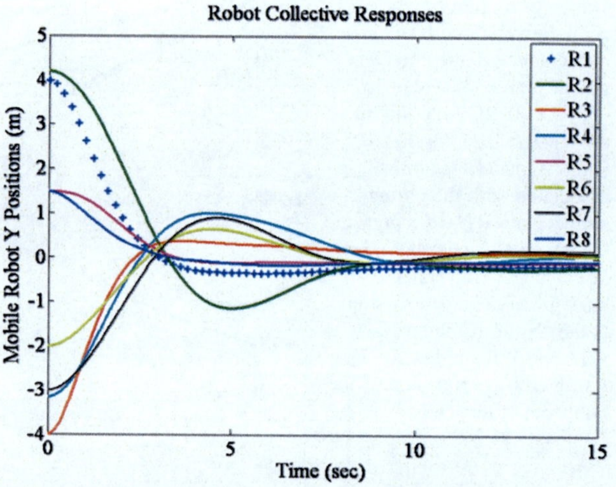

Fig. 7.13 Mobile robots transient responses for dissipative case Y-positions. Robinett III, R.D. and Wilson, D.G. [37], reprinted by permission of the publisher (Interscience Publishers)

mation and physical/information exergies in place of Shannon entropy. Exergy is described in detail in Chaps. 2, 3, 4, and references [7, 12, 13, 30]. From a communications point of view, Fisher Information is a measure of how well the receiver can estimate the message from the sender where as Shannon information/entropy is a measure of the sender's transmission efficiency over a communications channel [39–42]. Fisher Information [41] is defined as

$$
\begin{aligned}
I &= \int \left(\frac{\partial \ln \hat{p}(x)}{\partial x} \right)^2 \hat{p}(x)\, dx \\
&= \int \frac{1}{\hat{p}(x)} \left(\frac{\partial \hat{p}(x)}{\partial x} \right)^2 dx
\end{aligned}
$$

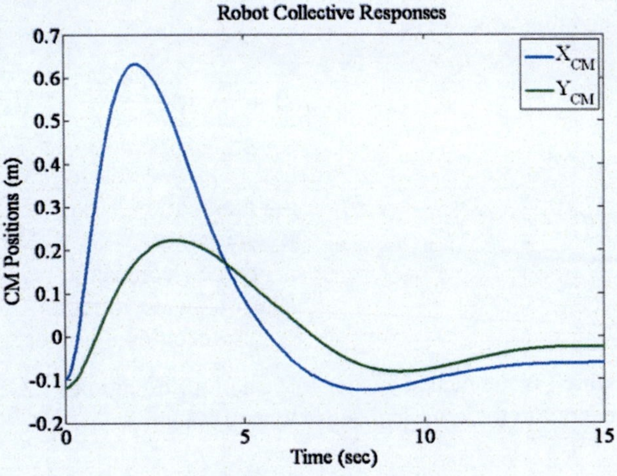

Fig. 7.14 Mobile robots collective center of mass transient responses for X-positions and Y-positions. Robinett III, R.D. and Wilson, D.G. [37], reprinted by permission of the publisher (Interscience Publishers)

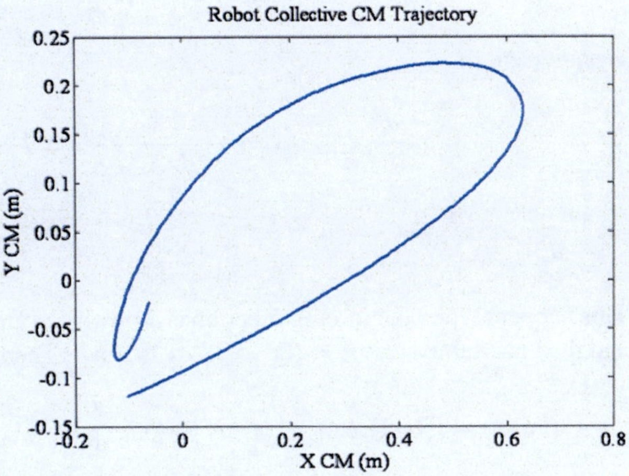

Fig. 7.15 Mobile robots collective center of mass phase plane response. Robinett III, R.D. and Wilson, D.G. [37], reprinted by permission of the publisher (Interscience Publishers)

$$= 4 \int \left(\frac{\partial q(x)}{\partial x} \right)^2 dx, \tag{7.22}$$

where $q^2(x) = \hat{p}(x)$ is a "real amplitude" function of the probability density function $\hat{p}(x)$. Equation (7.22) can be interpreted as the "mean kinetic energy," and for the purposes of this discussion,

$$I = 4 \int \dot{q}^2 \, dt = 4 \int \frac{2}{m} \mathcal{T} \, dt, \tag{7.23}$$

where $\mathcal{T} = \frac{1}{2} m \dot{q}^2$.

One way to explain this interpretation of Fisher Information as the mean kinetic energy is to begin with quantum mechanics and utilize the "classical limit" (as in classical mechanics). Following the derivations in [84], the expectation of the momentum squared is

$$\langle p^2 \rangle = \frac{\bar{h}^2}{2m} \int \left| \frac{\partial \Psi(x,t)}{\partial x} \right|^2 dx, \tag{7.24}$$

where

p—momentum,
m—mass,
$\bar{h}$—Dirac's constant,
$\Psi(x,t)$—wave function,
$\langle\,\rangle$—expectation operator,

which is the scaled Fisher Information. Equation (7.24) becomes the mean kinetic energy in the *classical* sense when [84]

$$\frac{d}{dt}\langle p \rangle_t = m \frac{d^2}{dt^2}\langle x \rangle_t = -\left\langle \frac{d\mathcal{V}(x)}{dx} \right\rangle_t = \langle F(x) \rangle_t \tag{7.25}$$

is equivalent to

$$m \frac{d^2}{dt^2} x_{\text{classical}} = -\frac{d\mathcal{V}(x_{\text{classical}})}{dx_{\text{classical}}} = F(x_{\text{classical}}), \tag{7.26}$$

which occurs when

$$x_{\text{classical}} = \langle x \rangle.$$

This happens when the forcing function is a slowly varying function of its arguments and the uncertainty $(\Delta x)^2 = \langle (x - \langle x \rangle)^2 \rangle$ is small. Then

$$F(x) = F(\langle x \rangle) + (x - \langle x \rangle) F'(\langle x \rangle) + \frac{1}{2!}(x - \langle x \rangle)^2 F''(\langle x \rangle) + \cdots,$$

which leads to

$$\begin{aligned} \langle F(x) \rangle &\approx F(\langle x \rangle) + \langle x - \langle x \rangle \rangle F'(\langle x \rangle) \\ &\approx F(\langle x \rangle). \end{aligned}$$

I is the scaled integral of the mean kinetic energy portion of both the Lagrangian and the Hamiltonian and is referred to as the Fisher Data Information. The second part, the potential energy, $\mathcal{V}$, is referred to as the phenomenological or bound Fisher Information [41], J. The Lagrangian is

$$\mathcal{L} = \mathcal{T} - \mathcal{V}, \tag{7.27}$$

where

$$I - J = \frac{8}{m} \int [\mathcal{T} - \mathcal{V}]\, dt. \tag{7.28}$$

HSSPFC (see Chap. 4) can be utilized to demonstrate the equivalence between physical exergy and information exergy via the Hamiltonian that is the sum of the exergies. The Hamiltonian surface which determines the accessible states of the system can be shaped with a controller potential ($\bar{\mathcal{V}}_c$) which combines physical ($\bar{\mathcal{T}}$ and $\bar{\mathcal{V}}$) and information ($\bar{\mathcal{V}}_c$) exergies as

$$\bar{\mathcal{H}} = \bar{\mathcal{T}} + \bar{\mathcal{V}} + \bar{\mathcal{V}}_c,$$

where $\bar{\mathcal{T}} = \sum_{i=1}^{N} \frac{1}{m_i}\mathcal{T}_i$, and formalized further through bound Fisher Information [41] as

$$J = 8\int [\bar{\mathcal{V}} + \bar{\mathcal{V}}_c]\, dt \tag{7.29}$$

with

$$\bar{\mathcal{V}} = \sum_{i=1}^{N} \frac{1}{m_i}\mathcal{V}_i$$

and

$$\bar{\mathcal{V}}_c = \sum_{i=1}^{N} \frac{1}{m_i}\mathcal{V}_{c_i}.$$

Now, explicitly utilizing Fisher Information and defining the Fisher Lagrangian as

$$I - J = 8\int [\bar{\mathcal{T}} - (\bar{\mathcal{V}} + \bar{\mathcal{V}}_c)]\, dt = 8\int \bar{\mathcal{L}}\, dt, \tag{7.30}$$

one can tie the information and physical exergies to Fisher Information. The derivation of the equations of motion informs one about the functionality of the physical infrastructure (robots, sensors, etc.) versus the information-driven collective. The Fisher Lagrangian can be rewritten as the Fisher Hamiltonian which provides the *Fisher Information Equivalency* as

$$I + J = 8\int [\bar{\mathcal{T}} + (\bar{\mathcal{V}} + \bar{\mathcal{V}}_c)]\, dt = 8\int \bar{\mathcal{H}}\, dt, \tag{7.31}$$

which leads to the constraints of static and dynamic stability

$$\dot{I} + \dot{J} = 8[\bar{\mathcal{H}}] > 0 \quad \forall \mathbf{x}_i \neq \mathbf{x}_i^*,\ \dot{\mathbf{x}}_i \neq 0, \qquad \mathcal{V}_i(\mathbf{x}_i^*) + \mathcal{V}_{c_i}(\mathbf{x}_i^*) = 0,$$

and

$$\frac{1}{\tau_c}\int_0^{\tau_c} [\ddot{I} + \ddot{J}]\, dt = \frac{8}{\tau_c}\int_0^{\tau_c} [\dot{\bar{\mathcal{H}}}]\, dt < 0,$$

where

$$\bar{\mathcal{H}} = \sum_{i=1}^{N} \frac{1}{m_i}\mathcal{H}_i.$$

This relationship provides a direct connection between stability of, performance of, and information flow within the collective since the physical exergy flow is directly related to the information exergy (virtual potential) flow and the Fisher Information flow. Equation (7.31) provides an ideal optimization functional to generalize the idea of optimizing information flow under the constraints of stability and performance.

7.6 Chapter Summary

Chapter 7 has demonstrated the design of distributed decentralized controllers for a team of robots based on equilibrium thermodynamics, HSSPFC, and information theory. Initially, a collective kinematic update control law was designed based on HSSPFC and evaluated based on the minimum information associated with processing, memory, and communications. It was shown based on Shannon entropy that the sensor was limiting the performance of the team. These results were then extended to a collective kinetic controller to show equivalences between physical and information-based exergies from Fisher Information, control potentials for Hamiltonian surface shaping, and power flow control. The *Fisher Information Equivalency* can serve as an ideal optimization functional to measure and monitor both performance and stability of collective systems with respect to required physical and information resources. One specific aspect of this result was investigated for a multiagent system of robots. A distributed decentralized control law for collective robotic systems was developed based on HSSPFC and information theory. Stability boundaries and system performances were then determined with static and dynamic stabilities. A mobile robot collective plume tracing numerical simulation example demonstrated stable convergence to the source with this decentralized HSSPFC collective control architecture.

Chapter 8
Case Study #3: Nonlinear Aeroelasticity

8.1 Introduction

Researchers have been investigating limit cycle behavior for many different engineering fields. Specific applications that relate to the category of time-periodic systems include helicopter blades in forward flight, wind turbine blades, and airplane wing flutter, all of which Limit-Cycle Oscillations (LCO) may become present. The prediction and control of LCO in a system continues to be a challenge and an ongoing area of research. For example, Gopinath, Beran, and Jameson [16] explore various methods in the computation of time-periodic solutions for autonomous systems. The goal was to determine the range of applicability of models of varying fidelity to the numerical prediction of LCOs and related evaluations. A simple aeroelastic model of an airfoil with nonlinear structural coupling was used to show the efficacy of the procedure. Several researchers are investigating cyclic methods to compute limit-cycle oscillations for potentially large, nonlinear, systems of equations. One such method by Hall, Thomas, and Clark [17] introduces a harmonic balance technique for modeling unsteady nonlinear flows in turbomachinery. For the example presented in [17], a transonic front stage rotor of a high-pressure compressor was found to flutter in torsion but reached a stable limit cycle, demonstrating that strongly nonlinear flows can be modeled accurately with a small number of harmonics. Additional wing flutter LCO identification and control investigations by others are further discussed in the following references [18–22]. In more general mathematical descriptions of limit cycles, Sabatini [23] discusses a uniqueness theorem for limit cycles of a class of plane differential systems. The main result is applicable to second-order systems with dissipative terms which depend on both position and velocity. In addition, Carletti and Villari [24] consider the Liénard equation for which a sufficient condition to ensure the existence and uniqueness of limit cycles is given.

The goal of Chap. 8 is to demonstrate how limit cycles are the stability boundaries for linear and nonlinear aeroelastic control systems. The Poincaré–Bendixson Theorem is a good place to begin this quest. Boyce and DiPrima [25] provide some insight into the proof of this theorem in the form of Green's Theorem applied to a

R.D. Robinett III, D.G. Wilson, *Nonlinear Power Flow Control Design*, Understanding Complex Systems,
DOI 10.1007/978-0-85729-823-2_8,

line integral over a closed curve that equals zero:

$$\dot{x} = F(x, y), \qquad \dot{y} = G(x, y),$$
$$\oint_C [F(x, y)\,dy - G(x, y)\,dx] = \iint_R [F_x(x, y) + G_y(x, y)]\,dA = 0.$$

This line integral will be modified for Hamiltonian systems to determine the limit cycles resulting from power flow control. In particular, the work per cycle [26–29] defined by the line integral of the power flow

$$W_{\text{cyclic}} = \oint_\tau F \cdot \dot{x}\,dt = 0$$

is the modified form of choice since the time derivative of the Hamiltonian is the generalized power flow for natural systems [13].

A specific application of these concepts is the analysis of classical flutter. Classical flutter is a linear limit cycle that is a result of the coalescence of a bending mode and a torsional mode to produce a self-excited oscillation. Abramson [31] describes the existence of a flutter mode as follows:

> *We see now that negative work is done on the wing by part of the torsional motion, by the flexural motion, and by the elastic restoring forces; positive work is done on the wing by part of the torsional motion. The motion will maintain itself (the condition for flutter) when the net positive work just balances the dissipation of energy due to all the damping forces. The magnitude of the positive work done by the additional lift due to the twist is directly dependent upon the phase relationship between the coupled torsional and flexural motions . . .*

Chapter 8 develops the existence of limit cycles for aeroelastic systems based on power flows that leads to a balance between positive work and energy dissipation due to damping. This approach is generalized to nonlinear aeroelastic systems with nonlinear limit cycles where power flows are balanced over a cycle instead of point-by-point cancellation. This chapter will discuss the applicability of this aeroservelastic analysis to wind turbines that fall in the power range of 1.5 MW to 5 MW rated power [85, 86].

Chapter 8 is divided into four sections. Section 8.2 introduces a nonlinear stall flutter model and also discusses various aspects of additional nonlinearities plus the design and analysis of a PID controller based on HSSPFC. Nonlinear power flow control design allows the nonlinear dynamical system to be partitioned into three categories: generation, dissipation, and storage. By identifying the power flow over a cycle, the system stability and performance characteristics can be determined. In addition, linear flutter analysis is also discussed. Section 8.3 performs numerical simulations that are specific to a 5 MW wind turbine analysis. Finally, Sect. 8.4 summarizes the results with concluding remarks.

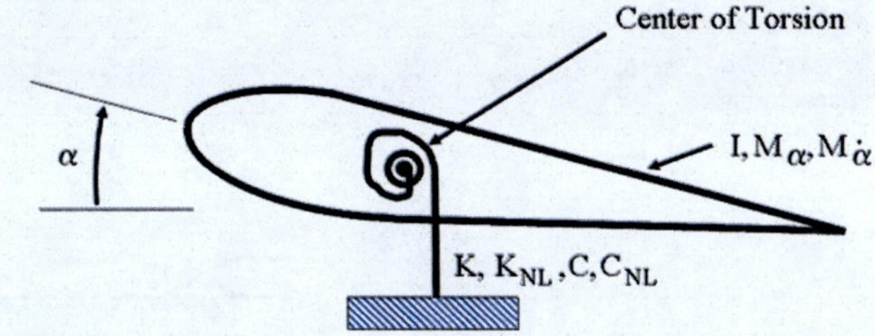

Fig. 8.1 Nonlinear flutter model

8.2 Nonlinear Stall Flutter Model

This section develops a single degree-of-freedom (DOF) model which is representative of the first torsional mode (on-the-order of 5.58 Hz) of a large wind turbine (5 MW) blade. The first torsion mode is the dominant feature which leads to the single DOF model. Figure 8.1 depicts the simplified nonlinear model, where K is the torsional stiffness, K_{NL} the nonlinear torsional stiffness, C the torsional damping, C_{NL} the nonlinear torsional damping, I the wing section torsional moment of inertia, and M_α, $M_{\dot{\alpha}}$ the aerodynamic moments. The equation of motion is derived from Lagrange's equation

$$\frac{d}{dt}\left(\frac{\partial \mathcal{L}}{\partial \dot{\alpha}}\right) - \frac{\partial \mathcal{L}}{\partial \alpha} = \mathcal{Q}_\alpha, \tag{8.1}$$

where

$$\begin{aligned}
&\mathcal{L} = \mathcal{T} - \mathcal{V},\\
&\mathcal{T} = \frac{1}{2} I \dot{\alpha}^2,\\
&\mathcal{V} = \frac{1}{2} K \alpha^2 + \frac{1}{4} K_{\mathrm{NL}} \alpha^4,\\
&\mathcal{Q}_\alpha = \mathcal{Q}_{\mathrm{damp}} + \mathcal{Q}_{\mathrm{aero}} + \mathcal{Q}_{\mathrm{control}},\\
&\mathcal{Q}_{\mathrm{damp}} = -C\dot{\alpha} - C_{\mathrm{NL}} \operatorname{sign}(\dot{\alpha}),\\
&\mathcal{Q}_{\mathrm{aero}} = M_\alpha(\alpha) + M_{\dot{\alpha}}(\dot{\alpha}, \alpha), \quad \text{and}\\
&\mathcal{Q}_{\mathrm{control}} = u = -K_P \alpha - K_I \int_0^t \alpha \, d\tau - K_D \dot{\alpha},
\end{aligned}$$

where $\mathcal{L}$ is the Lagrangian, $\mathcal{T}$ the kinetic energy, $\mathcal{V}$ the potential energy, and $\mathcal{Q}_\alpha$ the generalized forces. The controller input u consists of Proportional-Integral-Derivative (PID) control action where K_P is the proportional gain, K_I the integral gain, and K_D the derivative gain. The aerodynamic moments M_α and $M_{\dot{\alpha}}$ are

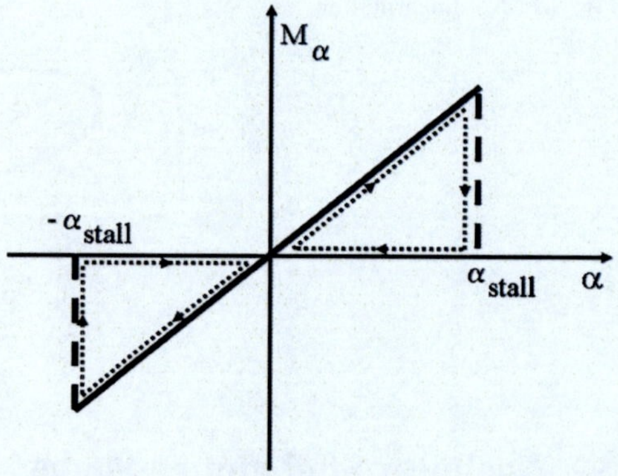

Fig. 8.2 Nonlinear hysteresis aerodynamic moment characteristic

generated based on the following nonlinear hysteresis logic (also shown visually in Fig. 8.2)

$$M_\alpha(\alpha) = \begin{cases} \hat{C}_{M_\alpha}\alpha & \text{for } |\alpha| < \alpha_{\text{stall}}, \\ 0 & \text{for } |\alpha| > \alpha_{\text{stall}}, \\ 0 & \text{for the return hysteresis cycle,} \end{cases}$$

and

$$M_{\dot{\alpha}}(\dot{\alpha}, \alpha) = \begin{cases} \hat{C}_{M_{\dot{\alpha}}}\dot{\alpha} & \text{for } |\alpha| < \alpha_{\text{stall}}, \\ 0 & \text{for } |\alpha| > \alpha_{\text{stall}}. \end{cases}$$

Applying (8.1) yields the equation of motion as

$$I\ddot{\alpha} + K\alpha + K_{\text{NL}}\alpha^3 = -C\dot{\alpha} - C_{\text{NL}}\text{sign}(\dot{\alpha}) + u + M_\alpha(\alpha) + M_{\dot{\alpha}}(\dot{\alpha}, \alpha),$$

which has many interesting properties given next. In the next four subsections the following regions will be investigated for: (i) the linear region, (ii) nonlinear stall flutter, (iii) further nonlinearities (such as cubic stiffness and Coulomb friction), and (iv) conventional PID control design.

8.2.1 Linear Region

For $|\alpha| < \alpha_{\text{stall}}$, the model is linear with $K_{\text{NL}} = C_{\text{NL}} = 0$ or

$$I\ddot{\alpha} + \left[C - \hat{C}_{M_{\dot{\alpha}}}\right]\dot{\alpha} + \left[K - \hat{C}_{M_\alpha}\right]\alpha = u,$$

which produces typical linear aeroelastic behavior. Divergence (statically unstable) occurs when

$$\hat{C}_{M_\alpha} \geq K \quad \text{for } u = 0,$$

where

$$\hat{C}_{M_\alpha} = K_{M_\alpha} q A = K_{M_\alpha} A\left(\frac{1}{2}\rho V^2\right)$$

with $A =$ cross-sectional area, $\rho =$ air density, $V =$ speed, and $K_{M_\alpha} =$ moment coefficient. Torsional flutter occurs when

$$\hat{C}_{M_{\dot\alpha}} \geq C \quad \text{for } K - \hat{C}_{M_\alpha} > 0 \quad \text{and} \quad u = 0,$$

where

$$\hat{C}_{M_{\dot\alpha}} = K_{M_{\dot\alpha}} q A d,$$

and $d =$ chord length.

8.2.2 Nonlinear Stall Flutter with Linear Dynamics

When the motion reaches $|\alpha| > \alpha_{\text{stall}}$, the model becomes nonlinear, where

$$I\ddot\alpha + C\dot\alpha + K\alpha = M_\alpha(\alpha) + M_{\dot\alpha}(\dot\alpha, \alpha)$$

with

$$\mathcal{H} = \frac{1}{2} I\dot\alpha^2 + \frac{1}{2} K\alpha^2,$$
$$\dot{\mathcal{H}} = [I\ddot\alpha + K\alpha]\dot\alpha = [-C\dot\alpha + M_\alpha(\alpha) + M_{\dot\alpha}(\dot\alpha, \alpha)]\dot\alpha,$$

which produces a limit cycle when

$$\oint_\tau [M_\alpha(\alpha) + M_{\dot\alpha}(\dot\alpha, \alpha)]\dot\alpha\, dt = \oint_\tau [C\dot\alpha]\dot\alpha\, dt.$$

The numerical results are given in Figs. 8.3–8.7 for each case. In Figs. 8.3 and 8.4 the limit cycle behavior of generation up to stall followed by dissipation until flow reattachment is clearly displayed. Figures 8.5–8.7 give the details of the three cases: stable (see Fig. 8.5), neutral stable (see Fig. 8.6), and unstable (see Fig. 8.7). The highly discontinuous moments are clearly presented in the figures. Also, the average power (slope of the energy curve) is shown to be equal and off-setting which leads to a limit cycle.

8.2.3 Nonlinear Stall Flutter with Nonlinear Dynamics

The nonlinear stall flutter can be further modified by adding the nonlinear stiffness and damping

$$I\ddot\alpha + C\dot\alpha + C_{\text{NL}}\text{sign}(\dot\alpha) + K\alpha + K_{\text{NL}}\alpha^3 = M_\alpha(\alpha) + M_{\dot\alpha}(\dot\alpha, \alpha)$$

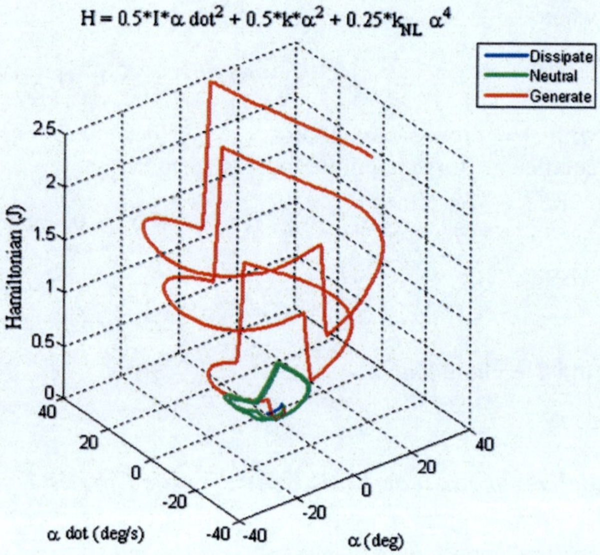

Fig. 8.3 Nonlinear stall flutter with linear dynamic results: 3D Hamiltonian. Robinett III, R.D. and Wilson, D.G. [85], reprinted with permission of the American Institute of Aeronautics and Astronautics

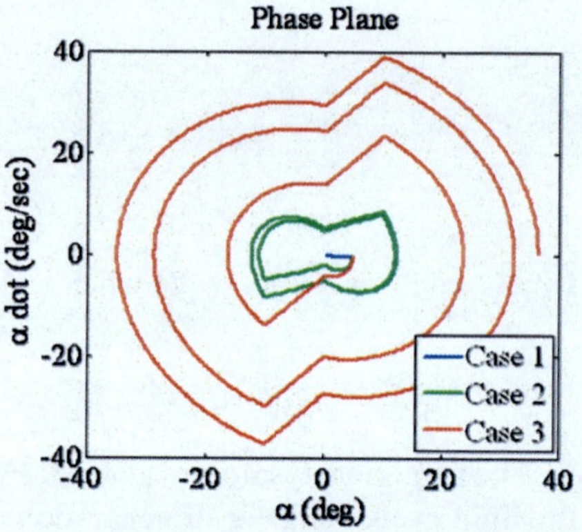

Fig. 8.4 Nonlinear stall flutter with linear dynamic results: phase plane plot. Robinett III, R.D. and Wilson, D.G. [85], reprinted with permission of the American Institute of Aeronautics and Astronautics

with

$$\mathcal{H} = \frac{1}{2}I\dot{\alpha}^2 + \frac{1}{2}K\alpha^2 + \frac{1}{4}K_{\text{NL}}\alpha^4,$$

$$\dot{\mathcal{H}} = \left[I\ddot{\alpha} + K\alpha + K_{\text{NL}}\alpha^3\right]\dot{\alpha} = [-C\dot{\alpha} - C_{\text{NL}}\text{sign}(\dot{\alpha}) + M_\alpha(\alpha) + M_{\dot{\alpha}}(\dot{\alpha}, \alpha)]\dot{\alpha},$$

which produces a limit cycle when

$$\oint_\tau [M_\alpha(\alpha) + M_{\dot{\alpha}}(\dot{\alpha}, \alpha)]\dot{\alpha}\,dt = \oint_\tau [C\dot{\alpha} + C_{\text{NL}}\text{sign}(\dot{\alpha})]\dot{\alpha}\,dt.$$

The numerical results are given in Figs. 8.8–8.12 for each case. The results are similar to the previous case except for the compression of the trajectories due to

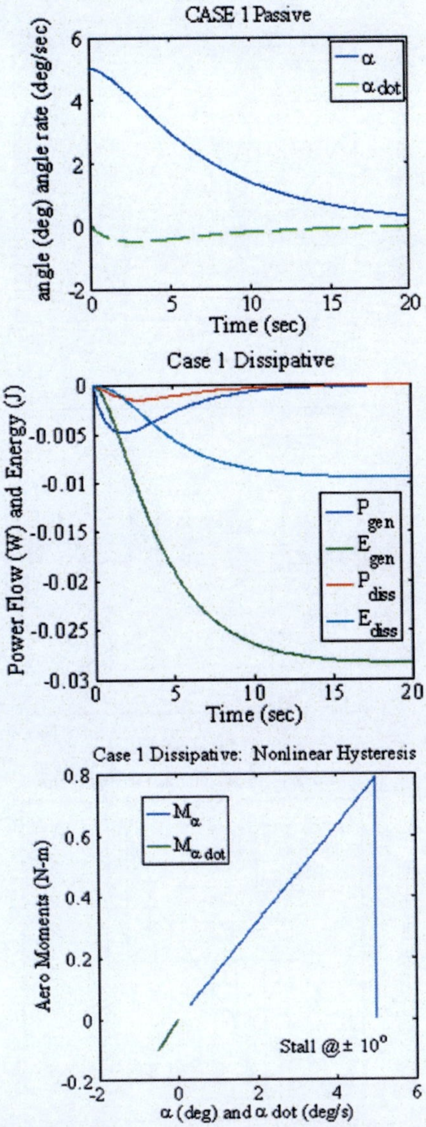

Fig. 8.5 Nonlinear stall flutter with linear dynamic results: angular responses, power and energy flow responses, and aero moment responses for Case 1 dissipative. Robinett III, R.D. and Wilson, D.G. [85], reprinted with permission of the American Institute of Aeronautics and Astronautics

nonlinear spring and nonlinear damping. Specifically, the cubic spring creates a new limit cycle at a higher energy state, while the nonlinear damping dissipates energy at a higher rate.

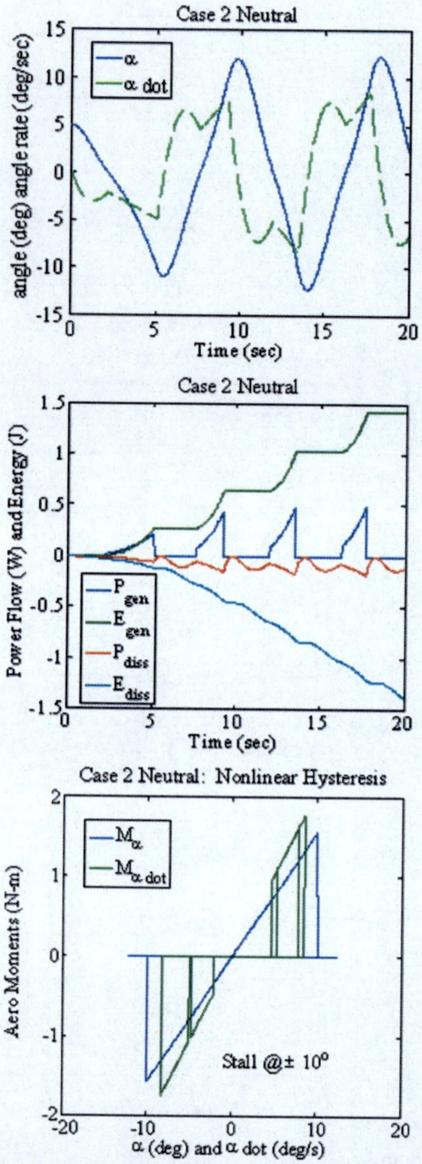

Fig. 8.6 Nonlinear stall flutter with linear dynamic results: angular responses, power and energy flow responses, and aero moment responses for Case 2 neutral. Robinett III, R.D. and Wilson, D.G. [85], reprinted with permission of the American Institute of Aeronautics and Astronautics

8.2.4 Controller Design

The nonlinear system can be modified by feedback control to meet several performance requirements. A PID controller is implemented to show the effects of feed-

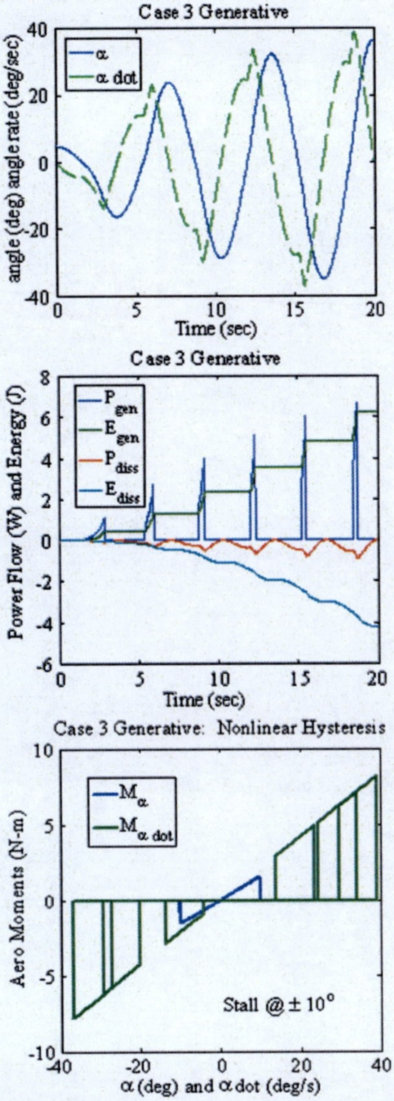

Fig. 8.7 Nonlinear stall flutter with linear dynamic results: angular responses, power and energy flow responses, and aero moment responses for Case 3 generative. Robinett III, R.D. and Wilson, D.G. [85], reprinted with permission of the American Institute of Aeronautics and Astronautics

back control. The model becomes

$$
\begin{aligned}
&I\ddot{\alpha} + [K + K_P]\alpha + K_{\mathrm{NL}}\alpha^3 \\
&\quad = -[C + K_D]\dot{\alpha} - C_{\mathrm{NL}}\mathrm{sign}(\dot{\alpha}) + M_\alpha(\alpha) + M_{\dot{\alpha}}(\dot{\alpha}, \alpha) - K_I \int_0^t \alpha\, d\tau
\end{aligned}
$$

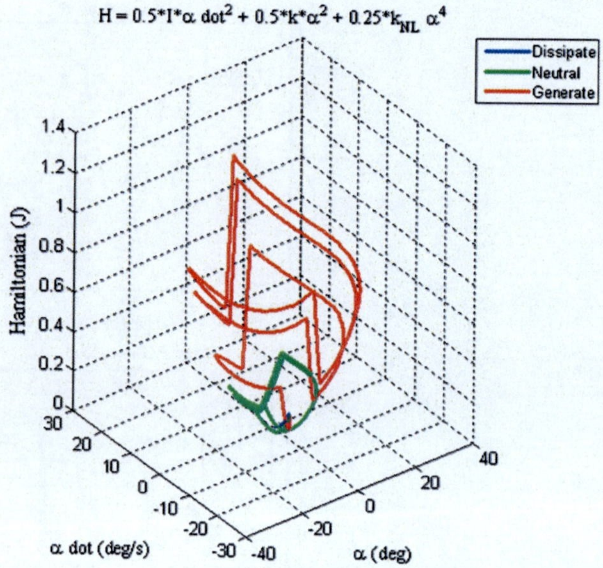

Fig. 8.8 Nonlinear stall flutter with nonlinear dynamic results: 3D Hamiltonian. Robinett III, R.D. and Wilson, D.G. [85], reprinted with permission of the American Institute of Aeronautics and Astronautics

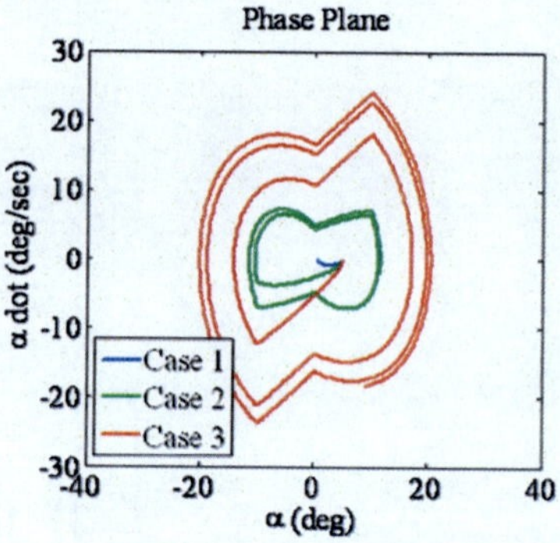

Fig. 8.9 Nonlinear stall flutter with nonlinear dynamic results: phase plane plot. Robinett III, R.D. and Wilson, D.G. [85], reprinted with permission of the American Institute of Aeronautics and Astronautics

with

$$\begin{aligned}\mathcal{H} &= \frac{1}{2}I\dot{\alpha}^2 + \frac{1}{2}[K + K_P]\alpha^2 + \frac{1}{4}K_{\mathrm{NL}}\alpha^4,\\ \dot{\mathcal{H}} &= \left[I\ddot{\alpha} + (K + K_P)\alpha + K_{\mathrm{NL}}\alpha^3\right]\dot{\alpha}\\ &= \left[-(C + K_D)\dot{\alpha} - C_{\mathrm{NL}}\mathrm{sign}(\dot{\alpha}) + M_\alpha(\alpha) + M_{\dot{\alpha}}(\dot{\alpha}, \alpha) - K_I\int_0^t \alpha\, d\tau\right]\dot{\alpha},\end{aligned}$$

which produces a limit cycle when

$$\oint_\tau \left[M_\alpha(\alpha) + M_{\dot{\alpha}}(\dot{\alpha}, \alpha) - K_I\int_0^t \alpha\, d\tau_1\right]\dot{\alpha}\, dt = \oint_\tau [(C + K_D)\dot{\alpha} + C_{\mathrm{NL}}\mathrm{sign}(\dot{\alpha})]\dot{\alpha}\, dt.$$

The numerical results are given in Figs. 8.13–8.17 for each case. The results are similar to the previous two example cases. The PID controller enables the trajecto-

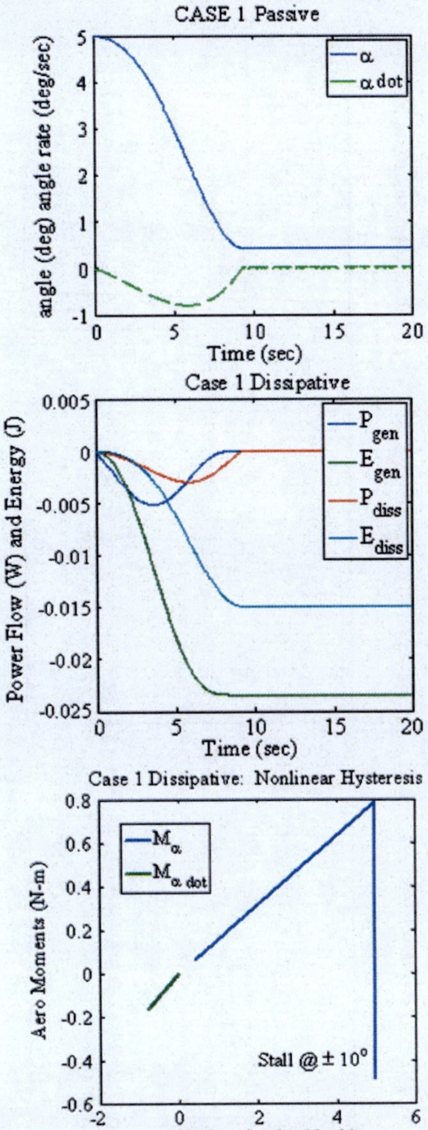

Fig. 8.10 Nonlinear stall flutter with nonlinear dynamic results: angular responses, power and energy flow responses, and aero moment responses Case 1 dissipative. Robinett III, R.D. and Wilson, D.G. [85], reprinted with permission of the American Institute of Aeronautics and Astronautics

ries to be shaped such that they are more smoothed, expanded, raised energy level and changed decay rate. This is due to the change in the Hamiltonian surface shape and the power flow.

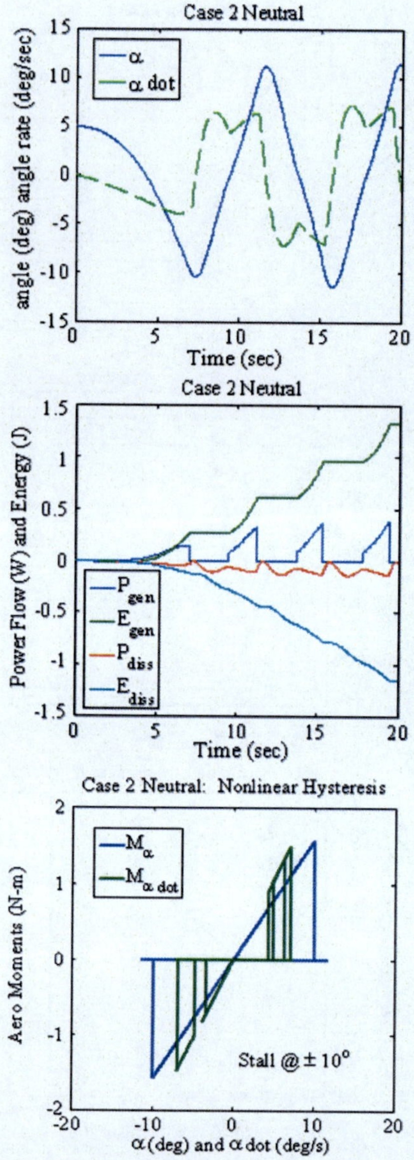

Fig. 8.11 Nonlinear stall flutter with nonlinear dynamic results: angular responses, power and energy flow responses, and aero moment responses Case 2 neutral. Robinett III, R.D. and Wilson, D.G. [85], reprinted with permission of the American Institute of Aeronautics and Astronautics

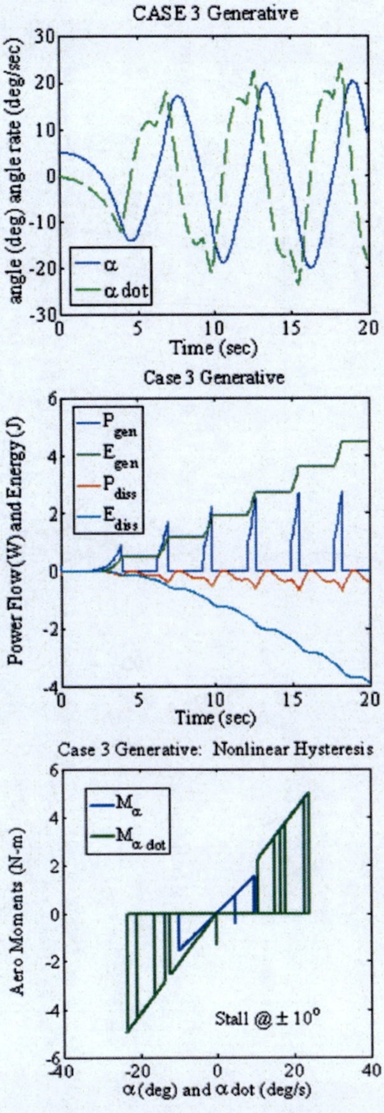

Fig. 8.12 Nonlinear stall flutter with nonlinear dynamic results: angular responses, power and energy flow responses, and aero moment responses Case 3 generative. Robinett III, R.D. and Wilson, D.G. [85], reprinted with permission of the American Institute of Aeronautics and Astronautics

8.3 Specific 5 MW Wind Turbine Control Design

Numerical simulations were performed for a representative 5 MW wind turbine blade first torsion mode (5.58 Hz). The previous sections have shown the general capabilities of the model and HSSPFC analyses. Three cases were considered. Case 1 represents a passivity PID control design, Case 2 represents a limit cycle oscilla-

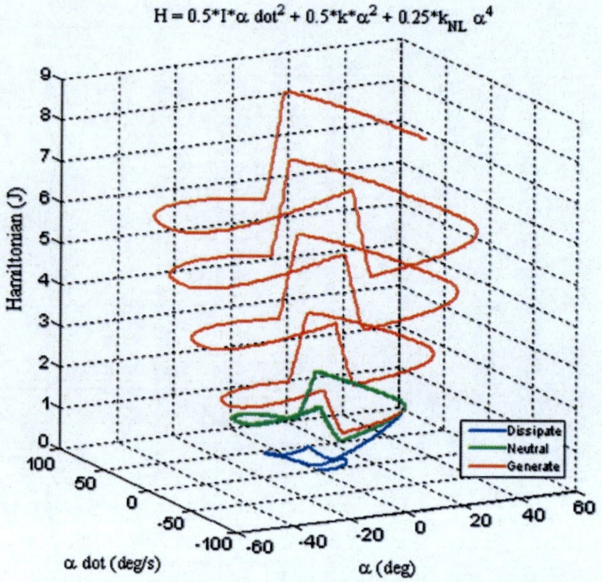

Fig. 8.13 Nonlinear stall Flutter with nonlinear dynamics and control system results: 3D Hamiltonian. Robinett III, R.D. and Wilson, D.G. [85], reprinted with permission of the American Institute of Aeronautics and Astronautics

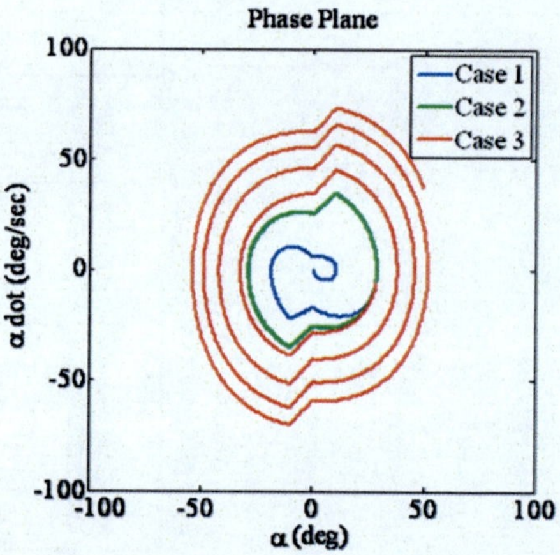

Fig. 8.14 Nonlinear stall flutter with nonlinear dynamics and control system results: phase plane plot. Robinett III, R.D. and Wilson, D.G. [85], reprinted with permission of the American Institute of Aeronautics and Astronautics

tion (LCO) for the system, and Case 3 is without control, where the aerodynamic loads drive the system unstable or generates more energy and power into the system then can be dissipated. These three cases are illustrated in a Hamiltonian 3D surface with the resulting paths given in Fig. 8.18. The corresponding phase plane plots are shown in Fig. 8.19.

Case 1 demonstrates a stable, damped response (see Fig. 8.20, top) that occurs when the power flows due to the system damping (linear, nonlinear) and derivative PID control action are greater than the power flows due to the nonlinear aerodynamic loads and the integral PID control action. Case 2 identifies the existence of a limit cycle oscillation (LCO) (see Fig. 8.20, middle) which results in nonlinear stall flutter when the power flows due to the derivative PID control action and damp-

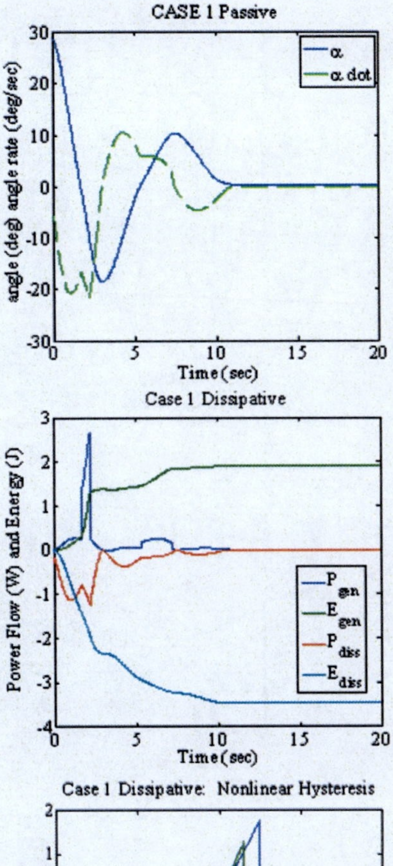

Fig. 8.15 Nonlinear stall flutter with nonlinear dynamics and control system results: angular responses, power and energy flow responses, and aero moment responses Case 1 dissipative. Robinett III, R.D. and Wilson, D.G. [85], reprinted with permission of the American Institute of Aeronautics and Astronautics

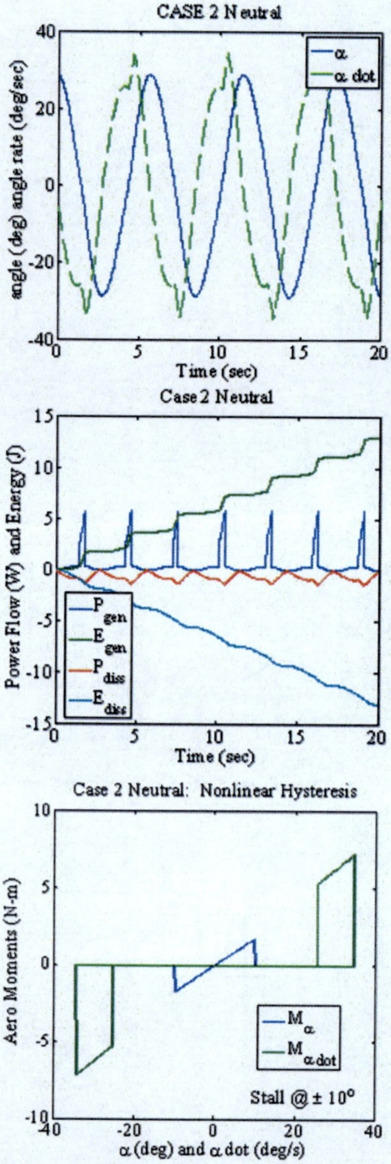

Fig. 8.16 Nonlinear stall flutter with nonlinear dynamics and control system results: angular responses, power and energy flow responses, and aero moment responses Case 2 neutral. Robinett III, R.D. and Wilson, D.G. [85], reprinted with permission of the American Institute of Aeronautics and Astronautics

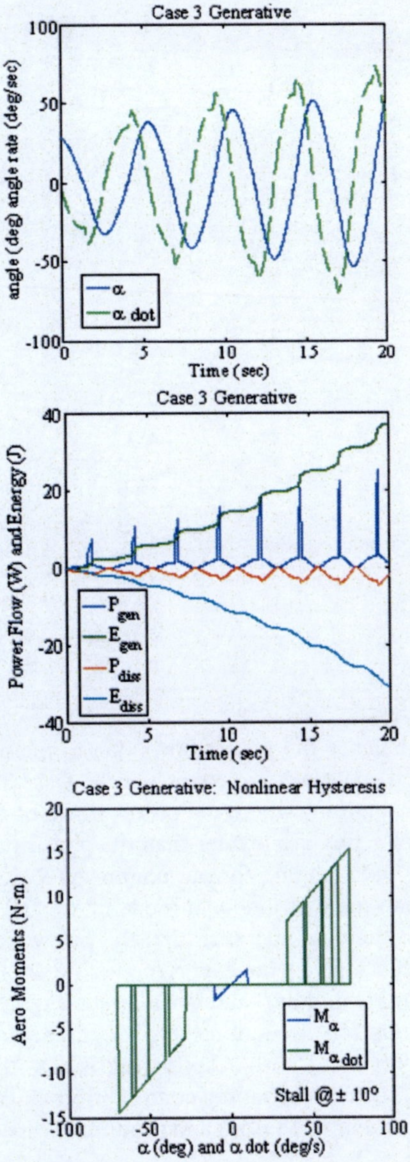

Fig. 8.17 Nonlinear stall flutter with nonlinear dynamics and control system results: angular responses, power and energy flow responses, and aero moment responses Case 3 generative. Robinett III, R.D. and Wilson, D.G. [85], reprinted with permission of the American Institute of Aeronautics and Astronautics

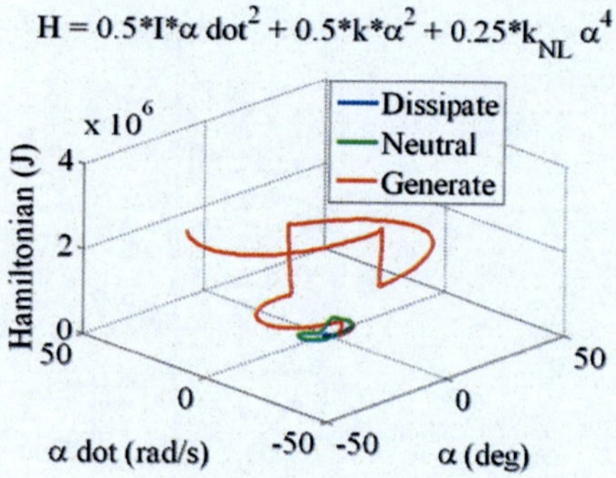

Fig. 8.18 Nonlinear stall flutter with nonlinear dynamics and PID control system results: 3D Hamiltonian trajectory paths. Robinett III, R.D. and Wilson, D.G. [86], reprinted by permission of the publisher (©2010 IEEE)

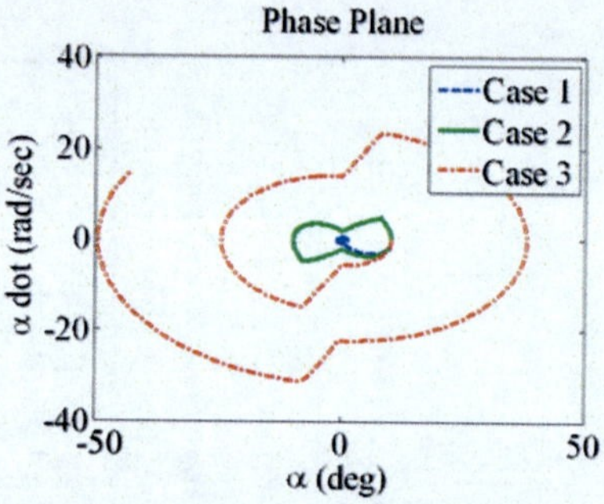

Fig. 8.19 Nonlinear stall flutter with nonlinear dynamics and PID control system results: phase plane plots. Robinett III, R.D. and Wilson, D.G. [86], reprinted by permission of the publisher (©2010 IEEE)

ing (linear, nonlinear) balance the power flows due to the nonlinear aerodynamic loads and the integral PID control action over a cycle. Case 3 illustrates an unstable response (see Fig. 8.20, bottom) where the power flows of the aerodynamic loads and integral PID control action are greater than the power flows due to the derivative PID control action and damping (linear, nonlinear). Notice that the limit cycle shows power flowing into the first torsional mode up to stall and then the decay of this energy state due to the damping such that the net work over a cycle is zero. These numerical results are given for energy/power flows; Case 1 (see Fig. 8.21, top), Case 2 (see Fig. 8.21, middle), and Case 3 (see Fig. 8.21, bottom). In addition, the numerical results for the nonlinear hysteretic aerodynamic moments are shown for Case 1 (see Fig. 8.22, top), Case 2 (see Fig. 8.22, middle), and Case 3 (see Fig. 8.22, bottom). The corresponding control effort and acceleration responses are provided for Case 1 in Fig. 8.23 (top) and (bottom), respectively.

8.4 Chapter Summary

HSSPFC was applied to a nonlinear stall flutter problem (dynamic stall) that approximates the first torsion mode of a large wind turbine blade (5 MW) under the influence of nonlinear aerodynamic loading. The methodology directly accommodated nonlinear structural and discontinuous aerodynamic models. The limit cycles were found by partitioning the power flows and identifying when the dissipation

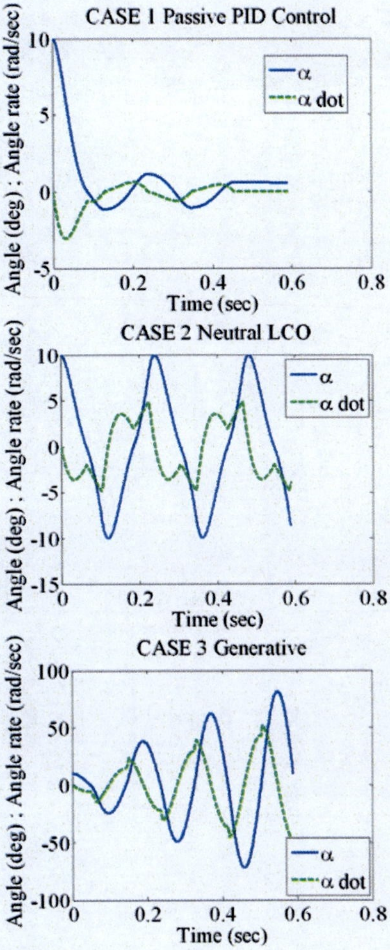

Fig. 8.20 Nonlinear stall flutter with nonlinear dynamics Case 1 dissipative PID control (*top*), Case 2 neutral LCO (*middle*), and Case 3 generative (*bottom*) transient numerical simulation results. Robinett III, R.D. and Wilson, D.G. [86], reprinted by permission of the publisher (©2010 IEEE)

and generation balanced over a cycle subject to the energy storage. The limit cycles were shown to be stability boundaries. The flutter suppression controllers were initially assessed by designing a PID controller. This initial step starts the process of how to design a nonlinear flutter/dynamic stall controller that could be incorporated into the conventional controllers for a typical 5 MW wind turbine in below-, at-, and above-rated power conditions. The closer the wind turbine can safely operate to dynamic stall, the greater the energy that can be generated.

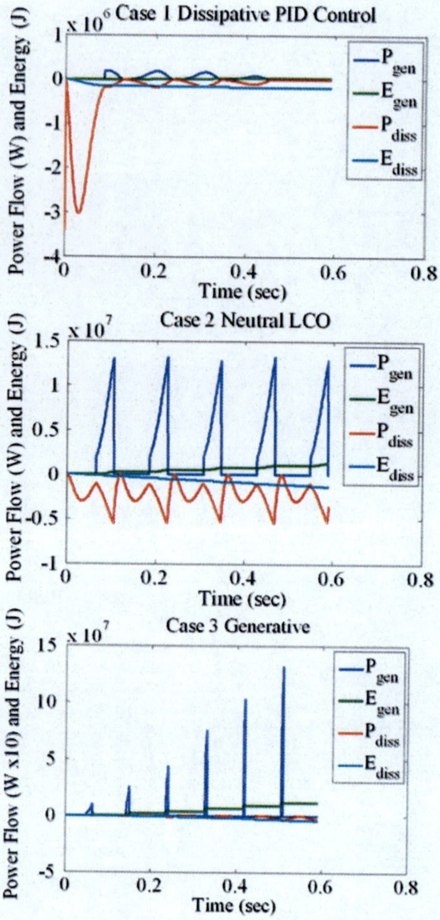

Fig. 8.21 Power and energy flow responses for Case 1 dissipative PID control (*top*), Case 2 neutral LCO (*middle*), and Case 3 generative (*bottom*) numerical simulation results. Robinett III, R.D. and Wilson, D.G. [86], reprinted by permission of the publisher (©2010 IEEE)

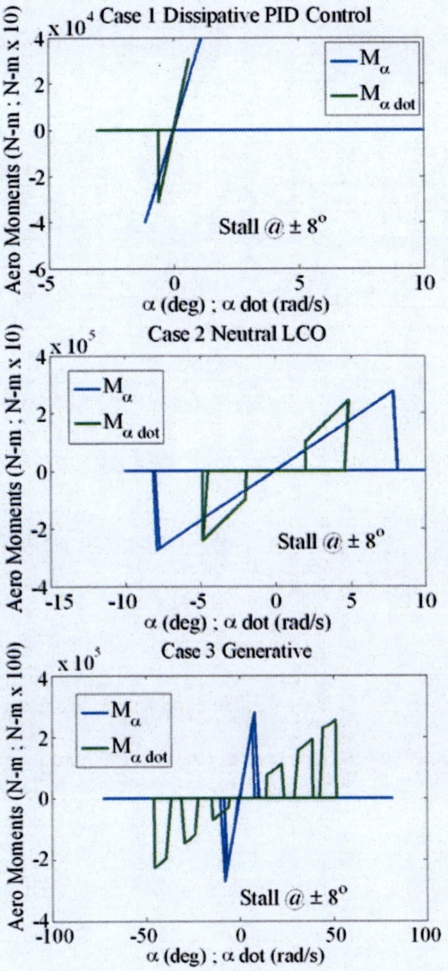

Fig. 8.22 Aero moment responses for Case 1 dissipative PID control (*top*), Case 2 neutral LCO (*middle*), and Case 3 generative (*bottom*) numerical simulation results. Robinett III, R.D. and Wilson, D.G. [86], reprinted by permission of the publisher (©2010 IEEE)

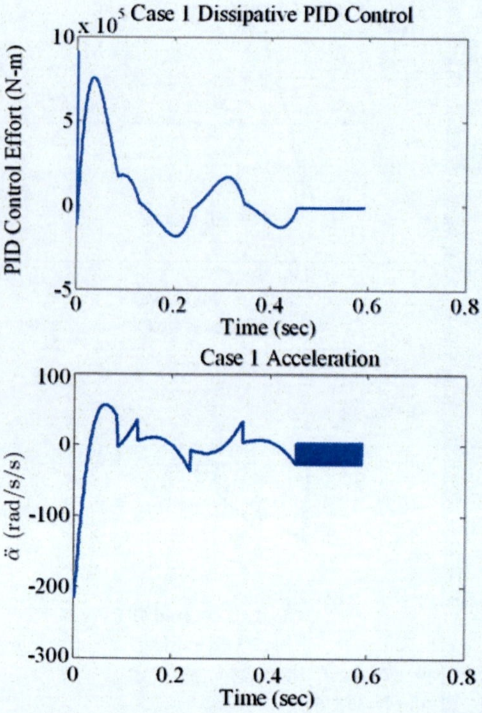

Fig. 8.23 Case 1 PID control effort (*top*) and acceleration (*bottom*) numerical simulation results. Robinett III, R.D. and Wilson, D.G. [86], reprinted by permission of the publisher (©2010 IEEE)

Chapter 9
Case Study #4: Fundamental Power Engineering

9.1 Introduction

Engineered power systems are rich in control and mathematical problems and will continue to require improved performance from control systems centered on physics-based principles and mathematics. Engineers are working to improve the performance and gain a deeper understanding of these power systems [87]. The need for increased efficiency has driven designers to develop more sophisticated control systems that lay a burden on traditional analysis tools and techniques. Further complications come from nonlinearities in the system that understanding comes only from simulation [87–89].

Power systems operate today with smaller stability margins and will need increased security plus reliability with improvements and developments in modeling, stability analysis approaches, and assessment tools [89]. Several researchers [90, 91] are investigating advanced architectures for the control and stability of multimachine power systems [92, 93]. In this chapter the goal is to introduce fundamental concepts that underpin nonlinear power flow control as it is applied to power engineering problems [94].

This chapter is divided into six sections. In Sect. 9.2 a power engineering application begins with a simple RLC network which shows the connections between power engineering and Hamiltonian formulations. Next Sect. 9.3 discusses the performance of an electric power grid system. Nonlinear power flow control is used to design both a robust PID and PID with adaptive control algorithms. Section 9.4 presents numerical simulations for a linear adaptive system. Section 9.5 presents numerical simulations for a nonlinear adaptive system. Finally, in Sect. 9.6 the key highlights and results are summarized with concluding remarks.

9.2 Power Engineering Application

In this section, an analog of the linear mass–spring–damper system, an RLC electrical network, is employed to explain the connection between HSSPFC control [7, 13,

R.D. Robinett III, D.G. Wilson, *Nonlinear Power Flow Control Design*, Understanding Complex Systems,
DOI 10.1007/978-0-85729-823-2_9,

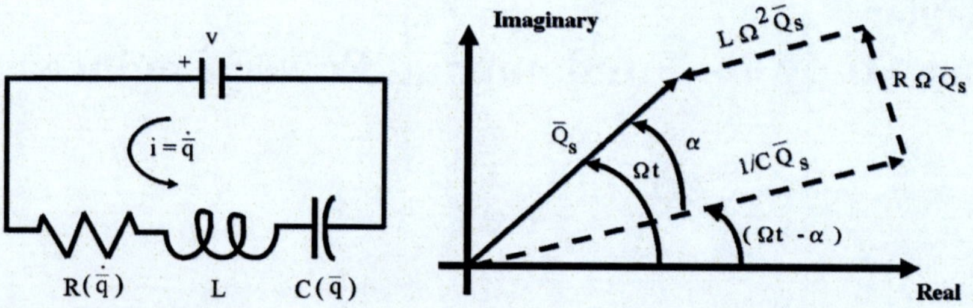

Fig. 9.1 Nonlinear RLC electric circuit (*left*) and RLC vector polygon (*right*). Robinett III, R.D. and Wilson, D.G. [94], reprinted by permission of the publisher (©2008 IEEE)

46] (also see Chap. 4) and power engineering [94]. The discussion begins by following the harmonic excitation development of a linear mass–spring–damper system by [68].

The first step is to derive the equations of motion using Lagrange's equations [68]

$$\frac{d}{dt}\left(\frac{\partial \mathcal{L}}{\partial \dot{q}_i}\right) - \frac{\partial \mathcal{L}}{\partial q_i} = Q_i,$$

where $\mathcal{L} = \mathcal{T} - \mathcal{V}$ is the Lagrangian, and $\delta W_i = Q_i \delta q_i$ is the virtual work of the ith nonconservative generalized force. For a mass–spring–damper system, the kinetic energy is $\mathcal{T} = \frac{1}{2}m\dot{x}^2$, the potential energy is $\mathcal{V} = \frac{1}{2}kx^2$, the Lagrangian is $\mathcal{L} = \frac{1}{2}m\dot{x}^2 - \frac{1}{2}kx^2$, and the virtual work is $\delta W = u\delta x - c\dot{x}\delta x$. The resulting equation of motion is

$$m\ddot{x} + c\dot{x} + kx = u. \tag{9.1}$$

For the RLC circuit in Fig. 9.1, left (with R, L, and C as constant linear parameters), the kinetic energy is $\mathcal{T} = \frac{1}{2}L\dot{\bar{q}}^2$, the potential energy is $\mathcal{V} = \frac{1}{2}\frac{1}{C}\bar{q}^2$, the Lagrangian is $\mathcal{L} = \frac{1}{2}L\dot{\bar{q}}^2 - \frac{1}{2}\frac{1}{C}\bar{q}^2$, and the virtual work is $\delta W = v\delta\bar{q} - R\dot{\bar{q}}\delta\bar{q}$, where L is the inductance, R is the resistance, C is the capacitance, $\bar{q}$ is the charge, $\dot{\bar{q}} = i$ is the current, and v is the applied voltage. The resulting equation of motion is

$$L\ddot{\bar{q}} + R\dot{\bar{q}} + \frac{1}{C}\bar{q} = v. \tag{9.2}$$

The second step is to solve (9.1) with a harmonic input [68] or

$$m\ddot{x} + c\dot{x} + kx = u_0 \cos \Omega t. \tag{9.3}$$

The steady-state response is $x_s = X_s \cos(\Omega t - \alpha)$, where X_s is the steady-state amplitude, and α is the phase angle of the steady-state response relative to the harmonic input. The solution of (9.3) is equivalent to determining X_s and α. The velocity and acceleration are $\dot{x}_s = -\Omega X_s \sin(\Omega t - \alpha)$ and $\ddot{x}_s = -\Omega^2 X_s \cos(\Omega t - \alpha)$, which upon substitution into (9.3) gives

$$-m\Omega^2 X_s \cos(\Omega t - \alpha) - c\Omega X_s \sin(\Omega t - \alpha) + kX_s \cos(\Omega t - \alpha) = u_0 \cos \Omega t. \tag{9.4}$$

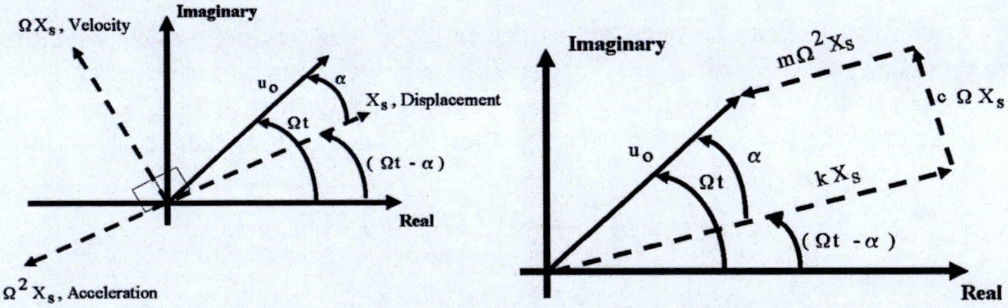

Fig. 9.2 Complex plane: rotating vectors (*left*) and force vector polygon (*right*). Robinett III, R.D. and Wilson, D.G. [94], reprinted by permission of the publisher (©2008 IEEE)

The terms in (9.4) can be presented in the complex plane as a set of rotating vectors (Fig. 9.2, left) and force vector polygon (Fig. 9.2, right). It is clear from Fig. 9.2 (right) and (9.4) that

$$u_0^2 = \left[kX_s - m\Omega^2 X_s\right]^2 + [c\Omega X_s]^2$$

and

$$\tan\alpha = c\Omega / \left[k - m\Omega^2\right], \tag{9.5}$$

which gives

$$\frac{X_s}{u_o} = 1/\left[\left(k - m\Omega^2\right)^2 + (c\Omega)^2\right]^{1/2}. \tag{9.6}$$

Now, it is time to relate the linear mass–spring–damper system to the RLC circuit. Equations (9.5) and (9.6) can be rewritten as

$$\begin{aligned} \frac{\bar{Q}_s}{v_0} &= 1 \Big/ \left[\left(\frac{1}{C} - L\Omega^2\right)^2 + (R\Omega)^2\right]^{1/2}, \\ \tan\alpha &= R\Omega \Big/ \left[\frac{1}{C} - L\Omega^2\right], \end{aligned} \tag{9.7}$$

where

$$\bar{q}_s = \bar{Q}_s \cos(\Omega t - \alpha) \tag{9.8}$$

and $v = v_0 \cos \Omega t$.

The third step is to relate (9.7) to power engineering. A goal of power engineering is to maximize the real power flow: match the frequency and phase of the applied voltage and resulting current. From Fig. 9.1 (right) it is clear that the goal is

$$\frac{1}{C}\bar{Q}_s = L\Omega^2 \bar{Q}_s \quad \Rightarrow \quad \Omega^2 = 1/LC,$$

which means that the RLC circuit should be driven at the undamped natural frequency and/or the inductance and capacitance should be balanced for 60 Hz.

Equation (9.8) can be rewritten as current $\dot{\bar{q}}_s = -\Omega \bar{Q}_s \sin(\Omega t - \alpha)$ which is $i_s = -I_s \sin(\Omega t - \alpha)$ for $I_s = \Omega \bar{Q}_s$. The phase angle α can be redefined in terms of the phase angle θ between the voltage vector and the current vector $\alpha = \pi/2 - \theta$, which produces $i_s = I_s \cos(\Omega t + \theta)$. Equation (9.7) can be rewritten as

$$\frac{I_s}{v_0} = 1 \bigg/ \left[\left(\frac{1}{C\Omega} - L\Omega \right)^2 + R^2 \right]^{1/2}, \tag{9.9}$$

which is $v_0 = Z I_s$ where $Z = [X^2 + R^2]^{1/2}$ is the magnitude of the complex impedance. The power factor is defined [49] as $\text{pf} = \cos\theta = P/VI$, and by convention, θ is positive when the voltage leads the current.

The goal of power engineering can be restated as $\theta = 0$, which happens when

$$\sin\theta = \frac{X}{Z} = \frac{X}{[X^2 + R^2]^{1/2}} = 0$$

with $X = 0 \Rightarrow \Omega^2 = 1/LC$.

The fourth step is to relate the Hamiltonian to the power factor. The Hamiltonian is defined as

$$\mathcal{H} = \sum_{j=1}^{n} \frac{\partial \mathcal{L}}{\partial \dot{q}_j} \dot{q}_j - \mathcal{L},$$

which is $\mathcal{H} = \mathcal{T} + \mathcal{V}$ for both the linear mass–spring–damper system and the RLC circuit. For the linear mass–spring–damper system, the Hamiltonian and the time derivative of the Hamiltonian are

$$\mathcal{H} = \frac{1}{2} m \dot{x}^2 + \frac{1}{2} k x^2,$$
$$\dot{\mathcal{H}} = [m\ddot{x} + kx]\dot{x} = [u - c\dot{x}]\dot{x} = u\dot{x} - c\dot{x}^2$$

and, for the RLC circuit,

$$\begin{aligned} \mathcal{H} &= \frac{1}{2} L \dot{\bar{q}}^2 + \frac{1}{2}\frac{1}{C} \bar{q}^2, \\ \dot{\mathcal{H}} &= \left[L\ddot{\bar{q}} + \frac{1}{C}\bar{q} \right] \dot{\bar{q}} = [v - R\dot{\bar{q}}]\dot{\bar{q}} = vi - Ri^2. \end{aligned} \tag{9.10}$$

The power engineering solution is "open-loop" since the applied voltage does not include explicit feedback terms. Although, the applied voltage and capacitance are "tuned" at a different update rate. The applied voltage and resulting current and charge are

$$v = v_0 \cos \Omega t,$$
$$i_s = I_s \cos(\Omega t + \theta),$$
$$\bar{q}_s = \frac{I_s}{\Omega} \sin(\Omega t + \theta),$$

which leads to the Hamiltonian

$$\mathcal{H} = \frac{1}{2} L I_s^2 \cos^2(\Omega t + \theta) + \frac{1}{2} \frac{I_s^2}{C\Omega^2} \sin^2(\Omega t + \theta) > 0$$

and the corresponding time derivative

$$\dot{\mathcal{H}} = v_0 I_s \cos \Omega t \cos(\Omega t + \theta) - R I_s^2 \cos^2(\Omega t + \theta).$$

The resulting constraint equations for power engineering are

$$\oint \left[v_0 I_s \cos \Omega t \cos(\Omega t + \theta) - R I_s^2 \cos^2(\Omega t + \theta) \right] dt = 0 \tag{9.11}$$

and

$$\oint [-L\Omega + 1/C\Omega] I_s \sin(\Omega t + \theta) \cos(\Omega t + \theta)\, dt = 0, \tag{9.12}$$

which implies

$$\Omega^2 = 1/LC,$$
$$v_0 = R I_s$$

for optimal power flow. For $\Omega^2 \neq 1/LC$, (9.11) and (9.12) are automatically satisfied, but the required applied voltage increases by $v_0 I_s \cos \theta = R I_s^2$. Figure 9.3 presents the 3D Hamiltonian (top) and phase plane plots (bottom) for optimal and suboptimal power flows. As can be seen from the suboptimal plots, part of the applied voltage is supporting the off-resonant storage system by varying the limit cycle trajectory across the Hamiltonian surface to produce the desired phase plane limit cycle [7, 13, 46].

9.3 Performance of Electric Power Grid System

A goal of the electric power grid is to distribute electricity from the source to the load with a power factor of 1 (see reference [49]). The power factor[1] is defined as the ratio of the real power to the apparent power and ranges from 0 to 1. Low power factor loads increase losses and energy costs in a power distribution system. The term VAR (volt-amperes reactive) is the unit of reactive power (or VAR support) and represents the power consumed by a reactive load. Power factor correction returns the power factor of an electric AC power transmission system to very near unity by switching in or out banks of capacitors or inductors which act to cancel the inductive or capacitive effects of the load. For example, the inductive effect of motor loads may be offset by locally connected capacitors. An active power factor corrector is

[1] *Power factor, VAR, power factor correction*, and *active power factor corrector* terms have been defined from the Wikipedia encyclopedia.

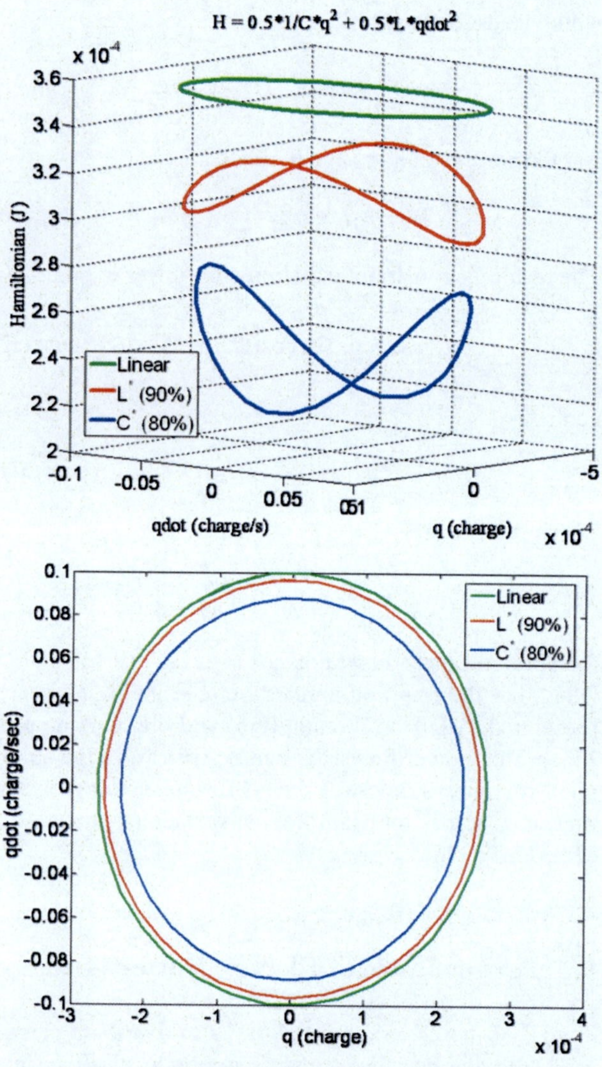

Fig. 9.3 The 3D Hamiltonian shows the linear, 10% variation in inductance, L or $L' = 90\%$ of L, and 20% variation in capacitance, C or $C' = 80\%$ of C (*top*) with the corresponding phase plane plots (*bottom*). Robinett III, R.D. and Wilson, D.G. [94], reprinted by permission of the publisher (©2008 IEEE)

a power electronic system that controls the amount of power drawn by a load in order to obtain a power factor as close to unity as possible. In the following section a simple RLC network that represents several key components in an electric power grid will be analyzed with respect to power flow.

A simple strategy to achieve a power factor of 1 is to have capacitor banks available to provide "VAR support" where needed on the grid, since most loads and generators are heavily inductive. A potentially more flexible approach is to use power electronics to provide proportional feedback to do real-time VAR support or

$$v = v_0 \cos \Omega t - K_p \bar{q},$$

which leads to

$$\begin{aligned}\mathcal{H} &= \frac{1}{2}L\dot{\bar{q}}^2 + \frac{1}{2}\left[\frac{1}{C} + K_P\right]\bar{q}^2,\\ \dot{\mathcal{H}} &= \left[L\ddot{\bar{q}} + \left(\frac{1}{C} + K_P\right)\bar{q}\right]\dot{\bar{q}} = \left[(v_0 \cos \Omega t)\dot{\bar{q}} - R\dot{\bar{q}}^2\right].\end{aligned}\tag{9.13}$$

Clearly, the cost of K_P (power electronics) versus $\frac{1}{C}$ (capacitor banks) can be evaluated with respect to exergy, exergy rate, flexibility, reliability, etc.

A more general, and possibly clearer, understanding of the trade-offs between information and physical exergies can be described by rewriting (9.13) as a tracking controller

$$\begin{aligned}\mathcal{H} &= \frac{1}{2}L(\dot{\bar{q}} - \dot{\bar{q}}_R)^2 + \frac{1}{2}\frac{1}{C}(\bar{q} - \bar{q}_R)^2,\\ \dot{\mathcal{H}} &= \left[L(\ddot{\bar{q}} - \ddot{\bar{q}}_R) + \frac{1}{C}(\bar{q} - \bar{q}_R)\right](\dot{\bar{q}} - \dot{\bar{q}}_R),\end{aligned}$$

where

$$L\ddot{\bar{q}} = -R\dot{\bar{q}} - \frac{1}{C}\bar{q} + v.$$

Next, modify $\mathcal{H}$ to account for K_P as

$$\begin{aligned}\hat{\mathcal{H}} &= \frac{1}{2}L(\dot{\bar{q}} - \dot{\bar{q}}_R)^2 + \frac{1}{2}\left[\frac{1}{C} + K_P\right](\bar{q} - \bar{q}_R)^2,\\ \dot{\hat{\mathcal{H}}} &= \left[L(\ddot{\bar{q}} - \ddot{\bar{q}}_R) + \left[\frac{1}{C} + K_P\right](\bar{q} - \bar{q}_R)\right](\dot{\bar{q}} - \dot{\bar{q}}_R)\\ &= [\Delta v + K_P(\bar{q} - \bar{q}_R) - R(\dot{\bar{q}} - \dot{\bar{q}}_R)](\dot{\bar{q}} - \dot{\bar{q}}_R),\end{aligned}$$

where

$$v = v_R + \Delta v,$$

$$v_R = L\ddot{\bar{q}}_R + R\dot{\bar{q}}_R + \frac{1}{C}\bar{q}_R.$$

If one designs v_R for $\omega^2 = \Omega^2 = 1/LC$, then $v_R\dot{\bar{q}}_R = R\dot{\bar{q}}_R^2$, which requires feedback control if $\omega^2 = 1/LC \neq \Omega^2$ to get a power factor of 1, then

$$\Delta v = -K_p(\bar{q} - \bar{q}_R) - K_I \int (\bar{q} - \bar{q}_R)\, dt - K_D(\dot{\bar{q}} - \dot{\bar{q}}_R).$$

Then by choosing $K_I = K_D = 0$ the control strategy becomes

$$\dot{\hat{\mathcal{H}}} = \left[L(\ddot{\bar{q}} - \ddot{\bar{q}}_R) + \left[\frac{1}{C} + K_P\right](\bar{q} - \bar{q}_R)\right](\dot{\bar{q}} - \dot{\bar{q}}_R) = -R(\dot{\bar{q}} - \dot{\bar{q}}_R)^2,$$

where

$$R\dot{\bar{q}}_R^2 = v_0 \cos \Omega t \dot{\bar{q}}_R$$

and

$$\frac{1}{\bar{C}} = \left[\frac{1}{C} + K_P\right].$$

In this example, the clear tradeoff is between a physical capacitor bank and a power flow control device. The capacitor is simple, reliable, and fixed capacity. The control device is flexible, adaptable, and presently more expensive and possibly less reliable. Also, there are two major issues with this approach that should be addressed in future studies:

1. Does it actually get closer to a power factor of 1 since one is expending exergy to run power electronics?
2. Does closing the control loop locally destabilize the collective grid especially when K_I and K_D are nonzero? In particular, does the controlled system performance result in a stable linear limit cycle (refer to Chaps. 3 and 4).

9.4 Linear Adaptive Power Engineering

One way to improve performance is to introduce an adaptive PID controller. For comparison, the nonadaptive (robust) PID control does not include the parameter updates required by robust control (see Chap. 5). Given the general RLC network, the dynamics of (9.2) produces (9.10). An adaptive control strategy can be defined by investigating a more general tracking controller composed of

$$\mathcal{H} = \mathcal{T} + \mathcal{V} = \frac{1}{2}L\left(\dot{\bar{q}} - \dot{\bar{q}}_R\right)^2 + \frac{1}{2}\frac{1}{C}(\bar{q} - \bar{q}_R)^2.$$

Next, modify $\mathcal{H}$ to account for additional storage or capacitance control as $\mathcal{H} = \mathcal{T} + \mathcal{V} + \mathcal{V}_{\text{cap}}$, where $\mathcal{V}_{\text{cap}} = \frac{1}{2}K_{\text{cap}}(\bar{q} - \bar{q}_R)^2$. Next, add a general information control potential [7] associated with the adaptation parameters or $\mathcal{H} = \mathcal{T} + \mathcal{V} + \mathcal{V}_{\text{cap}} + \mathcal{V}_I$, where $\mathcal{V}_I = \frac{1}{2}\tilde{\Phi}^T \Gamma^{-1} \tilde{\Phi}$. This results in

$$\mathcal{H} = \frac{1}{2}L\left(\dot{\bar{q}} - \dot{\bar{q}}_R\right)^2 + \frac{1}{2}\left(\frac{1}{C} + K_{\text{cap}}\right)(\bar{q} - \bar{q}_R)^2 + \frac{1}{2}\tilde{\Phi}^T \Gamma^{-1} \tilde{\Phi}. \tag{9.14}$$

Then taking the derivative, rearranging, and simplifying gives

$$\dot{\mathcal{H}} = \left[-L\ddot{\bar{q}}_R - \frac{1}{C}\bar{q}_R + K_{\text{cap}}(\bar{q} - \bar{q}_R) - R\dot{\bar{q}} + v\right] + \tilde{\Phi}^T \Gamma^{-1} \dot{\tilde{\Phi}}. \tag{9.15}$$

The controller is selected as [7, 13]

$$v = v_{\text{ref}} + \Delta v \quad \text{and} \quad v_{\text{ref}} = \hat{L}\ddot{\bar{q}}_R + \hat{R}\dot{\bar{q}} + \frac{\hat{1}}{C}\bar{q}_R. \tag{9.16}$$

One wants to design v_R for $\omega^2 = 1/LC = \Omega^2$ (an ideal 60 Hz condition); then $v_{\text{ref}}\dot{\bar{q}}_R = R\dot{\bar{q}}_R^2$, which requires feedback control if $\omega^2 = 1/LC \neq \Omega^2$ to get a power factor of 1. Then select the feedback portion as

$$\Delta v = -K_{\text{cap}}(\bar{q} - \bar{q}_R) - K_{\text{gen}} \int_0^t (\bar{q} - \bar{q}_R)\, d\tau - K_{\text{diss}}(\dot{\bar{q}} - \dot{\bar{q}}_R).$$

Next substitute the controller (9.16) into (9.15), simplify, and rearrange as

$$\begin{aligned} \dot{\mathcal{H}} = {} & \left[-K_{\text{gen}} \int_0^t (\bar{q} - \bar{q}_R)\, d\tau - K_{\text{diss}}(\dot{\bar{q}} - \dot{\bar{q}}_R)\right](\dot{\bar{q}} - \dot{\bar{q}}_R) \\ & + \tilde{\Phi}^T\left[Y^T(\dot{\bar{q}} - \dot{\bar{q}}_R) + \Gamma^{-1}\dot{\tilde{\Phi}}\right], \end{aligned} \tag{9.17}$$

where

$$Y\tilde{\Phi} = (\hat{L} - L)\ddot{\bar{q}}_R + \left(\frac{\hat{1}}{C} - \frac{1}{C}\right)\bar{q}_R + (\hat{R} - R)\dot{\bar{q}}$$

and

$$\begin{aligned} \tilde{\Phi}^T &= \begin{bmatrix}(\hat{L} - L) & \left(\frac{\hat{1}}{C} - \frac{1}{C}\right) & (\hat{R} - R)\end{bmatrix}, \\ \dot{\tilde{\Phi}}^T &= \dot{\hat{\Phi}}^T = \begin{bmatrix}\dot{\hat{L}} & \dfrac{\dot{\hat{1}}}{C} & \dot{\hat{R}}\end{bmatrix}, \\ Y &= \begin{bmatrix}\ddot{\bar{q}}_R & \bar{q}_R & \dot{\bar{q}}\end{bmatrix}. \end{aligned}$$

Next to identify the adaptive parameter update equations, the last term in (9.17) is set to zero, which yields

$$\begin{aligned} \dot{\hat{L}} &= -\gamma_1\ddot{\bar{q}}_R(\dot{\bar{q}} - \dot{\bar{q}}_R), \\ \frac{\dot{\hat{1}}}{C} &= -\gamma_2\bar{q}_R(\dot{\bar{q}} - \dot{\bar{q}}_R), \\ \dot{\hat{R}} &= -\gamma_3\dot{\bar{q}}(\dot{\bar{q}} - \dot{\bar{q}}_R). \end{aligned} \tag{9.18}$$

Initially, for perfect parameter matching, the following nonlinear stability boundary concludes from (9.17):

$$\oint_\tau \left[-K_{\text{gen}} \int_0^t (\bar{q} - \bar{q}_R)\, d\tau'\right](\dot{\bar{q}} - \dot{\bar{q}}_R)\, dt < \oint_\tau \left[K_{\text{diss}}(\dot{\bar{q}} - \dot{\bar{q}}_R)^2\right] dt, \tag{9.19}$$

which results in an asymptotically stable (passivity [95]) solution.

Numerical simulations were performed with variations in $L = 77.4$ mH, $C = 100$ μF ($\neq 60$ Hz), and $R = 9$ Ω. Both a robust PID and PID with adaptive controllers were investigated for the linear RLC network. The tracking results for the charge and current along with the tracking errors are given in Fig. 9.4 for PID and in Fig. 9.5 for PID with adaptive controllers, respectively. In addition, the power flow

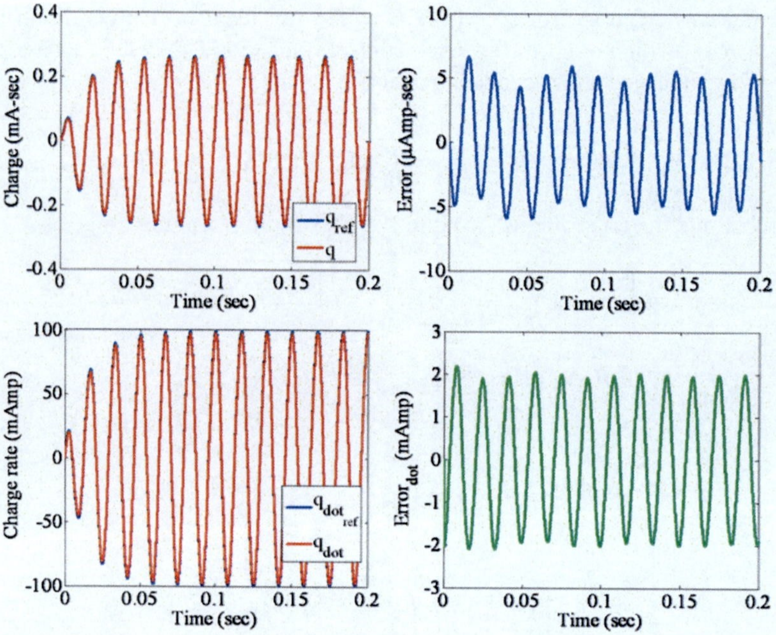

Fig. 9.4 Linear RLC network with PID controller: charge and charge error responses along with charge-rate and charge-rate error responses. Robinett III, R.D. and Wilson, D.G. [94], reprinted by permission of the publisher (©2008 IEEE)

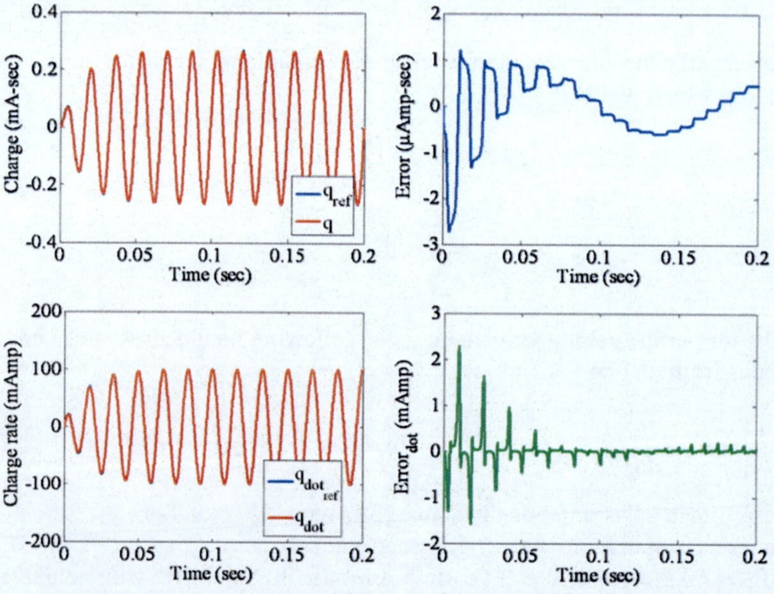

Fig. 9.5 Linear RLC network with PID/adaptive controller: charge and charge error responses along with charge-rate and charge-rate error responses. Robinett III, R.D. and Wilson, D.G. [94], reprinted by permission of the publisher (©2008 IEEE)

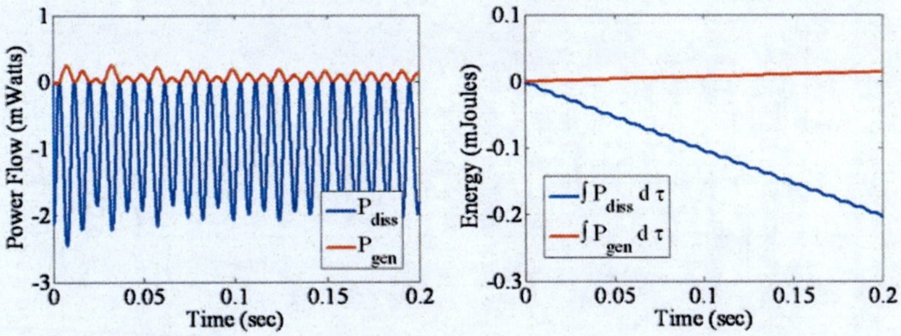

Fig. 9.6 Linear RLC network with PID controller: power flow and energy responses. Robinett III, R.D. and Wilson, D.G. [94], reprinted by permission of the publisher (©2008 IEEE)

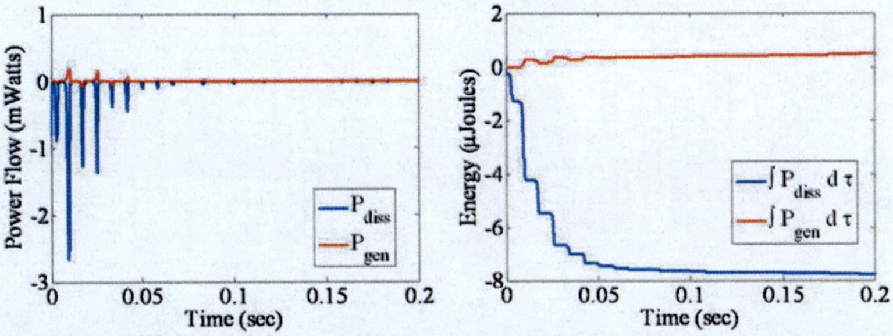

Fig. 9.7 Linear RLC network with PID/adaptive controller: power flow and energy responses. Robinett III, R.D. and Wilson, D.G. [94], reprinted by permission of the publisher (©2008 IEEE)

and energy consumed by the PID and PID with adaptive controllers are shown in Fig. 9.6 and Fig. 9.7, respectively.

The adaptive PID control versus the robust PID control alone is analogous to information flow versus physical power flow. In this case, the adaptive PID controller with information flow is less demanding on the system than the PID controller alone. For the commanded voltage input, $v = v_{\text{ref}} + \Delta v$, the signal is composed of the PID controller feedback signal Δv and the reference voltage v_R computed from the adaptive controller. The ideal voltage signal v_{ideal} is given for the ideal linear RLC circuit where $\omega = \Omega$. The responses are shown in Fig. 9.8 for both the PID (left) and PID with adaptive (right) controllers, respectively. The estimates for the adaptive controller reference input parameters are given for the linear model in Fig. 9.10 (left column) and include $\hat{L}$, $\hat{1}/C$, and $\hat{R}$.

9.5 Nonlinear Adaptive Power Engineering

This section describes the extension of power engineering using nonlinear power flow (exergy/entropy) control to nonlinear power engineering. The investigation begins by modifying a problem described by [95]. The RLC circuit in Fig. 9.1 (left)

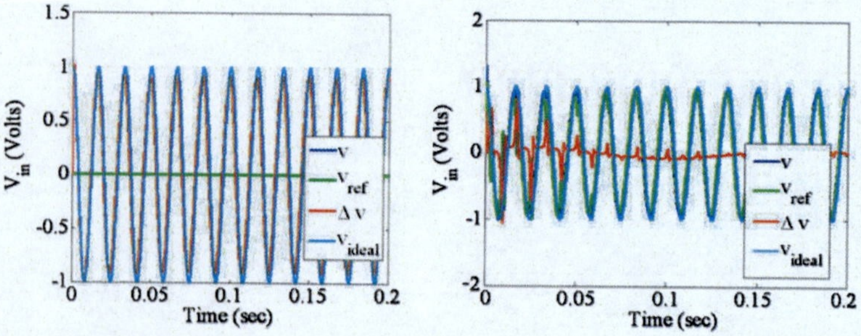

Fig. 9.8 Linear RLC network with PID and PID/adaptive controller: voltage input responses. Robinett III, R.D. and Wilson, D.G. [94], reprinted by permission of the publisher (©2008 IEEE)

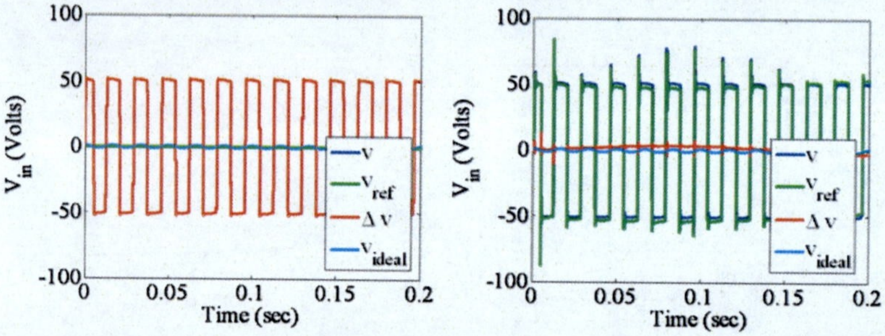

Fig. 9.9 Nonlinear RLC network with PID and PID/adaptive controller: voltage input responses. Robinett III, R.D. and Wilson, D.G. [94], reprinted by permission of the publisher (©2008 IEEE)

shows nonlinear resistance and capacitance elements. The nonlinear equation of motion is

$$L\ddot{\bar{q}} + R(\dot{\bar{q}}) + C(\bar{q}) = v,$$

where

$$\begin{aligned} \dot{\bar{q}}R(\dot{\bar{q}}) &> 0 \quad \text{for } \dot{\bar{q}} \neq 0, \\ \bar{q}C(\bar{q}) &> 0 \quad \text{for } \bar{q} \neq 0. \end{aligned}$$

The Hamiltonian is

$$\mathcal{H} = \frac{1}{2}L\dot{\bar{q}}^2 + \int_0^{\bar{q}} C(y)\,dy,$$

and the time derivative is

$$\dot{\mathcal{H}} = [L\ddot{\bar{q}} + C(\bar{q})]\dot{\bar{q}} = [v - R(\dot{\bar{q}})]\dot{\bar{q}} = vi - R(i)i.$$

The goal is to design an applied voltage that creates a desired limit cycle (e.g., 60 Hz) $\oint[L\ddot{\bar{q}}\dot{\bar{q}} + C(\bar{q})\dot{\bar{q}}]\,dt = 0$, where the storage terms are matched and $\oint[vi - R(i)i]\,dt = 0$, and vi is maximally used by the load.

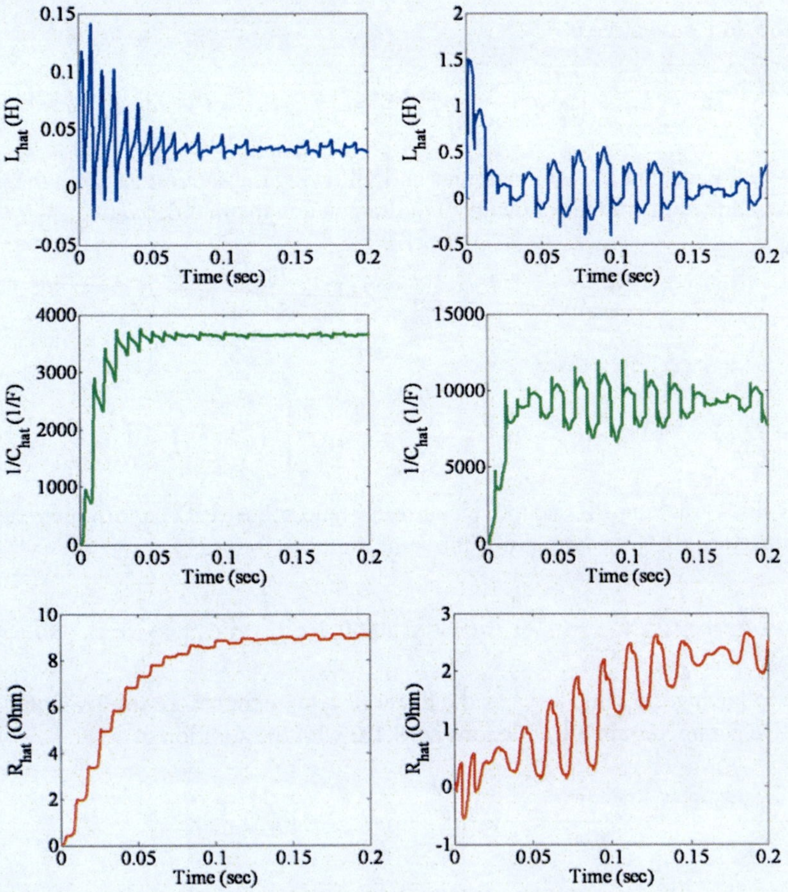

Fig. 9.10 Linear (*left column*) and nonlinear (*right column*) RLC adaptive controllers: $\hat{L}$, $\hat{\frac{1}{C}}$, and $\hat{R}$ responses. Robinett III, R.D. and Wilson, D.G. [94], reprinted by permission of the publisher (©2008 IEEE)

An interesting example is to modify the linear RLC network to include nonlinear capacitance (Duffing oscillator) and nonlinear resistance (Coulomb friction-like) defined as

$$L\ddot{\bar{q}} + \left[\frac{1}{C}\bar{q} + \frac{1}{C_{\mathrm{NL}}}\bar{q}^3\right] = v - R\dot{\bar{q}} - R_{\mathrm{NL}}\mathrm{sign}(\dot{\bar{q}}).$$

The Hamiltonian is

$$\mathcal{H} = \frac{1}{2}L\dot{\bar{q}}^2 + \left[\frac{1}{2}\frac{1}{C}\bar{q}^2 + \frac{1}{4}\frac{1}{C_{NL}}\bar{q}^4\right],$$

and the time derivative is

$$\dot{\mathcal{H}} = \left[L\ddot{\bar{q}} + \left(\frac{1}{C}\bar{q} + \frac{1}{C_{\mathrm{NL}}}\bar{q}^3\right)\right]\dot{\bar{q}} = [v - R\dot{\bar{q}} - R_{\mathrm{NL}}\mathrm{sign}(\dot{\bar{q}})]\dot{\bar{q}}.$$

Taking into account the nonlinearities and following the same steps from the previous section results in the following Hamiltonian and the first derivative:

$$\mathcal{H} = \frac{1}{2}L(\dot{\bar{q}} - \dot{\bar{q}}_R)^2 + \frac{1}{2}\left(\frac{1}{C} + K_{\mathrm{cap}}\right)(\bar{q} - \bar{q}_R)^2 + \frac{1}{4}\frac{1}{C_{\mathrm{NL}}}(\bar{q} - \bar{q}_R)^4 + \frac{1}{2}\tilde{\Phi}^T\Gamma^{-1}\tilde{\Phi} \tag{9.20}$$

and

$$\dot{\mathcal{H}} = \left[L(\ddot{\bar{q}} - \ddot{\bar{q}}_R) + \frac{1}{C}(\bar{q} - \bar{q}_R) + \frac{1}{C_{\mathrm{NL}}}(\bar{q} - \bar{q}_R)^3\right] \cdot (\dot{\bar{q}} - \dot{\bar{q}}_R) + \tilde{\Phi}^T\Gamma^{-1}\dot{\tilde{\Phi}}. \tag{9.21}$$

Again, a tracking PID adaptive controller (and robust PID control only case) is introduced as (9.16), where Δv is the same, and

$$v_{\mathrm{ref}} = \hat{L}\ddot{\bar{q}}_R + \widehat{\frac{1}{C}}\bar{q}_R + \widehat{\frac{1}{C_{\mathrm{NL}}}}\bar{q}_R^3 + \hat{R}\dot{\bar{q}} + \hat{R}_{\mathrm{NL}}\mathrm{sign}(\dot{\bar{q}}) + \widehat{\frac{1}{C_{\mathrm{NL}}}}(\bar{q}^3 - \bar{q}_R^3) + \widehat{\frac{1}{C_{\mathrm{NL}}}}(\bar{q} - \bar{q}_R)^3. \tag{9.22}$$

Then following the same steps as the linear development before results in the same adaptive parameter update equations in (9.18) with the addition of

$$\dot{\widehat{\frac{1}{C_{\mathrm{NL}}}}} = -\gamma_4[\bar{q}^3 - (\bar{q} - \bar{q}_R)^3](\dot{\bar{q}} - \dot{\bar{q}}_R),$$

$$\dot{\hat{R}}_{\mathrm{NL}} = -\gamma_5\,\mathrm{sign}(\dot{\bar{q}})(\dot{\bar{q}} - \dot{\bar{q}}_R).$$

This results in an asymptotically stable (passivity [95]) solution or the realization of (9.19).

Numerical simulations were performed with variations in L, C ($\neq$ 60 Hz), R (RLC parameters same as in the linear model), $C_{\mathrm{NL}} = 0.1$ μF, and $R_{\mathrm{NL}} = 50$ Ω. Both a robust PID and PID with adaptive controllers were investigated for the nonlinear RLC network. The tracking results for the charge and current along with the tracking errors are given in Fig. 9.11 for PID and Fig. 9.12 for PID with adaptive controllers, respectively. In addition, the power flow and energy consumed by the PID and PID with adaptive controllers are shown in Fig. 9.13 and Fig. 9.14, respectively. Once again, the adaptive PID controller with information flow is more efficient than the robust PID controller with physical flow alone. For the nonlinear RLC network, the commanded voltage input responses are shown for v, Δv, v_{ref}, and v_{ideal} in Fig. 9.9 for both the PID (left) and PID with adaptive (right) controllers, respectively. The adaptive control estimated reference input parameter responses for the nonlinear RLC network are given for $\hat{L}$, $\widehat{1/C}$, and $\hat{R}$ in Fig. 9.10 (right column) and for the nonlinear terms, $\widehat{1/C_{\mathrm{NL}}}$ and $\hat{R}_{\mathrm{NL}}$, in Fig. 9.15.

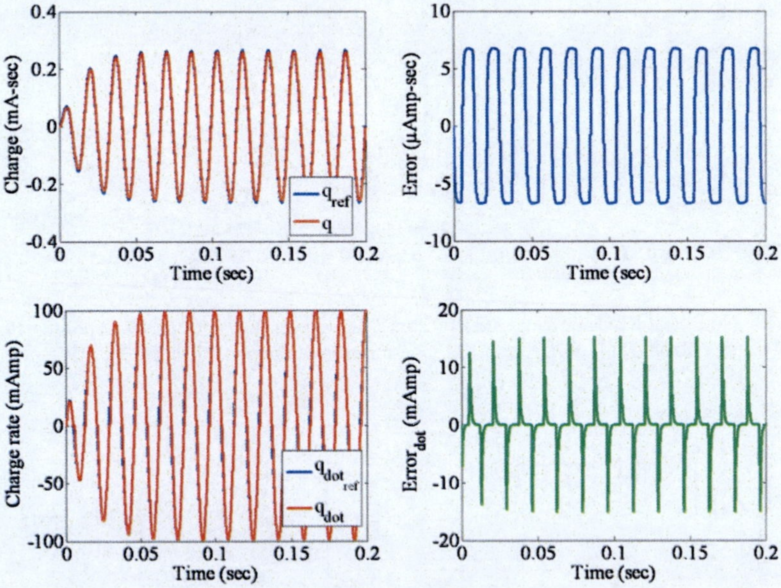

Fig. 9.11 Nonlinear RLC network with PID controller: charge and charge error responses along with charge-rate and charge-rate error responses. Robinett III, R.D. and Wilson, D.G. [94], reprinted by permission of the publisher (©2008 IEEE)

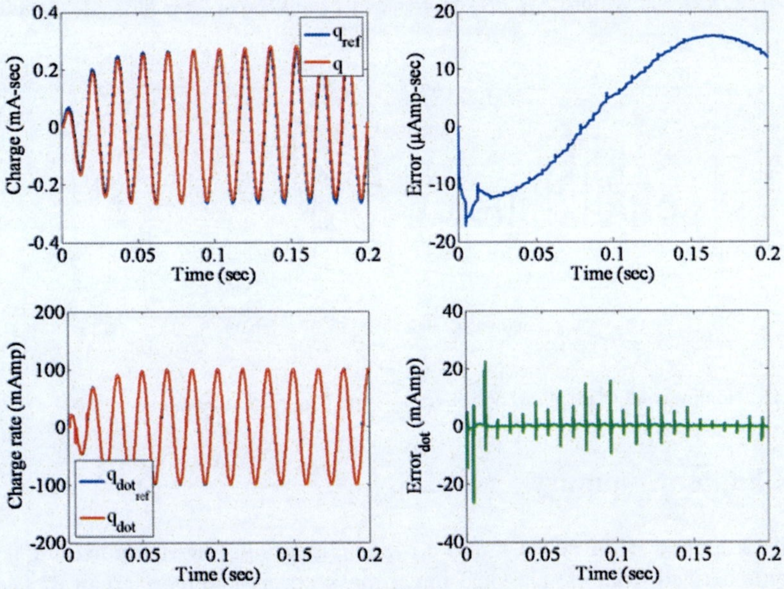

Fig. 9.12 Nonlinear RLC network with PID/adaptive controller: charge and charge error responses along with charge-rate and charge-rate error responses. Robinett III, R.D. and Wilson, D.G. [94], reprinted by permission of the publisher (©2008 IEEE)

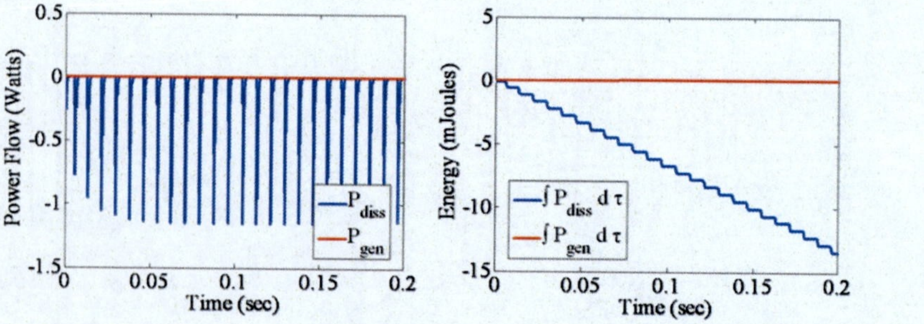

Fig. 9.13 Nonlinear RLC network with PID controller: power flow and energy responses. Robinett III, R.D. and Wilson, D.G. [94], reprinted by permission of the publisher (©2008 IEEE)

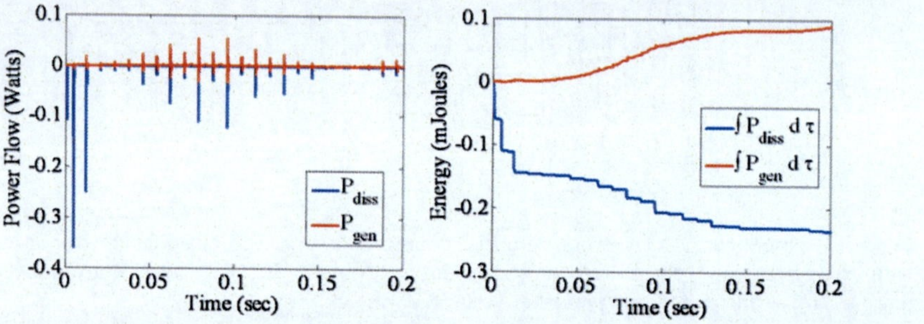

Fig. 9.14 Nonlinear RLC network with PID/adaptive controller: power flow and energy responses. Robinett III, R.D. and Wilson, D.G. [94], reprinted by permission of the publisher (©2008 IEEE)

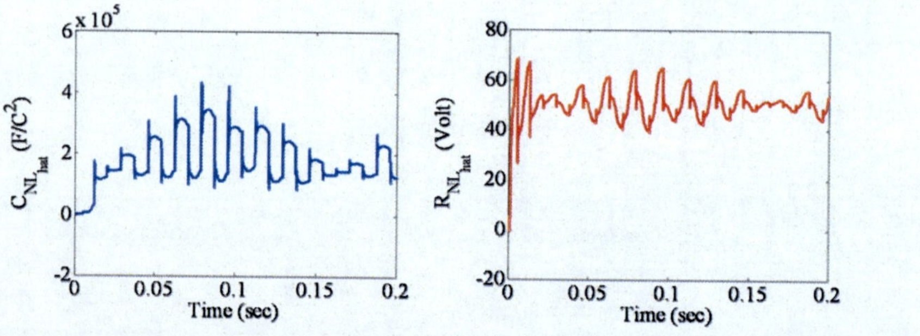

Fig. 9.15 Nonlinear RLC adaptive controller: $\widehat{\frac{1}{C_{\mathrm{NL}}}}$ and $\hat{R}_{\mathrm{NL}}$ responses

9.6 Chapter Summary

Chapter 9 applied HSSPFC design to power engineering. Power engineering terminology is derived from the classical linear mass–spring–damper and an RLC electrical network. The methodology is then used to design both robust PID and PID with adaptive control architectures for both a linear and nonlinear RLC dynamic network systems. Numerical simulations were performed to demonstrate the feasibility of the controller algorithms. In addition, the power flow and energy responses demonstrated the required energy consumption of each controller. The adaptive PID controller versions for both linear and nonlinear cases showed to be more efficient than just robust PID control alone. The main contribution of this chapter is to present HSSPFC design as it applies to power engineering and how it is enhanced through adaptive control.

Chapter 10
Case Study #5: Renewable Energy Microgrid Design

10.1 Introduction

Some of the most challenging problems the United States and other countries are facing is the integration of green renewable resources into existing aging Electric Power Grid (EPG) infrastructures. Many states within the US are faced with fast approaching deadlines due to Renewable Portfolio Standards (RPS) which are forcing the retrofit and patch in of renewables the best that can be done with existing options. Many of the proposed "Smart Grids" are simply overlaying information networks onto existing EPG infrastructures. What is needed is a paradigm shift in our current approach to the grid. At the heart of the EPG is the coordination and control of centralized dispatchable generation to meet customer loads via power engineering techniques. A new approach will be required to formally address the green grid of the future with distributed variable generation, buying and selling of power (bidirectional flow), and decentralization of the EPG.

Many researchers are attempting to address this problem. For example, the Solar Energy and Grid Integration Systems (SEGIS) program is trying to integrate large amounts of photovoltaic systems in the EPG.

> *Solar Energy Grid Integration Systems (SEGIS) concept will be key to achieving high penetration of photovoltaic (PV) systems into the utility grid. Advanced, integrated inverter/controllers will be the enabling technology to maximize the benefits of residential and commercial solar energy systems, both to the systems owners and to the utility distribution network as a whole. The value of the energy provided by these solar systems will increase through advanced communication interfaces and controls, while the reliability of electrical service, both for solar and non-solar customers, will also increase [96].*

The wind industry is attempting to add large amounts of wind power into the existing EPG without integrating additional storage. The Department of Energy (DOE) through its Energy Efficiency and Renewable Energy (EERE) Wind Program Office provides one way to do this through the report titled "20% Wind Energy by 2030" in [97].

The goal of Chap. 10 is to present a step toward addressing the integration of renewable resources into the EPG by applying a new nonlinear power flow control

R.D. Robinett III, D.G. Wilson, *Nonlinear Power Flow Control Design*, Understanding Complex Systems,
DOI 10.1007/978-0-85729-823-2_10, © Springer-Verlag London Limited 2011

technique to the analysis of the swing equations for renewable and conventional generators connected to the EPG while using a wind turbine as an example. The results of this research include the determination of the required performance of a proposed FACTS/Storage device to enable the maximum power output of a wind turbine while meeting the power system constraints on frequency and phase. The FACTS/Storage device is required to operate as both a generator and load (energy storage) on the power system in this design.

In Chap. 10, the swing equations for the renewable and conventional generators are formulated as a natural Hamiltonian system with externally applied nonconservative forces. A two-step process referred to as Hamiltonian Surface Shaping and Power Flow Control (HSSPFC) is used to analyze and design feedback controllers for the renewable and conventional generators system. This formulation extends previous results on the analytical verification of the Potential Energy Boundary Surface (PEBS) method to nonlinear control analysis and design. It justifies the decomposition of the system into conservative and nonconservative systems to enable a two-step, serial analysis and design procedure. In particular, this approach extends the work done by Vittal, Michel, and Fouad [98] by developing a formulation which applies to a larger set of Hamiltonian Systems that has Nearly Hamiltonian Systems as a subset.

The first step is to analyze the system as a conservative natural Hamiltonian system with no externally applied nonconservative forces. The Hamiltonian surface of the swing equations is related to the Equal-Area Criterion and the PEBS method to formulate the nonlinear transient stability problem by recognizing that the path of the system is constrained to the Hamiltonian surface. This formulation demonstrates the effectiveness of proportional feedback control to expand the stability region. Also, the two-step process directly includes nonconservative power flows in the analysis to determine the path of the system across the Hamiltonian surface to better determine the stability regions.

The second step is to analyze the system as a natural Hamiltonian system with externally applied nonconservative forces. The time derivative of the Hamiltonian produces the work/rate (power flow) equations that are used to ensure balanced power flows from the renewable and conventional generators to the loads. This step extends the work done by Alberto and Bretas [99] by developing a formulation which expands beyond the analysis of small perturbations of conservative Hamiltonian systems. The Melnikov number (see Chap. 11) for this class of systems is directly related to the balance of power flows for the stability (limit cycles) of natural Hamiltonian systems with externally applied nonconservative forces. The Second Law of Thermodynamics is applied to the power flow equations to determine the stability boundaries (limit cycles) of the renewable generators system. This enables design of feedback controllers that meet stability requirements while maximizing the power generation and flow to the load. Necessary and sufficient conditions for stability of renewable generators systems are determined based on the concepts of Hamiltonian systems, power flow, exergy (the maximum work that can be extracted from an energy flow) rate, and entropy rate.

Chapter 10 is divided into five sections. Section 10.2 presents the HSSPFC design for a typical One-Machine Infinite Bus (OMIB) system based on the Reduced

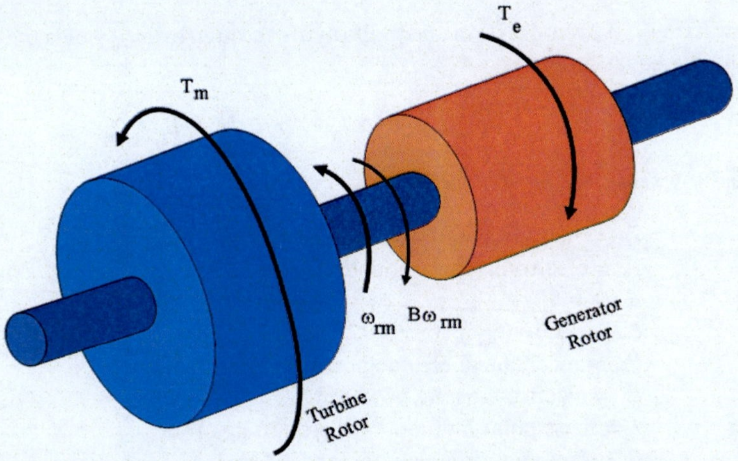

Fig. 10.1 Typical turbine-generator rotor system block diagram [101]. Wilson, D.G. and Robinett III, R.D. [102], reprinted by permission of the publisher (©2010 IEEE)

Network Model (RNM) technique [100]. Section 10.3 presents the HSSPFC applied to Unified Power Flow Controllers (UPFC) and renewable generators including numerical simulations. Section 10.4 presents the HSSPFC applied to a microgrid in islanded mode. Section 10.5 presents the HSSPFC applied to conventional and wind turbine 2MIB with UPFCs. Section 10.6 summarizes the chapter.

10.2 HSSPFC Design for a Typical OMIB System

The familiar OMIB model [101] based on the RNM technique [100] is employed to provide an illustrative example (refer to the turbine-generator system shown in Fig. 10.1). The power engineering model and equation of motion that best reflects the HSSPFC methodology are developed by starting with [100, 101]

$$T_m - T_e = J\dot{\omega}_{\mathrm{RM}} + B\omega_{\mathrm{RM}} \tag{10.1}$$

with

$$\omega_{\mathrm{RM}} = \frac{\omega}{N_p/2}, \qquad \omega = \omega_{\mathrm{ref}} + \dot{\delta}$$

or

$$T_m - T_e = \frac{2J}{N_p}\big(\dot{\omega}_{\mathrm{ref}} + \ddot{\delta}\big) + \frac{2B}{N_p}\big(\omega_{\mathrm{ref}} + \dot{\delta}\big). \tag{10.2}$$

Beginning with the reference torque equation

$$T_{m_{\mathrm{ref}}} - T_{e_{\mathrm{ref}}} = \frac{2J}{N_p}\dot{\omega}_{\mathrm{ref}} + \frac{2B}{N_p}\omega_{\mathrm{ref}} \tag{10.3}$$

and subtracting (10.3) from (10.2) and multiplying through by ω_{ref} yields the general OMIB power equation

$$P_m - P_e = (\Delta T_m - \Delta T_e)\omega_{\text{ref}} = \hat{J}\ddot{\delta} + \hat{B}\dot{\delta}, \tag{10.4}$$

which assumes $\omega_{\text{ref}} = \textit{constant}$ and $\omega_{\text{ref}} \gg \dot{\delta}$, where

$T_m =$ mechanical turbine torque in N m,
$T_e =$ electromagnetic counter torque in N m,
$\Delta T_m = T_m - T_{m_{\text{ref}}}$,
$\Delta T_e = T_e - T_{e_{\text{ref}}}$,
$P_m =$ mechanical turbine power in W,
$P_e =$ electromagnetic power in W,
$J =$ mass polar moment of inertia in kg m^2,
$B =$ damping torque coefficient in N m/s,
$N_p =$ number of generator poles

$$\hat{J} = \frac{2J}{N_p}\omega_{\text{ref}},$$

$$\hat{B} = \frac{2B}{N_p}\omega_{\text{ref}},$$

$\omega_{\text{RM}} =$ rotor shaft velocity in mechanical rad/s,
$\omega_{\text{ref}} =$ reference angular velocity in electrical rad/s of a revolving reference line which may or may not be constant, and
$\delta =$ power angle measured in electrical radians.

The first step in the HSSPFC design process is to recognize that the system is constrained to move on the Hamiltonian surface, the accessible phase space, which can be projected onto the phase plane. The Hamiltonian is the stored energy (exergy) of the system and is given (referring to the example OMIB from [100]) as

$$\begin{aligned} \mathcal{H} &= \mathcal{T} + \mathcal{V}, \\ \mathcal{H} &= \frac{1}{2}\hat{J}\dot{\delta}^2 + P_{\max}\big(1 - \cos(\delta)\big), \\ &\text{where} \\ \mathcal{T} &= \text{kinetic energy} = \frac{1}{2}\hat{J}\dot{\delta}^2, \\ \mathcal{V} &= \text{potential energy} = P_{\max}\big(1 - \cos(\delta)\big). \end{aligned} \tag{10.5}$$

The electrical power P_e is defined from the generalized potential $P_{\max}(1 - \cos(\delta))$. The equation of motion is determined from Lagrange's equation as

$$\frac{d}{dt}\left(\frac{\partial \mathcal{L}}{\partial \dot{\delta}}\right) - \frac{\partial \mathcal{L}}{\partial \delta} = Q,$$

where Q is the generalized force, and

$$\begin{aligned}
\mathcal{L} &= \mathcal{T} - \mathcal{V}, \\
\mathcal{L} &= \frac{1}{2}\hat{J}\dot{\delta}^2 - P_{\max}\big(1 - \cos(\delta)\big), \\
Q &= P_m + \Delta u.
\end{aligned} \tag{10.6}$$

Applying Lagrange's equations yields

$$\hat{J}\ddot{\delta} + P_{\max}\sin(\delta) = P_m + \Delta u - \hat{B}\dot{\delta}. \tag{10.7}$$

A nonlinear Proportional-Integral-Derivative (PID) control law Δu is defined from HSSPFC as

$$\Delta u = -K_P \sin(\delta - \delta_s) - K_D\dot{\delta} - K_I \int_0^t (\delta - \delta_s)\, d\tau_1, \tag{10.8}$$

where K_P, K_I, and K_D are positive PID controller gains, and δ_s is a set-point.

The time derivative of the Hamiltonian is the power flow or work-rate principle [15] (see Chap. 3)

$$\dot{\mathcal{H}} = \big[\hat{J}\ddot{\delta} + P_{\max}\sin(\delta) - P_m\big]\dot{\delta} = \big[\Delta u - \hat{B}\dot{\delta}\big]\dot{\delta} = \sum_{i=1}^{N} Q_i \dot{q}_i. \tag{10.9}$$

By adding the nonlinear PID control law both the energy storage and the power flow conditions have been modified. Substituting (10.8) into (10.9) yields the following:

$$\begin{aligned}
&\big[\hat{J}\ddot{\delta} + P_{\max}\sin\delta - P_m + K_P\sin(\delta - \delta_s)\big]\dot{\delta} \\
&\quad = \left[-K_I \int_0^t (\delta - \delta_s)\, d\tau_1 - (\hat{B} + K_D)\dot{\delta}\right]\dot{\delta}.
\end{aligned} \tag{10.10}$$

By sorting terms (storage, generation, and dissipation) the static stability condition becomes

$$\mathcal{H} = \frac{1}{2}\hat{J}\dot{\delta}^2 + (P_{\max} + K_P)\big(1 - \cos(\delta - \delta_s)\big) \tag{10.11}$$

with $\mathcal{H}$ being positive definite and

$$\delta_s = \sin^{-1}(P_m / P_{\max}). \tag{10.12}$$

The dynamic stability condition, referring to (10.10), is

$$\int_0^{\tau} [\hat{B} + K_D]\dot{\delta}^2\, dt > -\int_0^{\tau}\left[K_I \int_0^t (\delta - \delta_s)\, d\tau_1\right]\dot{\delta}\, dt. \tag{10.13}$$

Clearly, the nonlinear PID controller expands the region of stability by increasing the PEBS from $P_{\max}$ to $(P_{\max} + K_P)$ and enabling the system to respond more

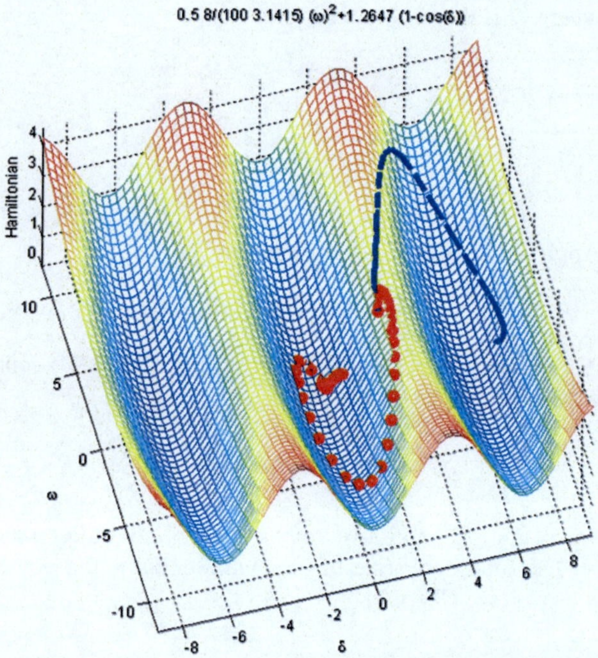

Fig. 10.2 Hamiltonian energy storage surface and power flow traversal path ($K_I = 0$, *blue dash*) which moves away from the operating point into third valley and ($K_I = 1.0$, *red circle*) returns to the steady-state operating point P_m in the middle valley). Wilson, D.G. and Robinett III, R.D. [102], reprinted by permission of the publisher (©2010 IEEE)

quickly by adding an integrator to the dissipator of reference [100]. The relationship between PEBS and Equal-Area Criterion is described in references [103, 104].

Figure 10.2 graphically shows, for the OMIB (all parameters are from [100]), the Hamiltonian (stored energy) surface and corresponding Hamiltonian rate (power flow) trajectory (red circle trace $K_I = 1.0$ and blue trace $K_I = 0.0$) constrained to traverse along the surface. The initial condition is close to the second peak. With no integral gain the trajectory tends to the unstable operating node in the third valley. With $K_I = 1.0$, the trajectory is brought back to the stable operating point. For larger values of K_I, the dynamic condition (10.13) is violated, and the trajectory will go unstable.

10.3 HSSPFC Applied to UPFCs and Renewable Generators

This section investigates power engineering models for UPFCs and renewable generators. First, define the Hamiltonian as

$$\mathcal{H} = \frac{1}{2}\hat{J}\dot{\delta}^2, \tag{10.14}$$

where the power flow or Hamiltonian rate becomes

$$\dot{\mathcal{H}} = \hat{J}\dot{\delta}\ddot{\delta} = \left[P_m - P_e - \hat{B}\dot{\delta}\right]\dot{\delta}. \tag{10.15}$$

Next, add the approximate power flows from the generator, mechanical controls, and UPFC [100]:

$$\begin{aligned} P_m &= P_{m_c} + u_m(\omega_{\text{ref}}), \\ P_e &= P_{e_c} \sin\delta + u_{e1} P_{e_c} \sin\delta - u_{e2} P_{e_c} \cos\delta. \end{aligned} \tag{10.16}$$

10.3.1 Example One—OMIB System with a UPFC

This example is an extension to the work done by Ghandhari [100] by developing a formulation which applies to a larger set of nonlinear control systems that has passivity controllers [95] as a subset of [56]. Starting with the reference torque (10.3) and substituting for the terms in (10.4) with (10.16) gives

$$\hat{J}\ddot{\delta} = -\hat{B}\dot{\delta} + P_{m_c} + u_m \omega_{\text{ref}} - P_{e_c}[(1 + u_{e_1}) \sin\delta - u_{e_2} \cos\delta].$$

Next, define the Hamiltonian as

$$\mathcal{H} = \frac{1}{2}\hat{J}\dot{\delta}^2;$$

then the derivative of the Hamiltonian becomes

$$\begin{aligned} \dot{\mathcal{H}} &= \hat{J}\ddot{\delta}\dot{\delta} \\ &= \big[-\hat{B}\dot{\delta} + P_{m_c} + u_m\omega_{\text{ref}} - P_{e_c}\big((1 + u_{e_1}) \sin\delta - u_{e_2} \cos\delta\big)\big]\dot{\delta}. \end{aligned} \tag{10.17}$$

Now assume that $u_m = 0$. Then

$$\hat{J}\ddot{\delta} + P_{e_c} \sin\delta - P_{m_c} = -\hat{B}\dot{\delta} - P_{e_c}[u_{e_1} \sin\delta - u_{e_2} \cos\delta]. \tag{10.18}$$

Next, select the following nonlinear PID control laws for the UPFC from HSSPFC:

$$\begin{aligned} u_{e_1} &= K_{P_e} \cos\delta_s + K_{D_e} \sin\delta\dot{\delta} + K_{I_e} \sin\delta \int_0^t \Delta\, d\tau, \\ u_{e_2} &= K_{P_e} \sin\delta_s - K_{D_e} \cos\delta\dot{\delta} - K_{I_e} \cos\delta \int_0^t \Delta\, d\tau, \end{aligned} \tag{10.19}$$

where $\Delta = \delta - \delta_s$. Finally, substituting (10.19) into (10.18) yields the following:

$$\begin{aligned} &\hat{J}\ddot{\delta} + [P_{e_c} \sin\delta - P_{m_c}] + P_{e_c} K_{P_e} \sin(\delta - \delta_s) \\ &\quad = -[\hat{B} + P_{e_c} K_{D_e}]\dot{\delta} - P_{e_c} K_{I_e} \int_0^t (\delta - \delta_s)\, d\tau. \end{aligned}$$

The static stability condition becomes

$$\mathcal{H} = \frac{1}{2}\hat{J}\dot{\delta}^2 + P_{e_c}(1 + K_{P_e})\big(1 - \cos(\delta - \delta_s)\big) \tag{10.20}$$

with $\mathcal{H}$ being positive definite and

$$\delta_s = \sin^{-1}(P_{m_c}/P_{e_c}).$$

The dynamic stability condition for a passively stable control design yields

$$\int_0^\tau [\hat{B} + P_{e_c} K_{D_e}]\dot{\delta}^2\,dt > -\int_0^\tau \left[P_{e_c} K_{I_e} \int_0^t (\delta - \delta_s)\,dt_1\right]\dot{\delta}\,dt. \qquad (10.21)$$

Clearly, the UPFC nonlinear PID controller expands the region of stability by increasing the PEBS from P_{e_c} to $P_{e_c}(1 + K_{P_e})$ and enabling the system to respond more quickly by adding an integrator to the dissipator of reference [100].

10.3.2 Example Two—Swing Equation for a Wind Turbine Connected to an Infinite Bus Through UPFC

The derivation starts with the wind turbine equations defined as

$$J_T\dot{\omega}_r = -K_T\omega_r + T_a - T_g; \qquad \omega_r = \dot{\theta}_r; \quad \delta = \theta_r - \theta_{\text{ref}}.$$

Then the Hamiltonian for the wind turbine becomes

$$\mathcal{H} = \frac{1}{2}J_T\dot{\delta}^2 = \frac{1}{2}J_T\left(\dot{\theta}_r - \dot{\theta}_{\text{ref}}\right)^2, \qquad (10.22)$$

and the corresponding Hamiltonian rate is

$$\dot{\mathcal{H}} = [-K_T\dot{\theta}_r + T_a - T_g - J_T\ddot{\theta}_{\text{ref}}]\dot{\delta},$$

where

$$\begin{aligned} T_a &= M\dot{\theta}_r^2, \\ T_g &= u_m + T_e, \\ u_m &= u_{m_{\text{ref}}} + \Delta u_m, \\ u_{m_{\text{ref}}} &= -J_T\ddot{\theta}_{\text{ref}} - K_T\dot{\theta}_{\text{ref}} + M\dot{\theta}_r^2, \end{aligned}$$

which lead to

$$\dot{\mathcal{H}} = [-K_T\dot{\delta} - \Delta u_m - T_e]\dot{\delta}. \qquad (10.23)$$

Next, select the following wind turbine generation and UPFC nonlinear controllers from HSSPFC:

$$\begin{aligned} \Delta u_m &= K_{P_m}\delta + K_{I_m}\int_0^t \delta\,d\tau + K_{D_m}\dot{\delta}, \\ T_e &= T_{e_{\max}}[\sin\delta + u_{e_1}\sin\delta - u_{e_2}\cos\delta], \end{aligned} \qquad (10.24)$$

where u_{e_1} and u_{e_2} are substituted from (10.19) with $\delta_s = 0$. Substituting Δu_m and T_e from (10.24) into (10.23) yields the following power flow equation:

$$\dot{\mathcal{H}} = \left[-(K_T + K_{D_m})\dot{\delta} - K_{P_m}\delta - K_{I_m}\int_0^t \delta\, d\tau - T_{e_{\max}}(\sin\delta + u_{e_1}\sin\delta - u_{e_2}\cos\delta)\right]\dot{\delta}$$

or

$$\dot{\mathcal{H}} = \Big[-(K_T + K_{D_m} + T_{e_{\max}} K_{D_e})\dot{\delta} - \big(K_{P_m}\delta + T_{e_{\max}}(1 + K_{P_e})\sin\delta\big) - \left(K_{I_m}\int_0^t \delta\, d\tau + T_{e_{\max}} K_{I_e}\int_0^t \delta\, d\tau\right)\Big]\dot{\delta}, \tag{10.25}$$

where again, as in the previous example, the static stability condition is that

$$\mathcal{H} = \frac{1}{2}J_T\dot{\delta}^2 + \frac{1}{2}K_{P_m}\delta^2 + T_{e_{\max}}(1 + K_{P_e})\big(1 - \cos(\delta)\big) \tag{10.26}$$

must be positive definite. For a passivity controller design, the terms in (10.25) need to be sorted into dissipators, generators, and storage terms over the cycle; the dissipators must be greater than the generators. The dynamic stability condition for a passively stable control design yields

$$\int_0^\tau [K_T + K_{D_m} + T_{e_{\max}} K_{D_e}]\dot{\delta}^2\, dt > -\int_0^\tau \left[K_{I_m}\int_0^t \delta\, dt_1 + T_{e_{\max}} K_{I_e}\int_0^t \delta\, dt_1\right]\dot{\delta}\, dt, \tag{10.27}$$

where the proportional terms are storage terms and contribute to the dynamic stability of the system as constraints.

Notice that the wind turbine controller is designed to maximize the output of the wind turbine while the UPFC controller is designed to smooth the output of the wind turbine to create $\omega_{\text{ref}} =$ constant. As a result, the UPFC performs as both a generator (firm up the wind turbine) and a load (storage excess power) to smooth out the peaks and valleys of the output from the wind turbine.

10.3.3 HSSPFC Applied to UPFC and Variable Generation

This section approximates the control designs of Sect. 10.3.2 by simulating a wind turbine with a constant P_{m_c} plus a random noise input [105]. This situation is handled by adding a feedforward control. Once again, the process begins with the Hamiltonian as

$$\mathcal{H} = \frac{1}{2}\hat{J}\dot{\delta}^2,$$

and then the derivative of the Hamiltonian becomes

$$\begin{aligned}\dot{\mathcal{H}} &= \hat{J}\ddot{\delta}\dot{\delta} \\ &= \big[-\hat{B}\dot{\delta} + P_{m_c}(t) + u_m\omega_{\text{ref}} - P_{e_c}\big((1+u_{e_1})\sin\delta - u_{e_2}\cos\delta\big)\big]\dot{\delta}.\end{aligned}$$

Now assume that $u_m = 0$. Then

$$\hat{J}\ddot{\delta} + P_{e_c}\sin\delta - P_{m_c}(t) = -\hat{B}\dot{\delta} - P_{e_c}[u_{e_1}\sin\delta - u_{e_2}\cos\delta]. \tag{10.28}$$

Next select the nonlinear PID control laws from HSSPFC given in (10.19). Finally, substituting (10.19) into (10.28) yields the following:

$$\begin{aligned}&\hat{J}\ddot{\delta} + P_{e_c}\sin\delta - P_{m_{\text{ref}}} + P_{e_c}K_{P_e}\sin(\delta - \delta_s) \\ &\quad = -[\hat{B} + P_{e_c}K_{D_e}]\dot{\delta} - P_{e_c}K_{I_e}\int_0^t (\delta - \delta_s)\,d\tau + P_{m_c}(t) - P_{m_{\text{ref}}}.\end{aligned}$$

The static stability condition becomes

$$\mathcal{H} = \frac{1}{2}\hat{J}\dot{\delta}^2 + P_{e_c}(1 + K_{P_e})\big(1 - \cos(\delta - \delta_s)\big) \tag{10.29}$$

with $\mathcal{H}$ being positive definite and

$$\delta_s = \sin^{-1}(P_{m_{\text{ref}}}/P_{e_c}).$$

The dynamic stability condition for a passively stable control design yields

$$\int_0^\tau [\hat{B} + P_{e_c}K_{D_e}]\dot{\delta}^2\,dt > -\int_0^\tau \bigg[P_{e_c}K_{I_e}\int_0^t (\delta - \delta_s)\,dt_1 + \big(P_{m_{\text{ref}}} - P_{m_c}(t)\big)\bigg]\dot{\delta}\,dt. \tag{10.30}$$

A feedforward control term is added to the UPFC controllers, (10.19), for u_{e_1} and u_{e_2} by

$$\begin{aligned}u_{e_1} &= u_{e_1} - \big[\big(P_{m_{\text{ref}}} - P_{m_c}(t)\big)/P_{\max}\big]\sin\delta, \\ u_{e_2} &= u_{e_2} + \big[\big(P_{m_{\text{ref}}} - P_{m_c}(t)\big)/P_{\max}\big]\cos\delta,\end{aligned} \tag{10.31}$$

where $P_{m_{\text{ref}}}$ is designed to emulate a constant input, $P_{m_c}(t)$ can become variable such as from wind or solar generation, and $P_{e_c} = P_{\max}$. In the next section these effects are explored numerically with simple wind turbine characteristics through a $P_{m_c}(t)$ variation.

10.3.4 Numerical Simulations

A numerical example presented next is based on an example in reference [100]. All of the parameters are given in Table 10.1. To demonstrate the controllers defined

Table 10.1 Numerical values for existing OMIB system from example in [100]

Parameter		Relationship and numerical values
$\hat{J}'$	$=$	$\frac{2H}{\omega_0} = \frac{8}{100\pi}$
$\hat{B}'$	$=$	$\frac{2}{\omega_0}$
P_m	$=$	1.1 (p.u)
$P_{\max}$	$=$	$bE'V$
b	$=$	$\frac{1}{x_L} = \frac{1}{0.85}$
E'	$=$	1.075 (p.u)
V	$=$	1.0 (p.u)
$\hat{J}'$	$\rightarrow$	scaled by $\omega_0 = 50\ \text{Hz} \cdot (2\pi)$
$\hat{B}'$	$\rightarrow$	scaled by $\omega_0 = 50\ \text{Hz} \cdot (2\pi)$

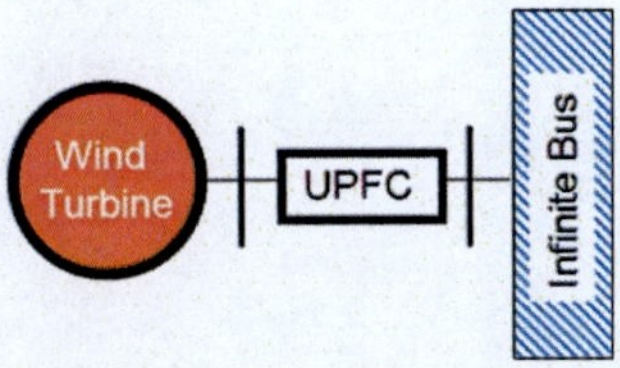

Fig. 10.3 One-machine infinite-bus model with UPFC and wind turbine generator. Wilson, D.G. and Robinett III, R.D. [105], reprinted by permission of the publisher (©2010 Energynautics GmbH)

in earlier sections, the OMIB model is modified to include the UPFC. In addition, simple wind turbine generator characteristics are investigated by allowing the P_m term to become time-varying in the swing equations. The OMIB wind turbine UPFC system is shown schematically in Fig. 10.3.

Initially, the OMIB is given a faulted initial condition that is away from the stable equilibrium point. This results in an unstable response, where the Hamiltonian (stored energy) surface (δ, ω) and corresponding Hamiltonian rate (power flow) trajectory (blue dashed trace along the surface) are presented in Fig. 10.4 (top). The first step in the HSSPFC process is to add the UPFC controller and increase the stable boundary region with the addition of K_{P_e}. This is defined and labeled as δ_1 responses. The next step is to define the dynamic stability and transient performance by adding K_{D_e} and K_{I_e}, defined and labeled as δ_2 responses. The stable response is demonstrated for δ_2 in Fig. 10.4 (bottom). The transient (top) and phase plane (bottom) responses for both δ_1 and δ_2 are given in Fig. 10.5, respectively. The corresponding power flow and energy responses for the UPFC δ_1 and δ_2 systems are illustrated in Fig. 10.6 (top).

As an initial investigation for simple wind turbine generator characteristics, a random input response for $P_{m_c}(t)$ was created, and a second-order filter (with a roll-off frequency of 2 Hz) was used in series to provide the effect for the wind turbine. In the following plots, δ_3 represents the feedback of the UPFC controller design, while δ_4 represents the addition of the feedforward control. The corresponding transient (top) and phase plane (bottom) responses are given in Fig. 10.7, respectively. The

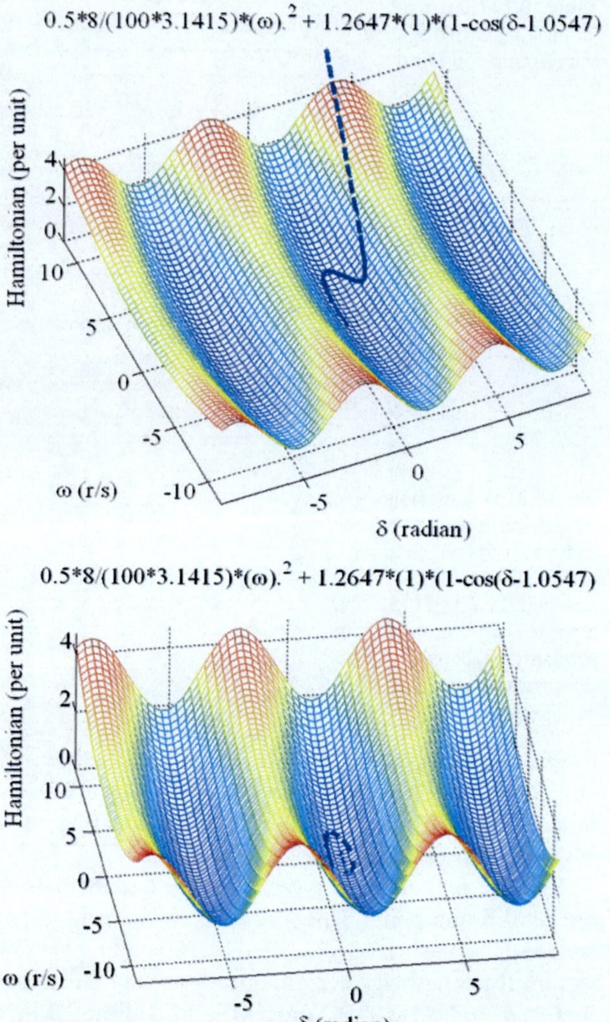

Fig. 10.4 OMIB Hamiltonian energy storage surface and power flow path where initially without any UPFC the machine goes unstable (*top*) and with the addition of UPFC the machine maintains stability and performance (*bottom*). Wilson, D.G. and Robinett III, R.D. [105], reprinted by permission of the publisher (©2010 Energynautics GmbH)

corresponding power flow and energy responses for the UPFC δ_3 and δ_4 systems are given in Fig. 10.6 (bottom). The addition of the feedforward control is to help $P_m(t)$ respond as a constant (see Fig. 10.8).

10.4 Microgrid in Islanded Mode

This section discusses and analyzes a microgrid in islanded mode which extends the results of [56]. This configuration is disconnected from the electric power grid (or infinite bus) as shown in Fig. 10.9. This control design attempts to emulate two decoupled conventional generators that can be modeled as OMIBs. The mathematical

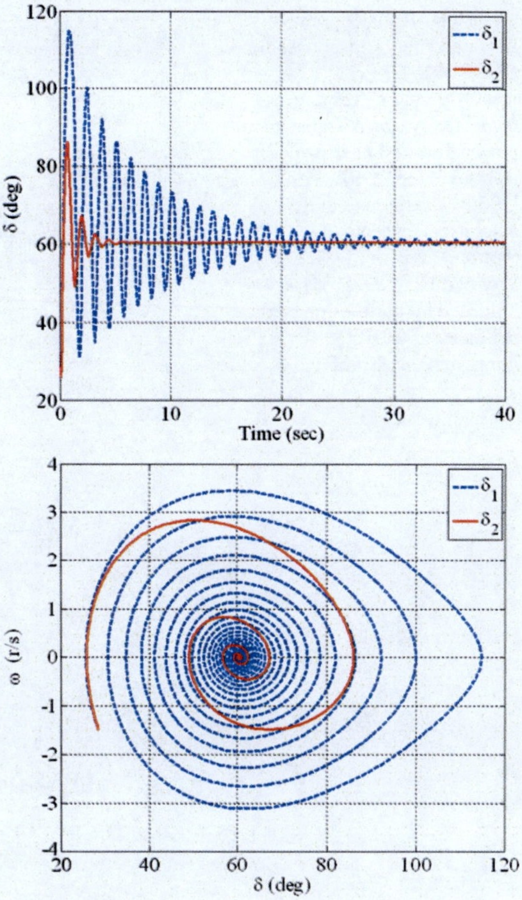

Fig. 10.5 Transient machine angle responses for OMIB with UPFC δ_1 response is increasing the static stability margin with K_{P_e}, and δ_2 response is for adding the dynamic stability and performance with K_{D_e} and K_{I_e} controller terms (*top*). Phase plane for OMIB with UPFC δ_1 response is increasing the static stability margin with K_{P_e}, and δ_2 response is for adding the dynamic stability and performance with K_{D_e} and K_{I_e} controller terms (*bottom*). Wilson, D.G. and Robinett III, R.D. [105], reprinted by permission of the publisher (©2010 Energynautics GmbH)

model equations [56, 100, 102] are

$$\begin{aligned}
\delta_{12} &= \delta_1 - \delta_2, \\
M_1\ddot{\delta}_1 &= P_{m_1} - P_{e_1} - D_1\dot{\delta}_1, \\
M_2\ddot{\delta}_2 &= P_{m_2} - P_{e_2} - D_2\dot{\delta}_2, \\
P_{e_1} &= C_{12}[(1+\hat{u}_{11})\sin\delta_{12} - \hat{u}_{12}\cos\delta_{12}], \\
P_{e_2} &= C_{12}[(1+\hat{u}_{11})\sin\delta_{21} - \hat{u}_{12}\cos\delta_{21}].
\end{aligned} \tag{10.32}$$

Substituting P_{e_i} for $i = 1, 2$ into the above equations yields

$$M_1\ddot{\delta}_1 = P_{m_1} - D_1\dot{\delta}_1 - C_{12}[(1+\hat{u}_{11})\sin\delta_{12} - \hat{u}_{12}\cos\delta_{12}], \tag{10.33}$$

$$M_2\ddot{\delta}_2 = P_{m_2} - D_2\dot{\delta}_2 - C_{12}[(1+\hat{u}_{11})\sin\delta_{21} - \hat{u}_{12}\cos\delta_{21}]. \tag{10.34}$$

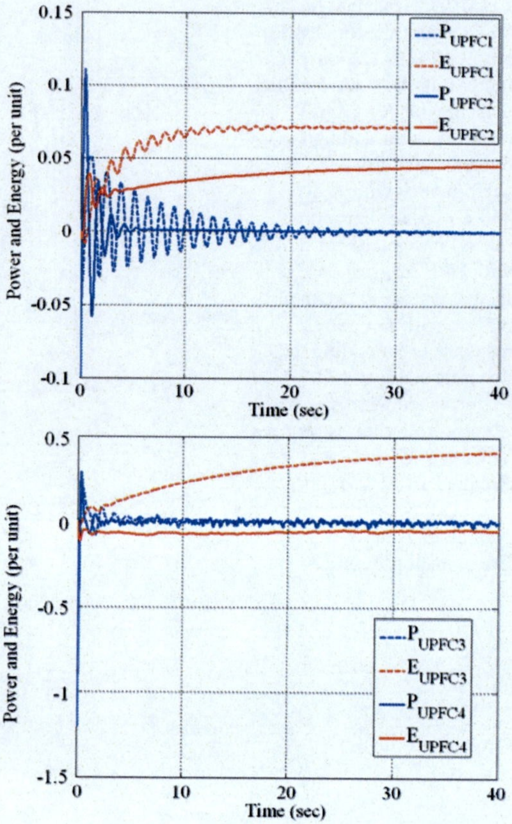

Fig. 10.6 Transient power flow and energy responses for OMIB with UPFC δ_1 responses and δ_2 responses, respectively (*top*). Transient power flow and energy responses for OMIB with UPFC δ_3 responses and δ_4 responses, respectively (*bottom*). Wilson, D.G. and Robinett III, R.D. [105], reprinted by permission of the publisher (©2010 Energynautics GmbH)

The matrix equation form of (10.33) and (10.34) is

$$\begin{bmatrix} M_1 & 0 \\ 0 & M_2 \end{bmatrix} \begin{Bmatrix} \ddot{\delta}_1 \\ \ddot{\delta}_2 \end{Bmatrix} + \begin{bmatrix} D_1 & 0 \\ 0 & D_2 \end{bmatrix} \begin{Bmatrix} \dot{\delta}_1 \\ \dot{\delta}_2 \end{Bmatrix} - \begin{Bmatrix} P_{m_1} \\ P_{m_2} \end{Bmatrix}$$
$$= -C_{12} \begin{bmatrix} \sin\delta_{12} & -\cos\delta_{12} \\ \sin\delta_{21} & -\cos\delta_{21} \end{bmatrix} \begin{Bmatrix} 1+\hat{u}_{11} \\ \hat{u}_{12} \end{Bmatrix}$$
$$= -C_{12} \begin{bmatrix} \sin\delta_{12} & -\cos\delta_{12} \\ -\sin\delta_{12} & -\cos\delta_{12} \end{bmatrix} \begin{Bmatrix} 1+\hat{u}_{11} \\ \hat{u}_{12} \end{Bmatrix} = -C_{12} \begin{Bmatrix} u_1 \\ u_2 \end{Bmatrix}. \quad (10.35)$$

The desired controller can be determined from

$$\begin{Bmatrix} 1+\hat{u}_{11} \\ \hat{u}_{12} \end{Bmatrix} = \begin{Bmatrix} \frac{1}{2\sin\delta_{12}}[u_1 - u_2] \\ \frac{-1}{2\cos\delta_{12}}[u_1 + u_2] \end{Bmatrix}. \quad (10.36)$$

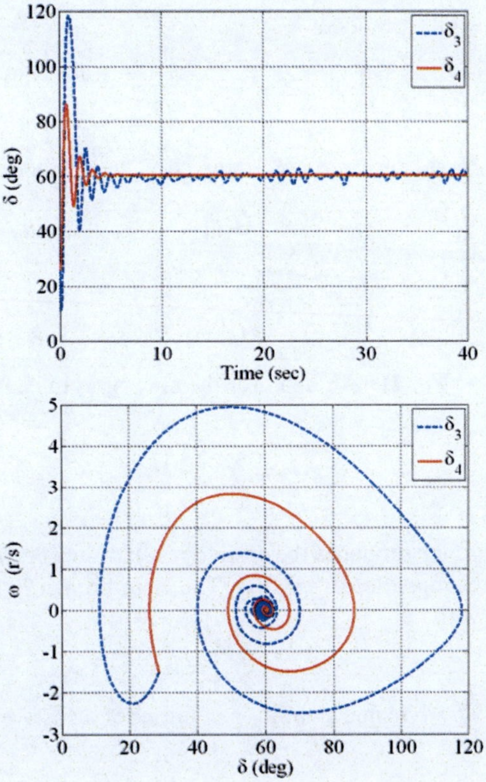

Fig. 10.7 Transient machine angle responses for OMIB with UPFC PID control only δ_3 response, and δ_4 response is for adding feedforward control in addition to feedback control (*top*). Phase plane for OMIB with UPFC δ_3 response is with UPFC PID control only, and δ_4 response is for adding feedforward control in addition to feedback control (*bottom*). Wilson, D.G. and Robinett III, R.D. [105], reprinted by permission of the publisher (©2010 Energynautics GmbH)

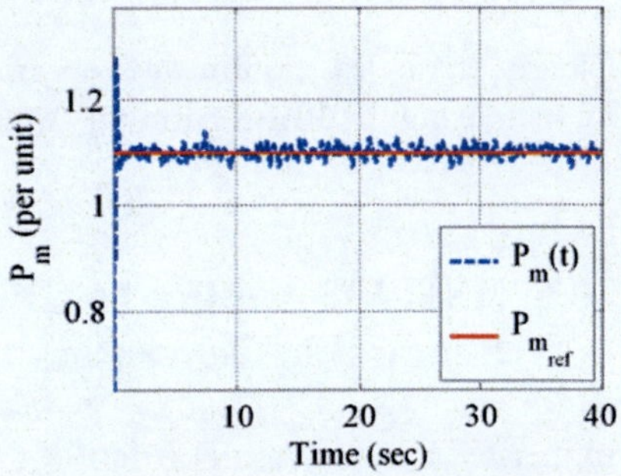

Fig. 10.8 Constant P_m reference signal compared to stochastic or random P_m response. Wilson, D.G. and Robinett III, R.D. [105], reprinted by permission of the publisher (©2010 Energynautics GmbH)

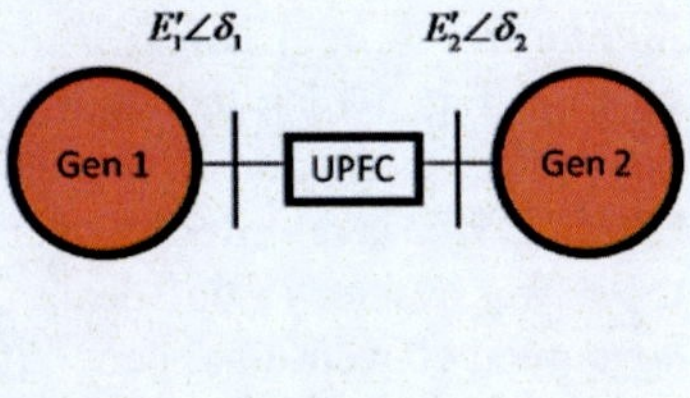

Fig. 10.9 Two-machine microgrid model with UPFC in islanded mode

A simple decoupling conroller is

$$\begin{aligned} u_1 &= K_P \sin 2\delta_1, \\ u_2 &= K_P \sin 2\delta_2 \end{aligned} \tag{10.37}$$

which produces

$$\begin{aligned}\hat{u}_{11} &= K_P \cos(\delta_1 + \delta_2) - 1,\\ \hat{u}_{12} &= -K_P \sin(\delta_1 + \delta_2).\end{aligned} \tag{10.38}$$

Next, substituting into the UPFC controllers and simplifying gives

$$M_1\ddot{\delta}_1 + D_1\dot{\delta}_1 + C_{12}K_P \sin 2\delta_1 - P_{m_1} = 0 \tag{10.39}$$

and

$$M_2\ddot{\delta}_2 + D_2\dot{\delta}_2 + C_{12}K_P \sin 2\delta_2 - P_{m_2} = 0. \tag{10.40}$$

The Hamiltonian for the microgrid in islanded mode can be summarized as

$$\mathcal{H} = \sum_{i=1}^{2}\left[\frac{1}{2}M_i\dot{\delta}_i^2 + C_{12}K_P[1 - \cos(2\delta_i - 2\delta_{i_s})]\right] \tag{10.41}$$

which produces the desired result of two decoupled conventional generators that can be modelled as OMIBs. The Hamiltonian $\mathcal{H}$ is positive definite and statically stable with

$$2\delta_{i_s} = \sin^{-1}\big(P_{m_i}/(C_{12}K_P)\big), \quad i = 1, 2. \tag{10.42}$$

The dynamic stability condition for a passively stable control design yields

$$\int_0^{\tau} D_i\dot{\delta}_i^2\, dt > 0, \quad i = 1, 2. \tag{10.43}$$

Clearly, the UPFC nonlinear Proportional controllers expand the region of stability by increasing the PEBS from C_{12} to $C_{12}K_P$.

10.5 HSSPFC Applied to UPFCs for Conventional and Renewable Generators

In this section, the results from Sect. 10.4 and references [56, 102, 105] are combined to enable conventional and wind turbine generators to be connected to an infinite bus through a FACTS/storage devices based on UPFCs (see configuration given in Fig. 10.10) to maximize energy capture of a wind turbine while meeting the power system constraints on frequency and phase for grid-tied applications. The UPFC device is required to operate as both a generator and load (energy storage) on the power system in this design. Also, this control design attempts to emulate two decoupled conventional generators that can be modeled as OMIBs.

Initially, the system is partitioned and analyzed as two subsystems. First, the right-hand side of Fig. 10.10, Generator 2 with a UPFC ($C_{12}, \hat{u}_{11}$, and $\hat{u}_{12}$) connected to a wind turbine and a UPFC (C_{23}, u_{21}, and u_{22}) connected to the infinite

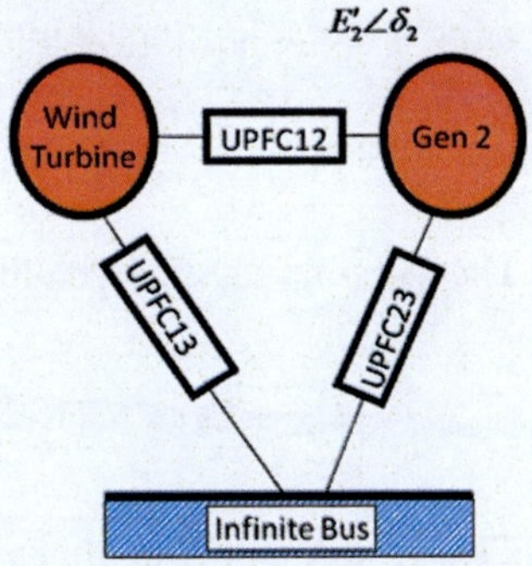

Fig. 10.10 Microgrid with UPFCs and Conventional and Renewable Generators Connected to Grid (Infinite Bus). Wilson, D.G. and Robinett III, R.D. [102], reprinted by permission of the publisher (©2010 IEEE)

bus is investigated by applying the RNM technique [100]. This results in

$$M_2\ddot{\delta}_2 = P_{m_2} - P_{e_2} - D_2\dot{\delta}_2. \tag{10.44}$$

The UPFC control and load takes the form [100]

$$P_{e_2} = C_{23}[(1+u_{21})\sin\delta_2 - u_{22}\cos\delta_2] + C_{12}[(1+\hat{u}_{11})\sin\delta_{21} - \hat{u}_{12}\cos\delta_{21}], \tag{10.45}$$

and substituting (10.45) into (10.44) yields

$$\begin{aligned} M_2\ddot{\delta}_2 = P_{m_2} - D_2\dot{\delta}_2 - C_{12}[(1+\hat{u}_{11})\sin\delta_{21} - \hat{u}_{12}\cos\delta_{21}] \\ - C_{23}[(1+u_{21})\sin\delta_2 - u_{22}\cos\delta_2]. \end{aligned} \tag{10.46}$$

The following nonlinear PID control laws are selected from HSSPFC for the conventional and wind turbine UPFCs as

$$\begin{aligned} u_{i1} &= K_{P_i}\cos\delta_{i_s} + K_{D_i}\sin\delta_i\dot{\delta}_i + K_{I_i}\sin\delta_i\int_0^t(\Delta_i)\,d\tau - 1, \\ u_{i2} &= K_{P_i}\sin\delta_{i_s} - K_{D_i}\cos\delta_i\dot{\delta}_i - K_{I_i}\cos\delta_i\int_0^t(\Delta_i)\,d\tau, \end{aligned} \tag{10.47}$$

where, $\Delta_i = \delta_i - \delta_{i_s}$, $i = 1$ is the wind turbine UPFC (C_{13}, u_{11}, and u_{12}), and $i = 2$ is the Generator 2 UPFC (C_{23}, u_{21}, and u_{22}). Note that $\hat{u}_{11}$ and $\hat{u}_{12}$ are given in Sect. 10.4 by (10.38). Finally, substituting, for $i = 2$, (10.47) into (10.46) and simplifying yields

$$\begin{aligned} M_2\ddot{\delta}_2 + C_{12}K_P\sin 2\delta_2 - P_{m_2} + C_{23}K_{P_2}\sin(\delta_2 - \delta_{2_s}) \\ = -[D_2 + C_{23}K_{D_2}]\dot{\delta}_2 - \left[C_{23}K_{I_2}\int_0^t(\delta_2 - \delta_{2_s})\,dt_1\right]. \end{aligned} \tag{10.48}$$

The static stability condition becomes

$$\mathcal{H} = \frac{1}{2}M_2\dot{\delta}_2^2 + C_{12}K_P[1-\cos(2\delta_2 - 2\delta_{2_s})] + C_{23}K_{P_2}[1-\cos(\delta_2 - \delta_{2_s})] \tag{10.49}$$

with $\mathcal{H}$ being positive definite and

$$2\delta_{2_s} = \sin^{-1}(P_{m_2}/C_{12}K_P).$$

The dynamic stability condition for a passively stable control design yields

$$\int_0^\tau [D_2 + C_{23}K_{D_2}]\dot{\delta}_2^2\, dt > -\int_0^\tau \left[C_{23}K_{I_2}\int_0^t (\delta_2 - \delta_{2_s})\, dt_1\right]\dot{\delta}_2\, dt. \quad (10.50)$$

Clearly, the UPFC nonlinear PID controller expands the region of stability by increasing the PEBS from C_{23} to $(C_{23}K_{P_2} + C_{12}K_P)$ and enabling the system to respond more quickly by adding an integrator to the dissipator of reference [100].

The wind turbine (refer to the left-hand side of Fig. 10.10) torque equation is defined as

$$J_T\ddot{\theta}_r = -K_T\dot{\theta}_r + T_a - T_g, \quad (10.51)$$

where

$$\begin{aligned}
T_a &= M\dot{\theta}_r^2 \\
&= \text{aerodynamic torque from the wind}, \\
T_g &= u_m + T_e \\
&= \text{wind turbine control and electric network torque}, \\
u_m &= u_{m_{\text{ref}}} + \Delta u_m \\
&= \text{reference torque and PID tracking controller}, \\
u_{m_{\text{ref}}} &= -J_T\ddot{\theta}_{\text{ref}} - K_T\dot{\theta}_{\text{ref}} + M\dot{\theta}_r^2 \\
&= \text{reference torque model}, \\
\delta_1 &= \theta_r - \theta_{\text{ref}} \\
&= \text{tracking error}, \\
\Delta u_m &= K_{P_m}\delta_1 + K_{I_m}\int_0^t \delta_1\, d\tau_1 + K_{D_m}\dot{\delta}_1 \\
&= \text{PID tracking controller}, \\
T_e &= C_{13}[(1+u_{11})\sin\delta_1 - u_{12}\cos\delta_1] + C_{12}[(1+\hat{u}_{11})\sin\delta_{12} - \hat{u}_{12}\cos\delta_{12}] \\
&= \text{UPFC electrical controller}.
\end{aligned}$$

Note that u_{11} and u_{12} are defined in (10.47). Also, $\hat{u}_{11}$ and $\hat{u}_{12}$ are previously defined in Sect. 10.4 by (10.38).

Next, following the same sequence used for Generator 2, the wind turbine model (10.51) is updated sequentially with all the defined variables and terms from the previous list, along with the UPFCs from (10.47) to give

$$\begin{aligned}
&J_T\ddot{\delta}_1 + K_{P_m}\delta_1 + C_{13}K_{P_1}\sin\delta_1 + C_{12}K_P\sin 2\delta_1 \\
&\quad = -[K_T + K_{D_m} + C_{13}K_{D_1}]\dot{\delta}_1 - \left[K_{I_m}\int_0^t \delta_1\, d\tau_1 + C_{13}K_{I_1}\int_0^t \delta_1\, d\tau_1\right]. \quad (10.52)
\end{aligned}$$

By examining the storage terms in (10.52) for the wind turbine, the Hamiltonian (stored energy) and static stability condition gives

$$\mathcal{H} = \frac{1}{2} J_T \dot{\delta}_1^2 + \frac{1}{2} K_{P_m} \delta_1^2 + C_{12} K_P (1 - \cos 2\delta_1) + C_{13} K_{P_1} (1 - \cos \delta_1), \quad (10.53)$$

which is positive definite, and

$$\delta_{1_s} = 0.$$

Once again examining (10.52) for the wind turbine and taking the derivative of (10.53), the corresponding Hamiltonian rate terms are determined. The Hamiltonian rate equation is decoupled, and the dynamic stability conditions have been partitioned for each subcomponent. For a passivity controller design, the terms in (10.52) are sorted into dissipators, generators, and storage terms over the cycle; the dissipators must be greater than the generators for a passive stable design. The dynamic stability condition for the wind turbine (10.52) and for a passively stable control design yields

$$\int_0^\tau [K_T + K_{D_m} + C_{13} K_{D_1}] \dot{\delta}_1^2 \, dt > - \int_0^\tau \left[(K_{I_m} + C_{13} K_{I_1}) \int_0^t \delta_1 \, d\tau_1 \right] \dot{\delta}_1 \, dt. \quad (10.54)$$

Notice that the wind turbine controller is designed to maximize the output of the wind turbine, while the UPFC controller is designed to smooth the output of the wind turbine to create $\omega_{\text{ref}} = \text{constant}$. This would be considered a sequential controller design process. As a result, the UPFC performs as both a generator (firm up the wind turbine) and a load (storage excess power) to smooth out the peaks and valleys of the output from the wind turbine.

10.6 Chapter Summary

The swing equations for renewable and conventional generators connected to the EPG and microgrid were developed. A wind turbine was used as an example of a renewable generator. The swing equations for the renewable generators were formulated as a natural Hamiltonian system with externally applied nonconservative forces. HSSPFC was used to analyze and design feedback controllers for the renewable and conventional generators system. This formulation extended previous results on the analytical verification of the Potential Energy Boundary Surface (PEBS) method to nonlinear control analysis and design. This justifies the decomposition of the system into conservative and nonconservative systems to enable a two-step, serial analysis and design procedure. Necessary and sufficient conditions for stability of renewable and conventional generators systems were determined based on the concepts of Hamiltonian systems, power flow, exergy rate, and entropy.

Several nonlinear control design examples were used to demonstrate the HSSPFC technique including the OMIB system with a UPFC. The nonlinear PID feedback controller design for the OMIB system with a UPFC was shown to be an extension of the Control Lyapunov Function approach. Also, the swing equation for a wind turbine connected to an infinite bus through a UPFC was used to determine the required performance of the UPFC to enable the maximum power output of a wind turbine while meeting the power system constraints on frequency and phase. The UPFC connected to the wind turbine is required to operate as both a generator and load (energy storage) on the power system in this design. Numerical simulations for four separate conditions of the renewable microgrid design were reviewed.

Finally, HSSPFC was used to design controllers for multiple machines in grid-tied and islanded (microgrid) configurations. The first HSSPFC controller design was implemented through a UPFC for an islanded microgrid with two conventional generators. The goal of the control design was to enable the two coupled generators to operate as two decoupled One-Machine Infinite-Bus systems. The second HSSPFC controller design was implemented through UPFCs for conventional and wind turbine generators connected to an infinite bus (EPG). For the wind turbine, an additional constraint on the UPFC was implemented to maximize energy capture of the wind turbine while meeting the power system constraints on frequency and phase for grid-tied applications. The UPFC device was required to operate as both a generator and load (energy storage) on the power system in this design. Also, this control design attempted to emulate two decoupled conventional generators that can be modeled as OMIBs.

Chapter 11
Case Study #6: Robotic Manipulator Control Design

11.1 Introduction

Many real-world problems require enhanced performance while ensuring stability of nonlinear systems. In particular, the aerospace community is constantly striving toward higher performance fighter aircrafts that are intrinsically unstable and nonlinear such as the X-29 with a forward swept wing canard configuration. The X-29 is statically unstable (and dynamically unstable) without a stability augmentation systems (SAS) [106]. The X-29 SAS was designed with linear control tools and gain-scheduling. This solution limited the X-29's performance since the present day researchers recognize the value and are attempting to utilize highly nonlinear behavior to enhance performance of future aircraft including unmanned combat air vehicles (UCAVs). Clearly, there is a need to develop nonlinear control design tools that determine and take advantage of the stability boundaries of nonlinear systems by reducing the conservativeness of current state-of-the-art tools.

The goal of developing necessary and sufficient conditions for nonlinear systems has been an area of intense research for many years. The sufficient conditions for stability are well defined in terms of the Lagrange–Dirichlet stability theorem [107, 108], Lyapunov's stability theorems [51, 95, 109], and the absolute stability problem (passivity theorems) [95]. In fact, many of these tools [110] are making progress in developing constructive control law design procedures such as controlled Lagrangians [111], energy-shaping [112], and energy-balancing [112, 113].

Attempts to find the necessary conditions for stability, finding the on-set of instability, include the sufficient conditions for an unstable system, Lyapunov's converse theorems [107, 108] and the inversion of the Lagrange–Dirichlet theorem [107, 108]. The limitations of these theorems are defining the transition from stability to instability and the requirements of conservative systems from the class C^2 or more restrictive [107, 108] to invert the Lagrange–Dirichlet theorem.

Chapter 11 defines the transition from stability to instability in the context of Lyapunov and dynamic stability (balanced power flow: a limit cycle) and in the context of static stability by utilizing the inversion of the Lagrange–Dirichlet theorem for systems of class C^2. The transition from stability to instability is defined in

R.D. Robinett III, D.G. Wilson, *Nonlinear Power Flow Control Design*, Understanding Complex Systems,
DOI 10.1007/978-0-85729-823-2_11, © Springer-Verlag London Limited 2011

Fig. 11.1 MIMO planar robot model. Robinett III, R.D. and Wilson, D.G. [66], reprinted by permission of the publisher (©2006 ASME)

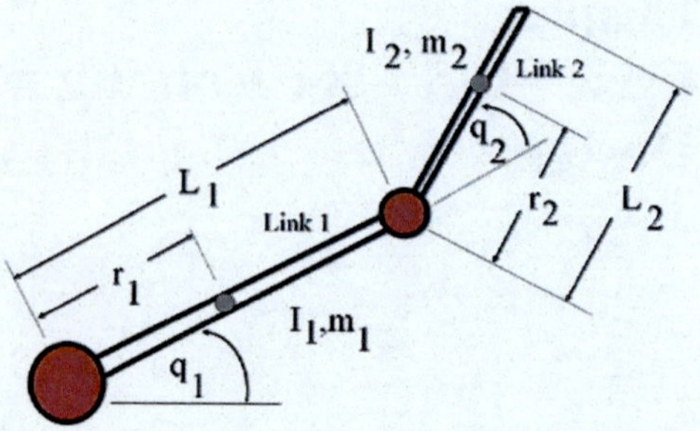

terms of the shape of the Hamiltonian surface and the power flow (work rate) by extending controlled Lagrangians [111], energy-shaping [112], and energy-balancing [112, 113] with exergy/entropy control [13] and static and dynamic stability [37]. The inversion of the Lagrange–Dirichlet theorem is used to formalize the concepts of static and dynamic stability in a nonlinear context. These extensions are a direct result of recognizing that the Hamiltonian is stored exergy and defines the accessible phase space of the system, the application of the Second Law of Thermodynamics to the power flow to determine the trajectory across the Hamiltonian surface, and the utilization of static and dynamic stability concepts to define a two-step design process.

This chapter is divided into four sections. Section 11.2 defines the equations of motion and their relationship to HSSPFC design. Section 11.3 develops a tracking controller utilizing HSSPFC and presents numerical results. Finally, in Sect. 11.4 the chapter results are summarized with concluding remarks.

11.2 Evaluation of the Equations of Motion

This section gives an illustrative example to demonstrate how to apply HSSPFC for a nonlinear mechanical system with gyroscopic or centripetal and Coriolis acceleration terms [66] that are normally associated with MIMO systems. Some of the early work performed by the authors and associated with single-axis systems are given in [30, 60, 114–116]. The equations of motion are derived with Lagrange's equations and evaluated with the Hamiltonian and its time derivative, power flow.

11.2.1 Two-Link Robot Model

Consider a planar two-link manipulator (see reference [117] for details) with two revolute joints in the vertical plane as shown in Fig. 11.1. Let m_i and L_i be the mass and length of link i, r_i be the distance from joint $(i-1)$ to the center of mass of link i, as indicated in the figure, and I_i be the moment of inertia of link i about the axis coming out of the page through the center of mass of link i. Referring to Fig. 11.1 the positions of each mass m_i, $i=1,2$, are given by

$$\begin{Bmatrix} x_{c_1} \\ y_{c_1} \end{Bmatrix} = \begin{Bmatrix} r_1 \cos q_1 \\ r_1 \sin q_1 \end{Bmatrix}$$

and

$$\begin{Bmatrix} x_{c_2} \\ y_{c_2} \end{Bmatrix} = \begin{Bmatrix} L_1 \cos q_1 + r_2 \cos(q_1+q_2) \\ L_1 \sin q_1 + r_2 \sin(q_1+q_2) \end{Bmatrix}.$$

The velocities are determined by differentiating the positions and resulting in

$$\mathbf{v}_{c_1} = \begin{Bmatrix} -r_1\dot{q}_1 \sin q_1 \\ r_1\dot{q}_1 \cos q_1 \end{Bmatrix}$$

and

$$\mathbf{v}_{c_2} = \begin{Bmatrix} -L_1\dot{q}_1 \sin q_1 - r_2(\dot{q}_1+\dot{q}_2)\sin(q_1+q_2) \\ L_1\dot{q}_1 \cos q_1 + r_2(\dot{q}_1+\dot{q}_2)\cos(q_1+q_2) \end{Bmatrix}.$$

The planar two-link robot presented in Fig. 11.1 has the following kinetic energy, potential energy, Lagrangian, and Hamiltonian:

$$\begin{aligned} \mathcal{T} &= \frac{1}{2}m_1\mathbf{v}_{c_1}\cdot\mathbf{v}_{c_1} + \frac{1}{2}I_1\dot{q}_1^2 + \frac{1}{2}m_2\mathbf{v}_{c_2}\cdot\mathbf{v}_{c_2} + \frac{1}{2}I_2(\dot{q}_1+\dot{q}_2)^2, \\ \mathcal{V} &= m_1 g y_{c_1} + m_2 g y_{c_2}, \\ \mathcal{L} &= \mathcal{T} - \mathcal{V}, \\ \mathcal{H} &= \mathcal{T} + \mathcal{V}, \end{aligned} \tag{11.1}$$

where g is the gravitational constant. The equations of motion are generated from Lagrange's equation

$$\frac{d}{dt}\left(\frac{\partial \mathcal{L}}{\partial \dot{q}_i}\right) - \frac{\partial \mathcal{L}}{\partial q_i} = Q_i, \quad i=1,2, \tag{11.2}$$

and the virtual work is

$$\delta W = \sum_{i=1}^{2} Q_i \delta q_i \quad \text{for } Q_i = u_i + f_i(q_i, \dot{q}_i), \tag{11.3}$$

where u_i is the ith control torque, and f_i is the ith frictional torque. Performing the indicated Lagrangian operations results in the dynamic equations of motion for the planar two-link robot

$$\mathbf{M(q)\ddot{q}} + \mathbf{C(q,\dot{q})\dot{q}} + \mathbf{F(q,\dot{q})} + \mathbf{G(q)} = \mathbf{u}, \tag{11.4}$$

where $\mathbf{q} = [q_1\ q_2]^T \in \mathcal{R}^2$ is the angular position vector, $\mathbf{u} = [\tau_1\ \tau_2]^T \in \mathcal{R}^2$ is the control torque, and

$$\mathbf{M}(\mathbf{q}) = \begin{bmatrix} \bar{m}_1 + \bar{m}_2 + 2\bar{m}_3 \cos q_2 & \bar{m}_2 + \bar{m}_3 \cos q_2 \\ \bar{m}_2 + \bar{m}_3 \cos q_2 & \bar{m}_2 \end{bmatrix},$$

$$\mathbf{C}(\mathbf{q}, \dot{\mathbf{q}}) = \begin{bmatrix} -\bar{m}_3 \dot{q}_2 \sin q_2 & -\bar{m}_3(\dot{q}_1 + \dot{q}_2) \sin q_2 \\ \bar{m}_3 \dot{q}_1 \sin q_2 & 0 \end{bmatrix},$$

$$\mathbf{F}(\mathbf{q}, \dot{\mathbf{q}}) = \begin{bmatrix} c_1 \dot{q}_1 + c_{NL_1} \operatorname{sign}(\dot{q}_1) \\ c_2 \dot{q}_2 + c_{NL_2} \operatorname{sign}(\dot{q}_2) \end{bmatrix},$$

$$\mathbf{G}(\mathbf{q}) = \begin{bmatrix} \bar{m}_4 g \cos q_1 + \bar{m}_5 g \cos(q_1 + q_2) \\ \bar{m}_5 g \cos(q_1 + q_2) \end{bmatrix},$$

$$\begin{aligned}
\bar{m}_1 &= m_1 r_1^2 + I_1 + (m_2 + m_p) L_1^2, \\
\bar{m}_2 &= m_2 r_2^2 + I_2 + m_p L_2^2, \\
\bar{m}_3 &= m_2 L_1 r_2 + m_p L_2^2, \\
\bar{m}_4 &= m_1 r_2 + (m_2 + m_p) L_1, \\
\bar{m}_5 &= m_2 r_2 + m_p L_2,
\end{aligned}$$

where $\mathbf{M}(\mathbf{q})$ is a positive definite matrix, $\mathbf{C}(\mathbf{q}, \dot{\mathbf{q}})$ is a skew symmetric matrix, $\mathbf{F}(\mathbf{q}, \dot{\mathbf{q}})$ is a linear and nonlinear friction vector with friction coefficients c_i and c_{NL_i}, $i = 1, 2$, $\mathbf{G}(\mathbf{q})$ is a gravitational force vector given by $\mathbf{G}(\mathbf{q}) = \partial \mathcal{V}/\partial \mathbf{q}$, and m_p is the payload mass.

11.2.2 Evaluation of the Hamiltonian Surface Shaping

The first step in the HSSPFC design process is to recognize that the system is constrained to move on the Hamiltonian surface, the accessible phase space, which can be projected onto the phase plane.

Following the processes in Chap. 4, a statically neutral system is not typically identified separately, but it has important physical significance for mechanical systems. For the two-link robot model operating in a horizontal plane ($\mathbf{G}(\mathbf{q}) = \mathbf{0}$), the system is statically neutral stable with two rigid body modes, $\mathcal{V}(\mathbf{q}) = 0\ \forall \mathbf{q}$, which leads to a singular stiffness operator and no preferred configuration. The two-link robot model is unstable without proportional feedback control. For the two-link robot operating in the vertical plane, the system is only statically stable about the robot hanging vertically straight down without feedback control. Static stability is a

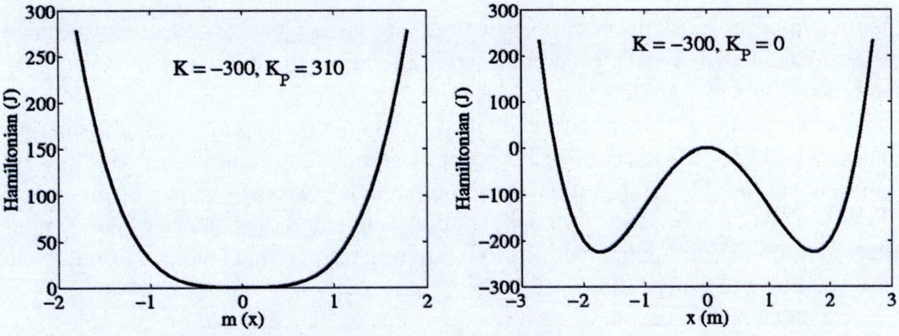

Fig. 11.2 Static stability to bifurcation of an equilibrium point. Robinett III, R.D. and Wilson, D.G. [66], reprinted by permission of the publisher (©2006 ASME)

necessary condition for stability, but not sufficient, and it limits the performance of the system. Returning to the X-29 example, this airplane was designed to be longitudinally statically unstable, and the degree of static stability was generated with the SAS depending upon the desired level of maneuverability (dog fighting, level flight, landing).

Controlled Lagrangians [111], energy-shaping [112], and energy-balancing [112, 113] can be used to construct a feedback controller that meets the sufficient conditions for stability. However, these tools do not recognize the importance of the Hamiltonian surface. Basically, any proportional feedback controller that derives from a C^2 function (and some C^1 functions) meets the requirements of static stability and can be used to increase performance by reducing the stability margin and even driving the system unstable for a portion of the path. A simple example is

$$\begin{aligned}\mathcal{V}(\mathbf{q}) = {} & \frac{1}{2}K_{P_1}(q_1 - q_{1_R})^2 + \frac{1}{2}K_{P_2}(q_2 - q_{2_R})^2 + \frac{1}{4}K_{NL_1}(q_1 - q_{1_R})^4 \\ & + \frac{1}{4}K_{NL_2}(q_2 - q_{2_R})^4, \end{aligned} \tag{11.5}$$

where

$$\mathbf{u} = \begin{Bmatrix} -K_{P_1}(q_1 - q_{1_R}) - K_{NL_1}(q_1 - q_{1_R})^3 \\ -K_{P_2}(q_2 - q_{2_R}) - K_{NL_2}(q_2 - q_{2_R})^3 \end{Bmatrix}. \tag{11.6}$$

An example of reducing the stability margin and bifurcating the statically stable equilibrium point occurs when K_{P_i} is changed from $K_{P_i} > 0$ to $K_{P_i} = 0$ and to $K_{P_i} < 0$. Figure 11.2 presents the results for K_{P_1}.

11.2.3 Evaluation of Power Flow

The second step in the HSSPFC design process is to identify the Hamiltonian as stored exergy, take the time derivative, and apply the Second Law of Thermodynamics in order to partition the power flow into three types: (i) the energy storage rate of change, (ii) power generation, and (iii) power dissipation [13, 37, 46, 50], and then follow the processes in Chap. 4.

A dynamically stable system is equivalent to energy-shaping [112] and energy-balancing [112, 113] except for the generator terms that do not meet passivity requirements, the line integral that is used to calculate average values of the power flows (i.e., AC power, discontinuous functions, etc.), and the balance of power generation to power dissipation subject to the power storage that leads to a limit cycle as a stability boundary [13, 37, 46].

A dynamically unstable system is equivalent to the converse of Lyapunov stability with the addition of the line integral. A dynamically neutral stable system is not typically identified separately, but it has important physical significance for mechanical systems, especially in aeroelasticity [118]. Dynamically neutral stability is the on-set of a limit cycle oscillation. Also, this equation plays an important role in determining the preservation of heteroclinic orbits [99]. The Melnikov number [99, 119] is defined as

$$L = \Delta\mathcal{H} = \int_{-\infty}^{\infty} \dot{\mathcal{H}}\,dt = \int_{0}^{\tau} \dot{\mathcal{H}}\,dt = 0, \tag{11.7}$$

which implies a zero change of energy over a heteroclinic orbit that is preserved.

For the structured robot dynamics, defined earlier, the Hamiltonian rate is

$$\dot{\mathcal{H}} = \dot{\mathbf{q}}^T[\mathbf{M}\ddot{\mathbf{q}} + \mathbf{G}(\mathbf{q})] = \dot{\mathbf{q}}^T[-\mathbf{F}(\mathbf{q}, \dot{\mathbf{q}}) - \mathbf{C}(\mathbf{q}, \dot{\mathbf{q}}) + \mathbf{u}].$$

It can be noted that the gyroscopic or centripetal/Coriolis terms do no work over a cycle and therefore drop out [15]. In addition, for the two-link robot model operating in a horizontal plane, $\mathbf{G}(\mathbf{q}) = \mathbf{0}$, the time derivative of the Hamiltonian is the power flow/work rate [15] for natural systems,

$$\dot{\mathcal{H}} = \dot{\mathcal{T}} = \sum_{i=1}^{2}\{Q_i\dot{q}_i = [u_i - c_i\dot{q}_i - c_{NL_1}\,\mathrm{sign}(\dot{q}_i)]\dot{q}_i\}. \tag{11.8}$$

This system is statically neutral stable and dynamically unstable with no control inputs due to the rigid body modes. For a nonlinear PID regulator,

$$u_i = -K_{P_i}q_i - K_{P_{NL_i}}q_i^3 - K_{I_i}\int_0^t q_i\,dt - K_{D_i}\dot{q}_i - K_{D_{NL_i}}\,\mathrm{sign}(\dot{q}_i).$$

The power flows can be sorted as [13, 37, 46]

$$\begin{aligned}\dot{\mathcal{H}} &= \dot{\mathcal{T}} + \dot{\mathcal{V}}_c \\ &= \sum_{i=1}^{2}\left[-K_{I_i}\int_0^t q_i\,d\tau\right]\dot{q}_i\end{aligned}$$

$$+\sum_{i=1}^{2}\left[-(K_{D_i}+c_i)\dot{q}_i^2-(K_{D_{NL_i}}+c_{NL_i})\,\mathrm{sign}(\dot{q}_i)\dot{q}_i\right] \tag{11.9}$$

with storage terms

$$\int_0^{\tau_c}\dot{\mathcal{T}}\,dt+\int_0^{\tau_c}\sum_{i=1}^{2}\left[K_{P_i}q_i+K_{P_{NL_i}}q_i^3\right]\dot{q}_i\,dt=0,$$

generator terms

$$\frac{1}{\tau_c}\int_0^{\tau_c}\sum_{i=1}^{2}\left[-K_{I_i}\int_0^t q_i\,d\tau\right]\dot{q}_i\,dt>0,$$

dissipator terms

$$\frac{1}{\tau_c}\int_0^{\tau_c}\sum_{i=1}^{2}\left[-(K_{D_i}+c_i)\dot{q}_i^2-(K_{D_{NL_i}}+c_{NL_i})\,\mathrm{sign}(\dot{q}_i)\dot{q}_i\right]dt<0,$$

statically stable

$$\mathcal{V}_c(\mathbf{q})>0\quad\text{for}\quad\mathbf{q}\neq\mathbf{0},\qquad\mathcal{V}_c(\mathbf{0})=0,$$

dynamically stable for

$$\begin{aligned}\frac{1}{\tau_c}\int_0^{\tau_c}\dot{\mathcal{H}}\,dt&=\frac{1}{\tau_c}\int_0^{\tau_c}\left[\dot{\mathcal{T}}+\sum_{i=1}^{2}\left(K_{P_i}q_i+K_{P_{NL_i}}q_i^3\right)\dot{q}_i\right]dt\\&=\frac{1}{\tau_c}\int_0^{\tau_c}\left[\sum_{i=1}^{2}\left(-K_{I_i}\int_0^t q_i\,d\tau-(K_{D_i}+c_i)\dot{q}_i\right.\right.\\&\qquad\left.\left.-(K_{D_{NL_i}}+c_{NL_i})\,\mathrm{sign}(\dot{q}_i)\right)\dot{q}_i\right]dt\end{aligned}$$

subject to

$$\begin{aligned}&\int_0^{\tau_c}\left(K_{I_i}\int_0^t q_i\,d\tau\right)\dot{q}_i dt\\&\quad<\int_0^{\tau_c}\left((-K_{D_i}+c_i)\dot{q}_i-(K_{D_{NL_i}}+c_{NL_i})\,\mathrm{sign}(\dot{q}_i)\right)\dot{q}_i\,dt,\end{aligned}$$

dynamically unstable when

$$\int_0^{\tau_c}\left(K_{I_i}\int_0^t q_i\,d\tau\right)\dot{q}_i\,dt$$

$$> \int_0^{\tau_c} ((-K_{D_i} + c_i)\dot{q}_i - (K_{D_{NL_i}} + c_{NL_i}) \operatorname{sign}(\dot{q}_i))\dot{q}_i \, dt,$$

and dynamically neutral stable when

$$\int_0^{\tau_c} \left(K_{I_i} \int_0^t q_i \, d\tau \right) \dot{q}_i \, dt$$
$$= \int_0^{\tau_c} ((-K_{D_i} + c_i)\dot{q}_i - (K_{D_{NL_i}} + c_{NL_i}) \operatorname{sign}(\dot{q}_i))\dot{q}_i \, dt$$

for $i = 1, 2$.

For the two-link robot operating in the vertical plane, a feedforward computed torque controller can be used to account for the gravity terms while everything else remains the same. Note that the statically and dynamically stable controller reaches a stable equilibrium point that is a minimum energy state [112] and a maximum entropy state [13] since the dominant dissipator term is equal to the irreversible entropy production term, $T_0\dot{\mathcal{S}}_i$.

11.3 Tracking Controller: Perfect Parameter Matching

In this section a tracking controller is designed for perfect parameter matching and to demonstrate decoupling of the DOFs for the three cases of stable, neutral, and unstable conditions [66]. The PID controller is defined for each DOF as

$$\boldsymbol{\tau} = \hat{\mathbf{M}}\ddot{\mathbf{q}}_{\text{ref}} + \hat{\mathbf{C}}\dot{\mathbf{q}}_{\text{ref}} - \mathbf{K}_P\tilde{\mathbf{q}} - \mathbf{K}_I \int_0^t \tilde{\mathbf{q}} \, d\tau - \mathbf{K}_D\dot{\tilde{\mathbf{q}}}, \tag{11.10}$$

where $\mathbf{K}_P$, $\mathbf{K}_I$, and $\mathbf{K}_D$ are the proportional, integral, and derivative diagonal matrix controller gains, respectively. The errors are given by $\tilde{\mathbf{q}} = \mathbf{q}_{\text{ref}} - \mathbf{q}$ and $\dot{\tilde{\mathbf{q}}} = \dot{\mathbf{q}}_{\text{ref}} - \dot{\mathbf{q}}$ with subscript ref equaling reference input terms.

The tracking servo control design begins with picking a Lyapunov function/Hamiltonian based on the error energy

$$V = \Delta\mathcal{H} = \frac{1}{2}\dot{\tilde{\mathbf{q}}}^T\mathbf{M}\dot{\tilde{\mathbf{q}}} + \frac{1}{2}\tilde{\mathbf{q}}^T\mathbf{K}_P\tilde{\mathbf{q}}, \tag{11.11}$$

which is positive definite and statically stable. The time derivative is

$$\begin{aligned}\dot{V} &= \frac{d}{dt}\Delta\mathcal{H} \\ &= \Delta\dot{W} - T_0\Delta\dot{\mathcal{S}}_i = \sum_{j=1}^{N} \delta Q_j \delta\dot{q}_j - \sum_{l=N+1}^{M+N} \delta Q_l \delta\dot{q}_l \\ &= \dot{\tilde{\mathbf{q}}}^T\mathbf{M}\ddot{\tilde{\mathbf{q}}} + \frac{1}{2}\dot{\tilde{\mathbf{q}}}^T\dot{\mathbf{M}}\dot{\tilde{\mathbf{q}}} + \dot{\tilde{\mathbf{q}}}^T\mathbf{K}_P\tilde{\mathbf{q}},\end{aligned} \tag{11.12}$$

and upon substitution of the dynamic error equations, simplifications, and assumption of perfect parameter matching ($\hat{\mathbf{M}} = \mathbf{M}$, $\hat{\mathbf{C}} = \mathbf{C}$), one has

$$\dot{V} = -\dot{\tilde{\mathbf{q}}}^T \mathbf{K}_D \dot{\tilde{\mathbf{q}}} - \dot{\tilde{\mathbf{q}}}^T \mathbf{K}_I \int_0^t \tilde{\mathbf{q}}\, d\tau \tag{11.13}$$

that is subject to the following general necessary and sufficient conditions:

$$\begin{aligned} \left[T_0 \Delta \dot{S}_i\right]_{\text{ave}} &\geq 0 \quad \text{positive semi-definite, always true,} \\ \left[\Delta \dot{W}\right]_{\text{ave}} &\geq 0 \quad \text{positive semi-definite; exergy pumped in.} \end{aligned}$$

Similar arguments can be made for the corollaries as compared to the earlier subsection. The formulation is identical to the regulator problem with the exception that for the nonlinear case, additional cross-terms are accommodated by the identity $\dot{\tilde{\mathbf{q}}}^T [\dot{\mathbf{M}} - 2\mathbf{C}] \dot{\tilde{\mathbf{q}}} = 0$.

Next, the PID tracking control in terms of exergy generation and exergy dissipation is investigated. The derivative of the Lyapunov function/Hamiltonian (11.13) yields

$$\begin{aligned} \left[T_0 \Delta \dot{S}_i\right]_{\text{ave}} &= \dot{\tilde{\mathbf{q}}}^T \mathbf{K}_D \dot{\tilde{\mathbf{q}}}, \\ \left[\Delta \dot{W}\right]_{\text{ave}} &= -\dot{\tilde{\mathbf{q}}}^T \mathbf{K}_I \int_0^t \tilde{\mathbf{q}}\, d\tau, \\ \left[T_0 \Delta \dot{S}_{\text{rev}}\right]_{\text{ave}} &= \left[\dot{\tilde{\mathbf{q}}}^T \mathbf{M} \ddot{\tilde{\mathbf{q}}}\right]_{\text{ave}} + \left[\dot{\tilde{\mathbf{q}}}^T \mathbf{K}_P \tilde{\mathbf{q}}\right]_{\text{ave}} = 0. \end{aligned} \tag{11.14}$$

The first expression in (11.14), the derivative tracking control term, is identified as dissipative terms. The second expression in (11.14), the integral tracking control term, is identified as a generative term. In the final expression in (11.14), the proportional tracking control term is identified as a reversible term along with the inertial terms.

To determine the nonlinear stability boundary from the HSSPFC design, let

$$\left[\Delta \dot{W}\right]_{\text{ave}} = \left[T_0 \Delta \dot{S}_i\right]_{\text{ave}}.$$

Substituting the actual terms yields the following:

$$\begin{aligned} \left[K_{D_1} \dot{\tilde{q}}_1^2\right]_{\text{ave}} &= \left[-K_{I_1} \int_o^t \tilde{q}_1\, d\tau \dot{\tilde{q}}_1\right]_{\text{ave}}, \\ \left[K_{D_2} \dot{\tilde{q}}_2^2\right]_{\text{ave}} &= \left[-K_{I_2} \int_o^t \tilde{q}_2\, d\tau \dot{\tilde{q}}_2\right]_{\text{ave}}, \end{aligned} \tag{11.15}$$

which are the *nonlinear stability boundaries* per DOF. To best understand how the boundary is determined, concepts and analogies from electric AC power have been introduced earlier. Essentially, when the average power_{in} is equivalent to the average $\text{power}_{\text{dissipated}}$ over a cycle, then the system is operating at the stability boundary. Later, in the exergy and exergy rate responses for the nonlinear system, one may

Table 11.1 MIMO planar robot PID controller numerical values. Robinett III, R.D. and Wilson, D.G. [66], reprinted by permission of the publisher (©2006 ASME)

Case No.	K_{P_1} (N m)	K_{P_2} (N m)	K_{D_1} (N m/s)	K_{D_2} (N m/s)	K_{I_1} (N m/s)	K_{I_2} (N m/s)
1	250	100	7	2	8.8	2.75
2	250	100	7	2	880	275
3	250	100	3.5	0.267	880	275

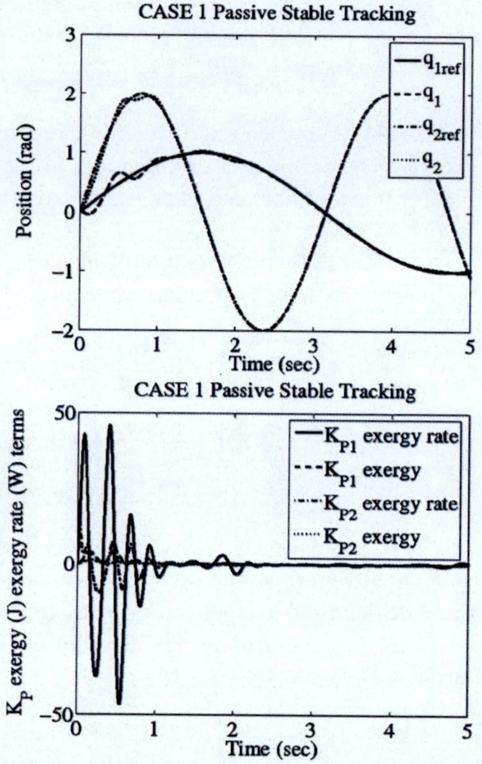

Fig. 11.3 Case 1: Robot MIMO tracking and reversible K_P exergy and exergy rate numerical results. Robinett III, R.D. and Wilson, D.G. [66], reprinted by permission of the publisher (©2006 ASME)

observe that the area under the curves for the exergy rate generation and the exergy rate dissipation are equivalent, and for the corresponding exergy responses, the slopes will be equal and opposite.

Numerical simulations are performed for three separate PID tracking control cases with the numerical values listed in Table 11.1.

The robot physical parameters are given as: link lengths; $L_1 = L_2 = 0.5$ m, center of gravity locations $r_1 = r_2 = 0.25$ m, moments of inertia $I_1 = I_2 = 0.1$ kg m^2, and masses $m_1 = 4.5$ kg and $m_2 = 2.5$ kg, respectively. The detailed dynamic model is given in [117].

The reference input signal is defined as $\mathbf{q}_{\text{ref}} = \mathbf{A}_{\text{in}} \sin \omega_{\text{in}} t$ with $\mathbf{A}_{\text{in}} = [1\ 2]$ rad and $\omega_{\text{in}} = [1\ 2]$ rad/s for each DOF, respectively. The MIMO robot system is initially at zero. For Case 1, the tracking position along with the reversible proportional tracking control exergy and exergy rate responses are given in Fig. 11.3 for

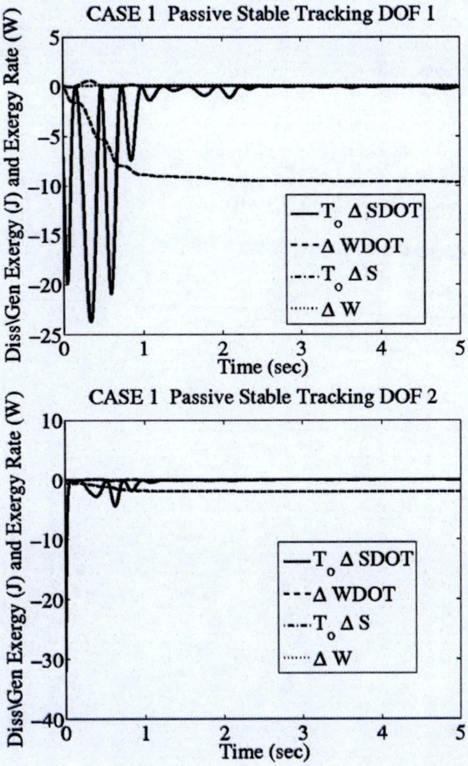

Fig. 11.4 Case 1: Robot MIMO tracking exergy and exergy rate numerical results. Robinett III, R.D. and Wilson, D.G. [66], reprinted by permission of the publisher (©2006 ASME)

both DOFs. Note that for over each cycle, the proportional tracking control exergy rate response is zero (valid for all three cases). Variations for dissipative and generative exergy and exergy rate responses are given in Fig. 11.4 for DOF one (left) and two (right), respectively. For Case 1, passive stable tracking is observed from the decaying response and exhibits a predominately larger perturbation dissipative term with respect to the perturbation generative term. The initial transient tracking position errors converge to the reference sinusoidal inputs. For Case 2, similar responses for the tracking position errors and reversible proportional tracking control exergy and exergy rate responses are presented in Fig. 11.5 for both DOFs. The variations for the dissipative and generative exergy and exergy rate responses are illustrated in Fig. 11.6 for DOF one (left) and two (right), respectively. This case demonstrates the neutrally stable tracking boundary or where the dissipative terms are equivalent and cancel the generative terms. For Case 3, similar responses are given in Figs. 11.7 and 11.8. Case 3 presents responses that are growing exponentially, since the generative term is greater than the dissipative term.

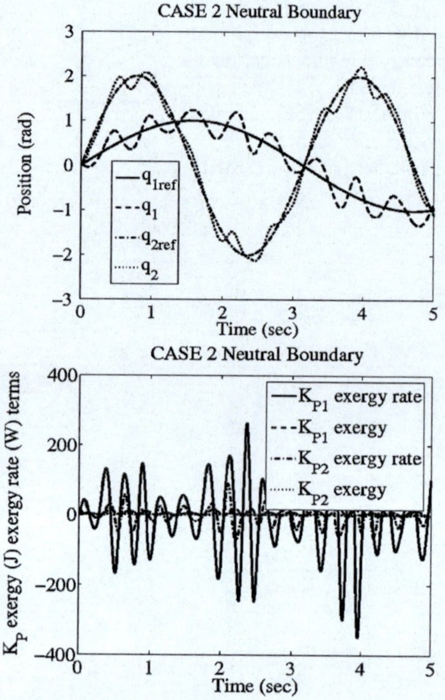

Fig. 11.5 Case 2: Robot MIMO tracking and reversible K_P exergy and exergy rate numerical results. Robinett III, R.D. and Wilson, D.G. [66], reprinted by permission of the publisher (©2006 ASME)

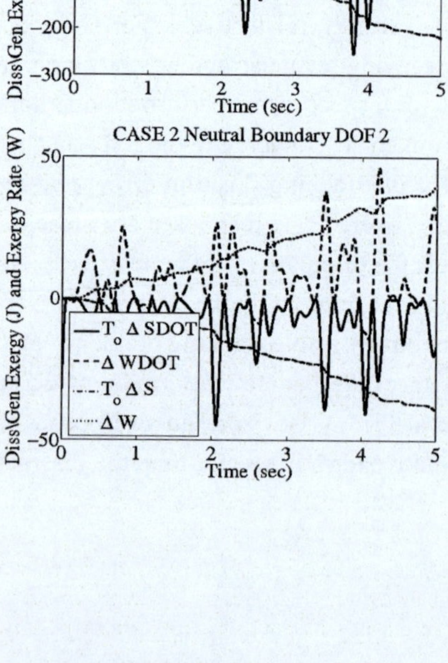

Fig. 11.6 Case 2: Robot MIMO tracking exergy and exergy rate numerical results. Robinett III, R.D. and Wilson, D.G. [66], reprinted by permission of the publisher (©2006 ASME)

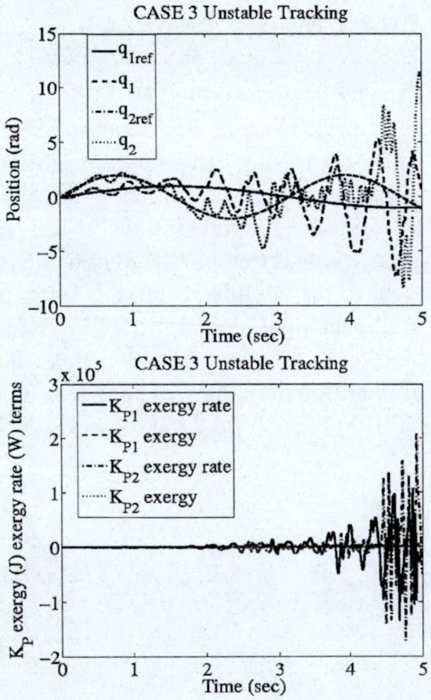

Fig. 11.7 Case 3: Robot MIMO tracking and reversible K_P exergy and exergy rate numerical results. Robinett III, R.D. and Wilson, D.G. [66], reprinted by permission of the publisher (©2006 ASME)

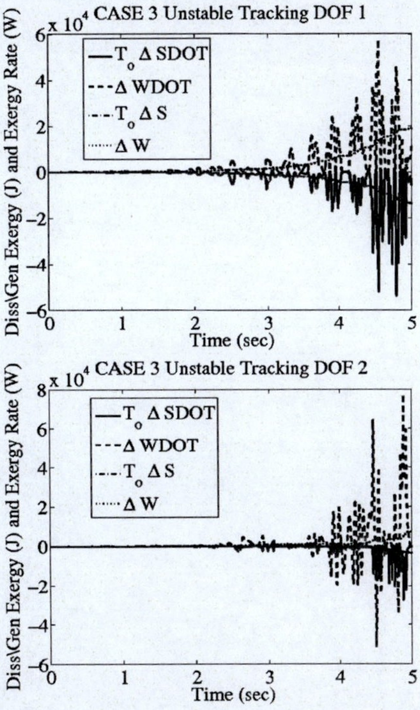

Fig. 11.8 Case 3: Robot MIMO tracking exergy and exergy rate numerical results. Robinett III, R.D. and Wilson, D.G. [66], reprinted by permission of the publisher (©2006 ASME)

11.4 Chapter Summary

Chapter 11 has applied HSSPFC to a nonlinear MIMO system with gyroscopic or centripetal/Coriolis acceleration terms. HSSPFC was demonstrated to be an extension of controlled Lagrangians, energy balancing, and energy shaping by developing necessary and sufficient conditions for stability of a class of nonlinear systems, Hamiltonian natural systems, based on the recognition that the Hamiltonian is stored exergy. It was demonstrated how the nonlinear dynamic stability constraint is equivalent to the Melnikov number for heteroclinic orbits. HSSPFC was used to design nonlinear regulator and tracking controllers with defined stability boundaries including limit cycles for a two-link robot. Also, the minimum energy state controller of energy-balancing was shown to be a maximum entropy state controller based on HSSPFC.

Chapter 12
Case Study #7: Satellite Reorientation Control

12.1 Introduction

Many mission objectives require that the spacecraft point in a desired direction [120]. Satellites used in communication and surveillance applications require that a part of the spacecraft point toward a desired location while another part, for example, the solar panel, point in another. To achieve the proper orientation and repositioning of antennas and solar panels, an attitude stabilization and control system must be properly designed for the spacecraft. There is a continued and active research interest in understanding the overall performance and stability issues that surround these applications. The large-angle attitude control problem of spacecraft has received much interest in recent decades. Many methods have been proposed to stabilize the attitude dynamics, some with respect to minimal time or minimal control energy performance requirements [121, 122]. Previous nonlinear control design methods for spacecraft attitude control include: nonlinear PID-like control [123], passivity-based control design [124], sliding mode control [95, 125–128], model reference adaptive control [129], adaptive dynamic inversion control [130], optimal Hamilton–Jacobi formulation control [131], quaternion feedback control [132, 133], nonlinear H_∞ control, and control Lyapunov function methods [134–137]. Most recently, the combination of quaternion feedback and nonlinear H_∞ control methods [138], by using a Hamilton–Jacobi general control Lyapunov function, accounted for the stability and robustness of the attitude control problems. The formulation [138] then led to a design framework suitable for Linear Matrix Inequality (LMI)-based design. The controller contained nonlinear feedback terms that enhanced design flexibility.

Real-world mechanical systems, such as spacecraft, present nonlinear behavior. In many cases simple linearization in modeling the system would not lead to satisfactory results. Traditionally, almost all modern control design is based on forcing the nonlinear systems to perform and behave like linear systems, thus limiting its maximum potential. Given this situation, it would be desirable to take advantage of the nonlinearities of the satellite system to enhance performance. Typical examples of system parameters currently used in nonlinear models of mechanical systems [139] are Coulomb damping and cubic stiffness. To gain insight into this new

R.D. Robinett III, D.G. Wilson, *Nonlinear Power Flow Control Design*,
Understanding Complex Systems,
DOI 10.1007/978-0-85729-823-2_12, © Springer-Verlag London Limited 2011

approach, nonlinear examples that include typical inherent nonlinearities and the full three-axis spacecraft attitude-control problem are reviewed. This HSSPFC approach begins to investigate the potential of designing inherently nonlinear control systems by better understanding the stability boundaries.

Chapter 12 presents HSSPFC design methodology applied to the spacecraft slewing problem [67]. Both necessary and sufficient conditions for stability are determined for a class of nonlinear systems by finding the stability boundaries, rigid body modes, and limit cycles which are a result of balancing power flows. The energy storage surface (Hamiltonian) is related to static stability and chosen as the Lyapunov candidate function. The time derivative of the energy storage surface is related to dynamic stability and defined in terms of the power flow into, the power dissipated within, and rate of change of stored energy in the system that is related to Lyapunov analysis.

Chapter 12 is divided into three sections. Section 12.1 is the introduction. Section 12.2 presents a nonlinear control design example, a MIMO three-axis slewing spacecraft that employs PID tracking control. Numerical simulations result in the demonstration of both performance and stability criteria for the example. Finally, Sect. 12.3 summarizes the results with concluding remarks.

12.2 Spacecraft Attitude Control Design

This section investigates a PID attitude control for MIMO three-axis spacecraft reorientation. Spacecraft generally have pointing requirements so that the satellite can slew and point at a desired target with a specified tolerance. In addition, many environmental effects can cause disturbances to the spacecraft. The attitude control system must be designed to achieve the specified performance requirements and account for disturbances while providing three-axis stability. This PID attitude control example designs a set of control laws to simultaneously account for both performance and stability, while including the spacecraft inherent nonlinearities, for a MIMO three-axis attitude control system (see Fig. 12.1). The spacecraft is considered a rigid body with the dynamical equations of motion given by

$$\mathbf{H}\dot{\boldsymbol{\omega}} = \mathbf{p} \times \boldsymbol{\omega} + \boldsymbol{\tau}$$

and the kinematics or the attitude of the rigid body described by the classical Euler angle representation

$$\dot{\mathbf{x}} = \mathbf{J}(\mathbf{x})\boldsymbol{\omega}.$$

For this analysis, a 3–2–1 or a (yaw, ϕ, pitch, θ, and roll, ψ) sequence is defined, where $\mathbf{x} = [\phi\ \theta\ \psi]^T$. For this Euler angle sequence, the kinematic relationship between the Euler angles and body angles is given as

$$\begin{Bmatrix} \omega_1 \\ \omega_2 \\ \omega_3 \end{Bmatrix} = \begin{bmatrix} -s\theta & 0 & 1 \\ s\psi c\theta & c\psi & 0 \\ c\psi c\theta & -s\psi & 0 \end{bmatrix} \begin{Bmatrix} \dot{\phi} \\ \dot{\theta} \\ \dot{\psi} \end{Bmatrix} \tag{12.1}$$

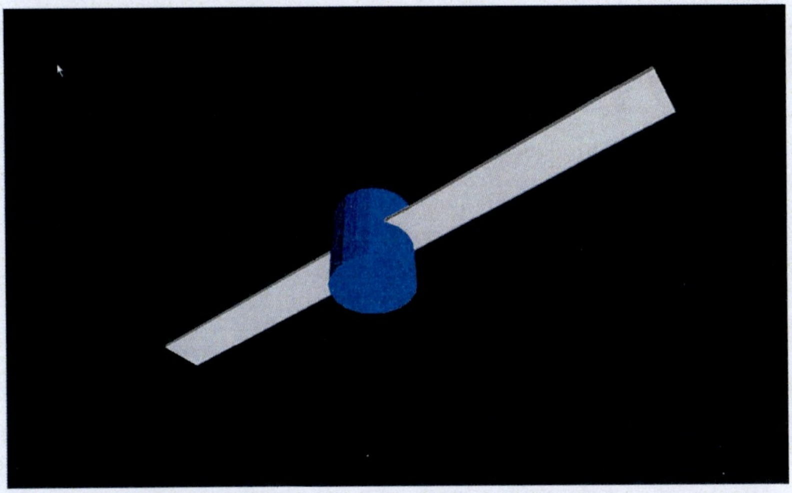

Fig. 12.1 General MIMO three-axis spacecraft system. Robinett III, R.D. and Wilson, D.G. [67], reprinted with the permission of the American Astronautical Society

with the inverse relationship [95] determined as

$$\begin{Bmatrix} \dot{\phi} \\ \dot{\theta} \\ \dot{\psi} \end{Bmatrix} = \frac{1}{c\theta} \begin{bmatrix} 0 & s\psi & c\psi \\ 0 & c\theta c\psi & -c\theta s\psi \\ c\theta & -s\theta s\psi & s\theta c\psi \end{bmatrix} \begin{Bmatrix} \omega_1 \\ \omega_2 \\ \omega_3 \end{Bmatrix}, \tag{12.2}$$

where the matrix is singular at $\theta = \pm\frac{\pi}{2}$, and c and s are shorthand for *cosine* and *sine* of a given angle. Both of these relationships in (12.1) and (12.2) can be written compactly as

$$\boldsymbol{\omega} = \mathbf{B}(\theta, \psi)\dot{\boldsymbol{\Theta}}$$

and

$$\dot{\boldsymbol{\Theta}} = \mathbf{J}(\theta, \psi)\boldsymbol{\omega}.$$

The momentum $\mathbf{p}$ can be expressed as a function of the Euler angles by observing that

$$\mathbf{p} = \mathbf{R}(\phi, \theta, \psi)\mathbf{p}^I,$$

where $\mathbf{p}^I$ is the inertial angular momentum, and the matrix $\mathbf{R}$ represents the coordinate transformation from the inertial frame to the spacecraft frame [95] and is given as

$$\mathbf{R}(\phi, \theta, \psi) = \begin{bmatrix} c\theta c\phi & c\theta s\phi & -s\theta \\ c\psi s\phi + s\psi s\theta c\phi & c\psi c\phi + s\psi s\theta s\phi & s\psi c\theta \\ s\psi s\phi + c\psi s\theta c\phi & -s\psi c\phi + c\psi s\theta s\phi & c\psi c\theta \end{bmatrix}.$$

Next, the system equations of motion are expressed and transformed into the $\mathbf{x}$ and $\dot{\mathbf{x}}$ space. Also note that the matrix $\mathbf{J}$ remains invertible in the range specified above.

Thus,

$$\mathbf{H}^*(\mathbf{x})\ddot{\mathbf{x}} + \mathbf{C}^*(\mathbf{x}, \dot{\mathbf{x}})\dot{\mathbf{x}} = \mathbf{F}, \tag{12.3}$$

where $\mathbf{H}^*$ and $\mathbf{C}^*$ are the inertial and centripetal/Coriolis matrices, and $\mathbf{F}$ is the external torque vector. By performing the implied transformations we have

$$\begin{aligned}
&\mathbf{F} = \mathbf{J}^{-T}\boldsymbol{\tau},\\
&\mathbf{H}^*(\mathbf{x}) = \mathbf{J}^{-T}\mathbf{H}(\mathbf{x})\mathbf{J}^{-1},\\
&\mathbf{C}^*(\mathbf{x}, \dot{\mathbf{x}}) = -\mathbf{J}^{-T}\mathbf{H}\mathbf{J}^{-1}\dot{\mathbf{J}}\mathbf{J}^{-1} - \mathbf{J}^{-T}[\mathbf{p}\times]\mathbf{J}^{-1},
\end{aligned}$$

where $[\mathbf{p}\times]$ refers to the skew-symmetric matrix for the vector $\mathbf{p}$.

The PID controller is defined in matrix/vector form as

$$\mathbf{F} = \mathbf{F}_{\text{ref}} + \delta\mathbf{F} = \hat{\mathbf{H}}^*\ddot{\mathbf{x}}_{\text{ref}} + \hat{\mathbf{C}}^*\dot{\mathbf{x}}_{\text{ref}} + \mathbf{K}_P\tilde{\mathbf{x}} + \mathbf{K}_I\int_0^t \tilde{\mathbf{x}}\,d\tau + \mathbf{K}_D\dot{\tilde{\mathbf{x}}}, \tag{12.4}$$

where $\mathbf{K}_P$, $\mathbf{K}_I$, and $\mathbf{K}_D$ are the proportional, integral, and derivative diagonal positive definite matrix controller gains, respectively. The errors are given by

$$\tilde{\mathbf{x}} = \mathbf{x}_{\text{ref}} - \mathbf{x}$$

and

$$\dot{\tilde{\mathbf{x}}} = \dot{\mathbf{x}}_{\text{ref}} - \dot{\mathbf{x}}$$

with subscript ref defined as the reference input terms.

The tracking attitude control design begins with defining a Hamiltonian/Lyapunov candidate function (that is positive definite for $\mathbf{H}^*(\mathbf{x})$, $\mathbf{K}_P$ positive definite) based on the error energy (error exergy) or

$$V = \Delta\mathcal{H} = \frac{1}{2}\dot{\tilde{\mathbf{x}}}^T\mathbf{H}^*\dot{\tilde{\mathbf{x}}} + \frac{1}{2}\tilde{\mathbf{x}}^T\mathbf{K}_P\tilde{\mathbf{x}} \tag{12.5}$$

where the proportional feedback ensures static stability. The time derivative is [13, 30, 67]

$$\begin{aligned}
\dot{V} &= \frac{d}{dt}\Delta\mathcal{H}\\
&= \Delta\dot{W} - T_0\Delta\dot{\mathcal{S}}_i = \sum_{j=1}^{N}\delta Q_j\delta\dot{q}_j - \sum_{l=N+1}^{M+N}\delta Q_l\delta\dot{q}_l\\
&= \dot{\tilde{\mathbf{x}}}^T\mathbf{H}^*\ddot{\tilde{\mathbf{x}}} + \frac{1}{2}\dot{\tilde{\mathbf{x}}}^T\dot{\mathbf{H}}^*\dot{\tilde{\mathbf{x}}} + \dot{\tilde{\mathbf{x}}}^T\mathbf{K}_P\tilde{\mathbf{x}}.
\end{aligned} \tag{12.6}$$

Next, from (12.3), solve for $\mathbf{H}^*\ddot{\tilde{\mathbf{x}}}$ by

$$\begin{aligned}
&\mathbf{H}^*\ddot{\mathbf{x}} = \mathbf{F} - \mathbf{C}^*\dot{\mathbf{x}},\\
&\mathbf{H}^*\ddot{\tilde{\mathbf{x}}} = -\mathbf{F} + \mathbf{H}^*\ddot{\mathbf{x}}_{\text{ref}} + \mathbf{C}^*\dot{\mathbf{x}}_{\text{ref}} - \mathbf{C}^*\dot{\tilde{\mathbf{x}}},
\end{aligned}$$

and substitute into (12.6) or

$$\dot{V} = \dot{\tilde{\mathbf{x}}}^T\left[-\mathbf{F} + \mathbf{H}^*\ddot{\mathbf{x}}_{\text{ref}} + \mathbf{C}^*\dot{\mathbf{x}}_{\text{ref}}\right] + \frac{1}{2}\dot{\tilde{\mathbf{x}}}^T\left[\dot{\mathbf{H}}^* - 2\mathbf{C}^*\right]\dot{\tilde{\mathbf{x}}} + \dot{\tilde{\mathbf{x}}}^T\mathbf{K}_P\tilde{\mathbf{x}}, \tag{12.7}$$

where the term $\dot{\mathbf{H}}^* - 2\mathbf{C}^*$ is skew symmetric and becomes identically zero [95]. Next, the attitude controller (12.4) is substituted into (12.7) or

$$\dot{V} = \dot{\tilde{\mathbf{x}}}^T\left[(\mathbf{H}^* - \hat{\mathbf{H}}^*)\ddot{\mathbf{x}}_{\text{ref}} + (\mathbf{C}^* - \hat{\mathbf{C}}^*)\dot{\mathbf{x}}_{\text{ref}} - \mathbf{K}_I\int_0^T \tilde{\mathbf{x}}\,d\tau - \mathbf{K}_D\dot{\tilde{\mathbf{x}}}\right].$$

Finally, by assuming perfect parameter matching or $\hat{\mathbf{H}}^* = \mathbf{H}^*$ and $\hat{\mathbf{C}}^* = \mathbf{C}^*$, we have

$$\dot{V} = -\dot{\tilde{\mathbf{x}}}^T\mathbf{K}_D\dot{\tilde{\mathbf{x}}} - \dot{\tilde{\mathbf{x}}}^T\mathbf{K}_I\int_0^t \tilde{\mathbf{x}}\,d\tau. \tag{12.8}$$

Next, the PID tracking control in terms of exergy generation and exergy dissipation is investigated. The derivative of the Hamiltonian/Lyapunov candidate function (12.8) yields [13, 30, 67]

$$\begin{aligned}
\left[T_0\Delta\dot{S}_i\right]_{\text{ave}} &= \frac{1}{\tau_c}\int_0^{\tau_c}\left[\dot{\tilde{\mathbf{x}}}^T\mathbf{K}_D\dot{\tilde{\mathbf{x}}}\right]dt,\\
\left[\Delta\dot{W}\right]_{\text{ave}} &= \frac{1}{\tau_c}\int_0^{\tau_c}\left[-\dot{\tilde{\mathbf{x}}}^T\mathbf{K}_I\int_0^t \tilde{\mathbf{x}}\,d\tau\right]dt,\\
\left[T_0\Delta\dot{S}_{\text{rev}}\right]_{\text{ave}} &= \left[\frac{1}{2}\frac{d}{dt}\dot{\tilde{\mathbf{x}}}^T\mathbf{H}\dot{\tilde{\mathbf{x}}}\right]_{\text{ave}} + \left[\dot{\tilde{\mathbf{x}}}^T\mathbf{K}_P\tilde{\mathbf{x}}\right]_{\text{ave}} = 0,
\end{aligned} \tag{12.9}$$

which is subject to the following conditions [30]:

$$\begin{aligned}
\left[T_0\Delta\dot{S}_i\right]_{\text{ave}} &\geq 0 \quad \text{positive semi-definite, always true}\\
&\qquad\ \ \text{by the second law of thermodynamics,}\\
\left[\Delta\dot{W}\right]_{\text{ave}} &\geq 0 \quad \text{positive semi-definite.}
\end{aligned}$$

The first row in (12.9) identifies the derivative tracking control as a power dissipator term. The second row in (12.9) identifies the integral tracking control term as a power generator term. The last row in (12.9) identifies the proportional tracking control term as a reversible (energy storage) term along with the inertial term.

To determine the nonlinear stability boundary from the HSSPFC design [13, 30, 67], let

$$\left[\Delta\dot{W}\right]_{\text{ave}} = \left[T_0\Delta\dot{S}_i\right]_{\text{ave}}.$$

Table 12.1 General MIMO three-Axis spacecraft PID control system gains. Robinett III, R.D. and Wilson, D.G. [67], reprinted with the permission of the American Astronautical Society

Case No.	K_{P_1} (N m)	K_{P_2} (N m/s)	K_{P_3} (N m/s)	K_{D_1} (kg m^2)	K_{D_2} (N m)	K_{D_3} (N m/s)	K_{I_1}	K_{I_2}	K_{I_3}
1	400.0	550.0	400.0	80.0	80.0	80.0	16.5	17.7	18.5
2	400.0	550.0	400.0	8.0	8.0	8.0	165.0	177.0	185.0
3	400.0	550.0	400.0	8.0	8.0	8.0	247.5	265.5	277.5

Substituting the actual terms yields the following:

$$\begin{aligned}
\left[K_{D_1}\dot{\tilde{x}}_1^2\right]_{\text{ave}} &= \left[-K_{I_1}\int_0^t \tilde{x}_1\,d\tau\,\dot{\tilde{x}}_1\right]_{\text{ave}}, \\
\left[K_{D_2}\dot{\tilde{x}}_2^2\right]_{\text{ave}} &= \left[-K_{I_2}\int_0^t \tilde{x}_2\,d\tau\,\dot{\tilde{x}}_2\right]_{\text{ave}}, \\
\left[K_{D_3}\dot{\tilde{x}}_3^2\right]_{\text{ave}} &= \left[-K_{I_3}\int_0^t \tilde{x}_3\,d\tau\,\dot{\tilde{x}}_3\right]_{\text{ave}},
\end{aligned} \tag{12.10}$$

which are the *nonlinear stability boundaries*, limit cycles, per DOF. To best understand how the boundary is determined, concepts and analogies from electric AC power were introduced in references [13, 30, 67]. Essentially, when the average power flowing into the system is equivalent to the average power dissipated within the system over the limit cycle, then the system is operating at the stability boundary. Later, in the exergy and exergy rate responses for the nonlinear system, one may observe that the area under the curves for the exergy rate generation and the exergy rate dissipation are equivalent, and for the corresponding exergy responses, the slopes will be equal and opposite.

Numerical simulations are performed for three separate PID tracking control cases with the numerical values listed in Table 12.1. The spacecraft dynamic model is given in [95], where the inertia matrix is

$$\mathbf{H} = \begin{bmatrix} 15 & 5 & 5 \\ 5 & 10 & 7 \\ 5 & 7 & 20 \end{bmatrix} \text{ kg m}^2,$$

and the constant inertial angular momentum is given as

$$\mathbf{p}^I = [1\ -1\ 0]^T.$$

The goal is to maneuver the spacecraft from an initial orientation $\mathbf{x}_i = [\phi\ \theta\ \psi]^T = [0°\ 0°\ 0°]$ to a final orientation of $\mathbf{x}_f = [\phi\ \theta\ \psi]^T = [70°\ 50°\ -60°]$ and returning. The initial and final reference velocities are zero. The sequence is then repeated. The corresponding reference acceleration pulses for all cases and each axis are given in Fig. 12.2.

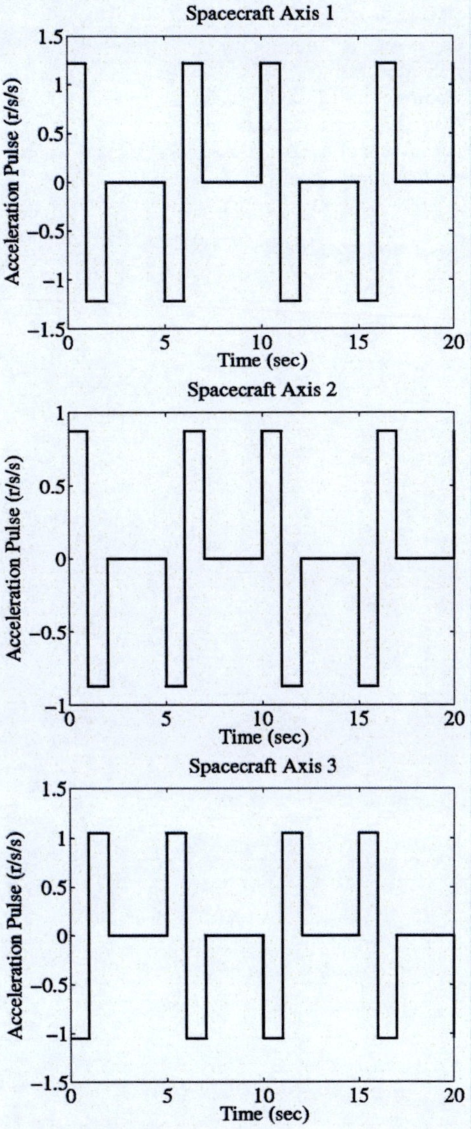

Fig. 12.2 Cases 1–3 reference acceleration pulse inputs. Robinett III, R.D. and Wilson, D.G. [67], reprinted with the permission of the American Astronautical Society

For Case 1, the tracking positions, velocities, and hub torques are presented in Fig. 12.3. The exergy and exergy-rate responses for each axis are illustrated in Fig. 12.4. For each axis, the dissipator terms are larger than the generator terms, resulting in stable passive responses. For Case 2, similar responses are given for position, velocity, and hub torque in Fig. 12.5. The exergy and exergy-rate responses are also shown in Fig. 12.6. For each axis, the dissipator terms are equivalent to the generator terms, resulting in neutrally stable responses and validation of the *nonlinear stability boundaries*, limit cycles, given in (12.10). For Case 3, similar responses

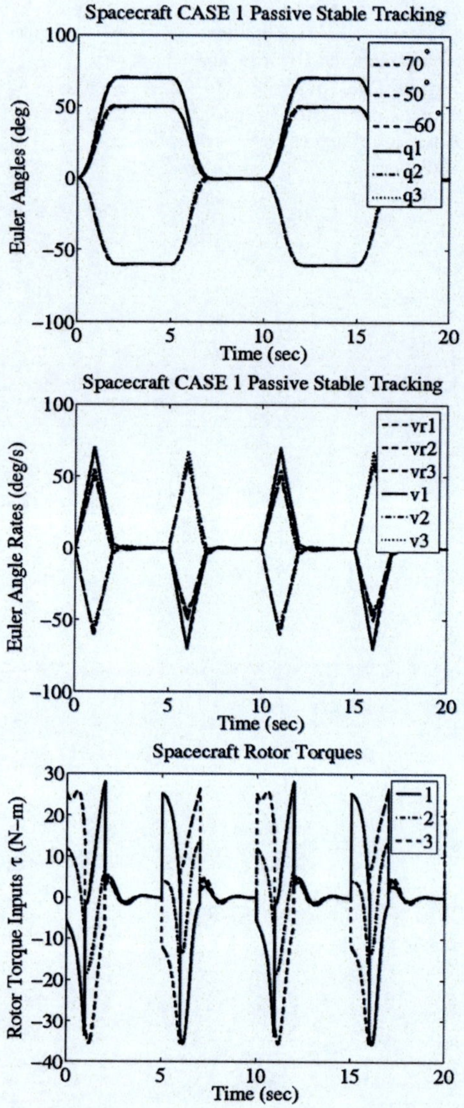

Fig. 12.3 Case 1 Euler angle, rate, and hub torque numerical results. Robinett III, R.D. and Wilson, D.G. [67], reprinted with the permission of the American Astronautical Society

are given for position, velocity, and hub torque in Fig. 12.7 along with exergy and exergy-rate responses for each axis in Fig. 12.8, respectively. This final case shows responses that are growing exponentially since the generator term is greater than the dissipator term.

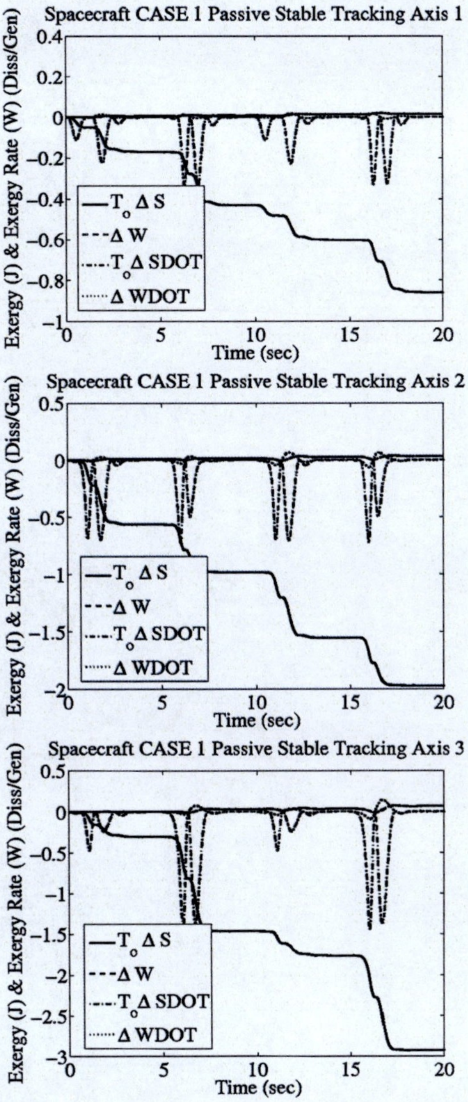

Fig. 12.4 Case 1 exergy and exergy rates for axes 1–3 numerical results. Robinett III, R.D. and Wilson, D.G. [67], reprinted with the permission of the American Astronautical Society

12.3 Chapter Summary

Chapter 12 presented HSSPFC control methodology applied to a slewing spacecraft problem. The HSSPFC methodology is applicable to a large class of nonlinear systems and demonstrated on a MIMO three-axis slewing spacecraft that employs PID tracking control with numerical simulation. These numerical results showed the stability boundaries (rigid-body modes and limit cycles) for each nonlinear system.

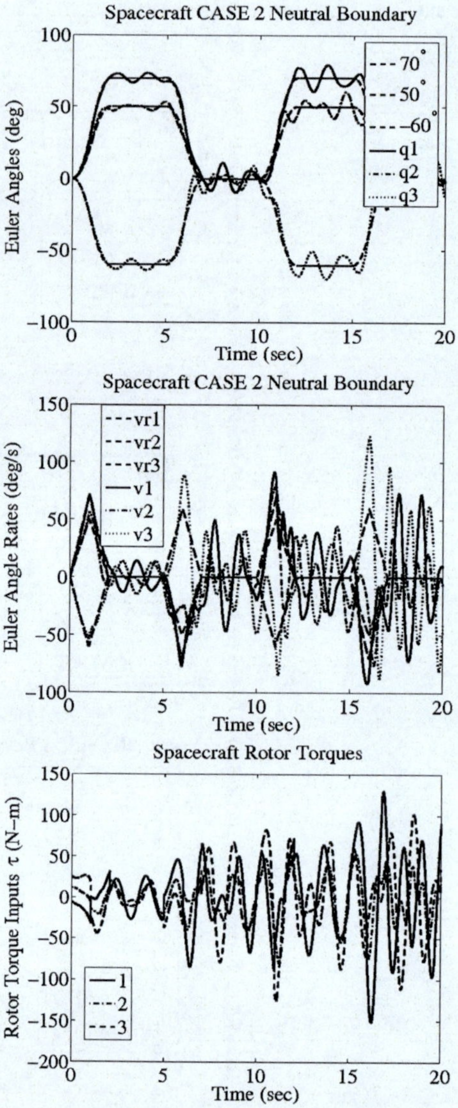

Fig. 12.5 Case 2 Euler angle, rate, and hub torque numerical results. Robinett III, R.D. and Wilson, D.G. [67], reprinted with the permission of the American Astronautical Society

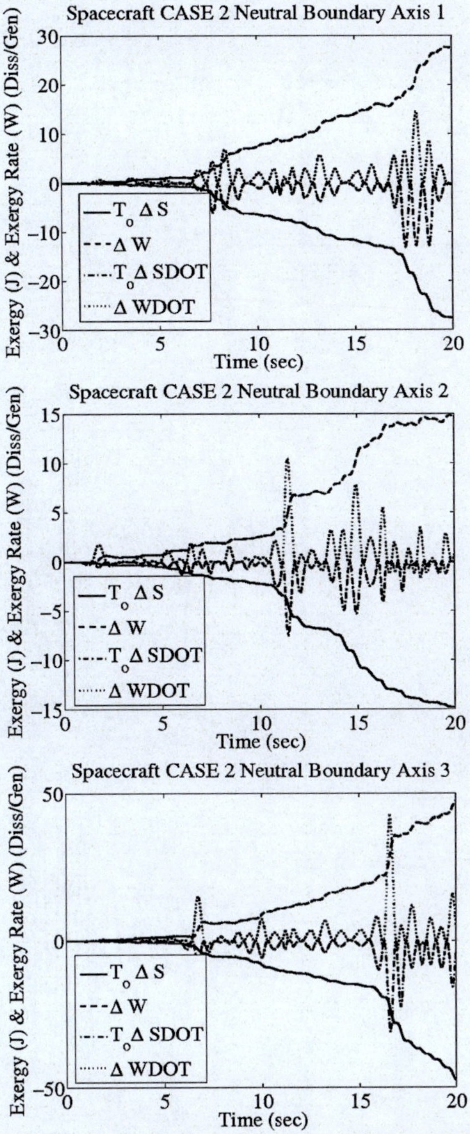

Fig. 12.6 Case 2 exergy and exergy rates for axes 1–3 numerical results. Robinett III, R.D. and Wilson, D.G. [67], reprinted with the permission of the American Astronautical Society

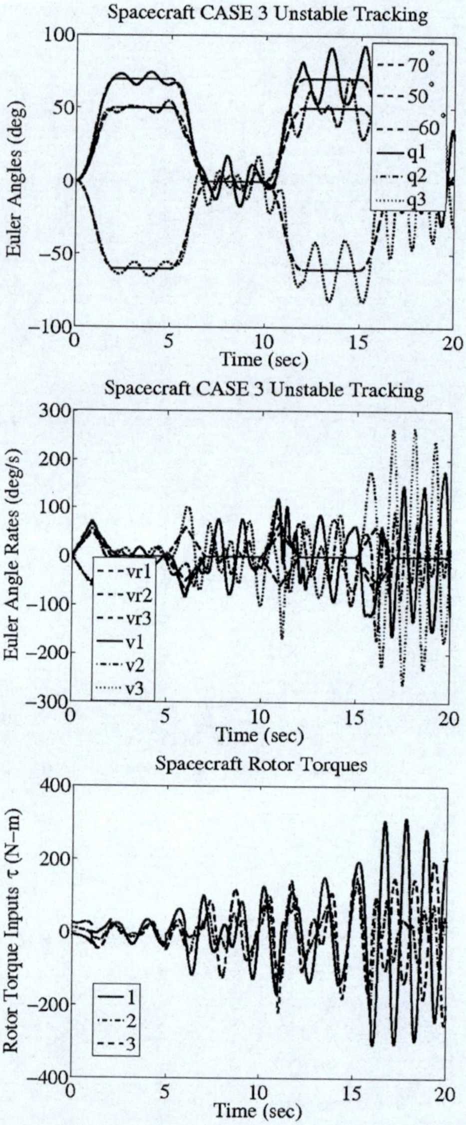

Fig. 12.7 Case 3 Euler angle, rate, and hub torque numerical results. Robinett III, R.D. and Wilson, D.G. [67], reprinted with the permission of the American Astronautical Society

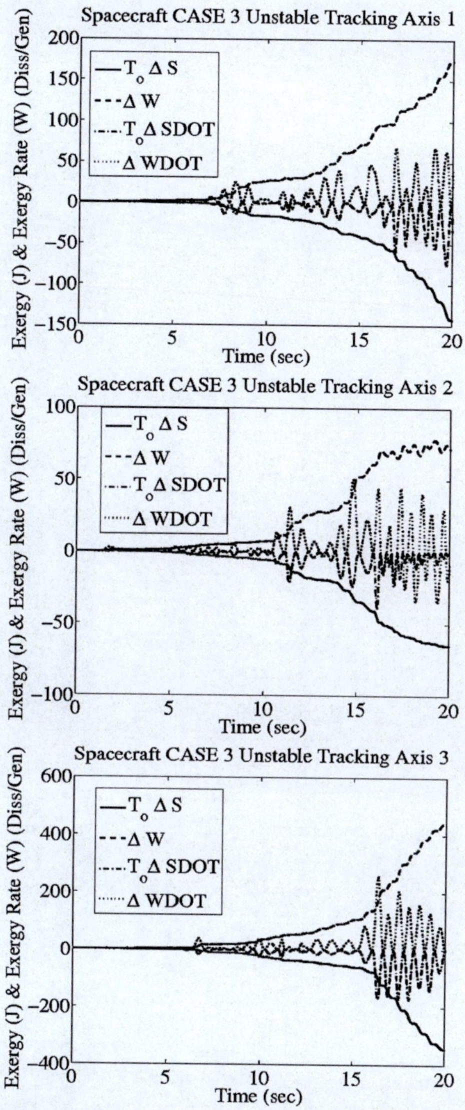

Fig. 12.8 Case 3 exergy and exergy rates for axes 1–3 numerical results. Robinett III, R.D. and Wilson, D.G. [67], reprinted with the permission of the American Astronautical Society

Chapter 13
Case Study #8: Wind Turbine Control Design

13.1 Introduction

Wind turbines are large complex dynamically flexible structures that must operate under very turbulent and unpredictable environmental conditions where efficiency and reliability are highly dependent upon a well-designed control strategy. Control strategies have been explored in the literature. Classical control strategies have been the mainstream for current wind turbine operations. More recently, modern control strategies based on linear time-invariant models have been investigated [140–143]. Robust control was studied in [144, 145], and nonlinear control has been proposed in [146–148]. The primary drawback is that these control techniques do not take into account the nonlinear effects associated with the wind turbine (with exception [148]). The objective of this chapter is to introduce an HSSPFC-based nonlinear/adaptive control methodology [13, 66] that takes into consideration the dynamics of the wind and wind turbine response.

Chapter 13 is divided into five sections. Section 13.2 defines the wind turbine dynamical model. Section 13.3 develops an adaptive control based on HSSPFC design. In Sect. 13.4 numerical simulation results are presented. Finally, in Sect. 13.5 the chapter is summarized with concluding remarks.

13.2 Wind Turbine Model

Typically, a variable-speed wind turbine aerodynamic power that is captured by the rotor is given [148] by

$$P_a = \frac{1}{2}\rho\pi R^2 C_P(\lambda, \beta)v^3,$$

where ρ is the air density, R is the rotor radius, C_P is the power coefficient that depends on the tip-speed ratio, λ, and the blade pitch angle, β, and v is the wind

R.D. Robinett III, D.G. Wilson, *Nonlinear Power Flow Control Design*, Understanding Complex Systems,
DOI 10.1007/978-0-85729-823-2_13,

speed. The tip-speed ratio is defined as

$$\lambda = \frac{\omega_r R}{v},$$

where ω_r is the rotor speed. The aerodynamic torque, T_a, can be identified from the power relationship

$$P_a = \omega_r T_a,$$

where the aerodynamic torque expression becomes

$$T_a = \frac{1}{2}\rho\pi R^3 C_T(\lambda, \beta)v^2.$$

The torque coefficient is related to the power coefficient by $C_T(\lambda, \beta) = C_P(\lambda, \beta)/\lambda$. The maximum theoretical power that can be extracted by the wind is defined by the Betz curve. Betz determined that the maximum power that can be extracted by the wind turbine without losses is $C_{P_{\max}} = 59\%$. The CART-2 (see Fig. 13.1, top, courtesy of NREL [143]) is a two-bladed, teetered, upwind, active-yaw wind turbine. The CART-2 is a machine used to test advanced control technology and is rated at 600-kW electrical power. For this study, the C_P versus λ for the NREL CART-2 machine is shown in Fig. 13.1 (bottom) with the corresponding characteristics given in Table 13.1. This power curve includes the losses.

The power coefficient $C_P(\lambda, \beta)$ has a unique maximum that corresponds to the maximum power production. In addition, for below rated power, the blade pitch angle, $\beta = \beta_{\text{opt}}$ such that the optimal value $\lambda = \lambda_{\text{opt}}$ will be tracked thus:

$$\omega_{r_{\text{ref}}} = \frac{\lambda_{\text{opt}}}{R} v.$$

In addition, the reference angular acceleration is determined as

$$\dot{\omega}_{r_{\text{ref}}} = \frac{\lambda_{\text{opt}}}{R} \dot{v},$$

where this development assumes that v and $\dot{v}$ are available and/or can be estimated. In this case a simple first-order estimator along with a derivative filter is used to generate these signals. The goal of the control system design is to maximize the wind power capture by adjusting the rotor speed ω_r to wind speed variation.

The wind turbine dynamics are given by the popular two-mass model [148] (see Fig. 13.2) as

$$J_r\dot{\omega}_r = T_a - T_{\text{ls}} - K_r\omega_r,$$

$$J_g\dot{\omega}_g = T_{\text{hs}} - K_g\omega_g - T_{\text{em}},$$

$$T_{\text{ls}} = K_{\text{ls}}(\omega_r - \omega_{\text{ls}}) + D_{\text{ls}}(\theta_r - \theta_{\text{ls}}),$$

where J_r is the rotor inertia, T_{ls} is the low-speed shaft torque, K_r is the rotor external damping, J_g is the generator inertia, T_{hs} is the high-speed shaft torque, K_g is the

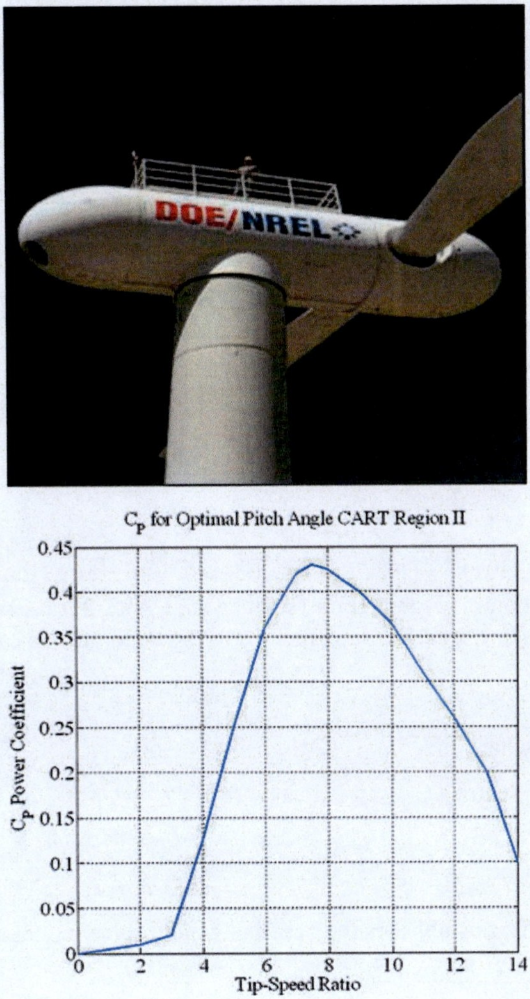

Fig. 13.1 NREL CART-2 Two-bladed wind turbine [143] (*top*) and Power curve for NREL CART-2 wind turbine (*bottom*). (Picture courtesy of NREL)

Table 13.1 NREL CART-2 [143] wind turbine characteristics

Rotor diameter	43.3 m
Gearbox ratio	43.165
Hub height	36.6 m
Generator electric power	600 kW
Maximum rotor torque	162 kN m
Optimal tip-speed ratio	7.7

generator external damping, T_{em} is the electromagnetic torque, ω_{ls} is the low-speed rotational speed, θ_r is the rotor angle, θ_{ls} is the low-speed angle, D_{ls} is the low-speed spring constant, and K_{ls} is the low-speed damping constant. The ideal gearbox ratio is defined as

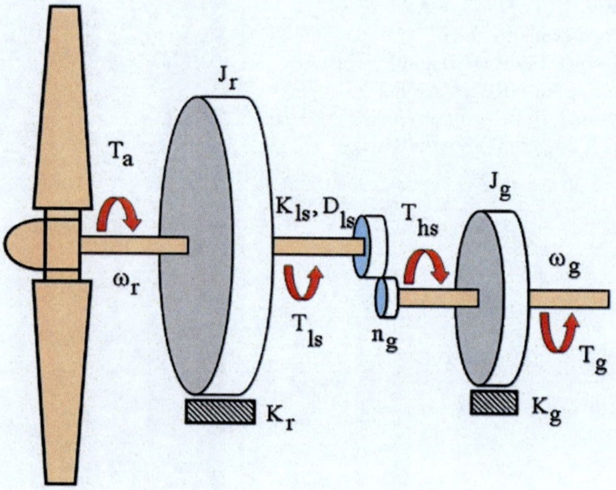

Fig. 13.2 Two mass wind turbine model

$$n_g = \frac{T_{ls}}{T_{hs}} = \frac{\omega_g}{\omega_{ls}}.$$

For a very stiff rotor (such as the CART-2) or assuming a perfectly rigid low-speed shaft, a single-mass model for the wind turbine can be defined as

$$J_T \dot{\omega}_r = T_a - K_T \omega_r - T_g, \tag{13.1}$$

where

$$J_T = J_r + n_g^2 J_g,$$
$$K_T = K_r + n_g^2 K_g,$$
$$T_g = n_g T_{em}.$$

This establishes the baseline wind turbine model used for the controller design.

13.3 Adaptive Power Flow Controller Design

The method known as the Indirect Speed Control (ISC) technique [149] is defined as

$$T_g = K_{opt} \omega_r^2 - K_T \omega_r,$$

where K_{opt} is calculated based on tracking along the optimal aerodynamic efficiency C_P curve. For this study, the ISC is considered the baseline controller that comparisons are made with the new adaptive power flow control design.

Following the nonlinear controller design methodology from references [13] and [66], the Hamiltonian is defined as

$$\mathcal{H} = \frac{1}{2} J_T (\omega_r - \omega_{r_{ref}})^2 + \frac{1}{2} \tilde{\Phi}^T \Gamma^{-1} \tilde{\Phi}. \tag{13.2}$$

From the wind turbine dynamic equation of motion (13.1) define

$$J_T\ddot{\theta}_r = M\dot{\theta}_r^2 - K_T\dot{\theta}_r - u,$$

where $\ddot{\theta}_r = \dot{\omega}_r$, $M\dot{\theta}_r^2 = T_a$, $\dot{\theta}_r = \omega_r$, and $u = T_g$. Next, define the control input as

$$u = u_{\text{ref}} + \Delta u,$$

where

$$u_{\text{ref}} = M\dot{\theta}_r^2 - \hat{K}_T\dot{\theta}_r - \hat{J}_T\ddot{\theta}_{\text{ref}}$$

with $\hat{K}_T$ and $\hat{J}_T$ representing the estimate of the parameter, and

$$\Delta u = K_D(\dot{\theta}_r - \dot{\theta}_{\text{ref}}),$$

where K_D is the derivative controller gain. The next step in the design is to take the derivative of the Hamiltonian (13.2) such that

$$\dot{\mathcal{H}} = \big[(\hat{J}_T - J_T)\ddot{\theta}_{\text{ref}} + (\hat{K}_T - K_T)\dot{\theta}_r - K_D(\dot{\theta}_r - \dot{\theta}_{\text{ref}})\big](\dot{\theta}_r - \dot{\theta}_{\text{ref}}) + \tilde{\Phi}^T\Gamma^{-1}\dot{\tilde{\Phi}}, \tag{13.3}$$

then recognize and identify

$$(\hat{J}_T - J_T)\ddot{\theta}_{\text{ref}} + (\hat{K}_T - K_T)\dot{\theta}_r = Y\tilde{\Phi},$$

and substituting into (13.3) and simplifying yields

$$\dot{\mathcal{H}} = -K_D(\dot{\theta}_r - \dot{\theta}_{\text{ref}})^2 + \tilde{\Phi}^T\big[Y^T(\dot{\theta}_r - \dot{\theta}_{\text{ref}}) + \Gamma^{-1}\dot{\tilde{\Phi}}\big], \tag{13.4}$$

where

$$\begin{aligned}
\tilde{\Phi}^T &= \big[(\hat{J}_T - J_T) \quad (\hat{K}_T - K_T)\big],\\
\dot{\tilde{\Phi}}^T &= \dot{\hat{\Phi}}^T = \big[\dot{\hat{J}}_T \quad \dot{\hat{K}}_T\big],\\
Y &= \big[\ddot{\theta}_{\text{ref}} \quad \dot{\theta}_r\big].
\end{aligned}$$

Next, set the last term in (13.4) to zero, and solving for the adaptive parameter update equations yields

$$\begin{aligned}
\dot{\hat{J}}_T &= -\gamma_1\ddot{\theta}_{\text{ref}}(\dot{\theta}_r - \dot{\theta}_{\text{ref}}),\\
\dot{\hat{K}}_T &= -\gamma_2\dot{\theta}_r(\dot{\theta}_r - \dot{\theta}_{\text{ref}},)
\end{aligned}$$

where γ_i, $i = 1, 2$, are the positive definite adaptation controller gains. Assuming that the parameters cancel, a passivity design then yields

$$\dot{\mathcal{H}} = -K_D(\dot{\theta}_r - \dot{\theta}_{\text{ref}})^2.$$

Note that for a formal passivity design [95], the Hamiltonian (13.2) would be selected as the positive definite Lyapunov function.

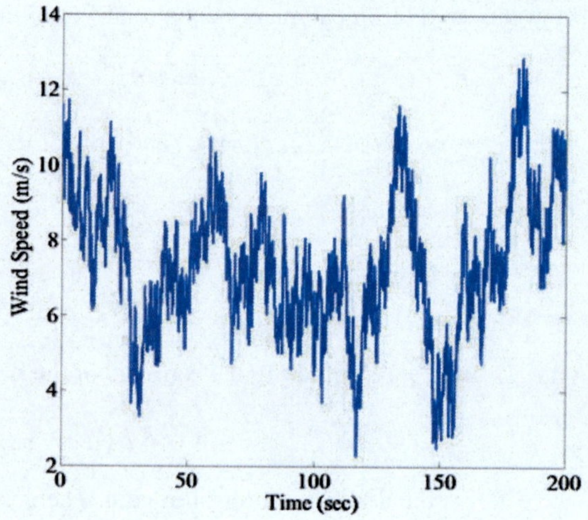

Fig. 13.3 Turbulent wind condition IEC NTM Type A 7 m/s

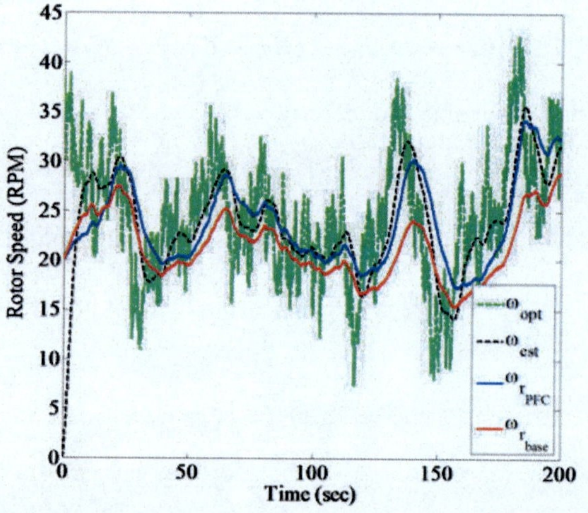

Fig. 13.4 Transient responses for rotor speed

13.4 Simple Model Simulation Results

As a numerical example, a 7-m/s turbulent wind condition [150] was used as the driving wind input as illustrated in Fig. 13.3. Both controller designs were run, and comparisons were made. In Fig. 13.4 the rotor speed responses are presented. The optimal rotor speed is exhibited in green, ω_{opt}, the estimated rotor speed in black, ω_{est}, the power flow control rotor speed is given in blue, ω_{PFC}, and the baseline ISC rotor speed is illustrated in red, ω_{base}. From the responses the power flow controller tracks the estimated rotor speed more closely than the baseline controller. Hence, in Fig. 13.5, where the electrical power generated is shown, the power flow control generates on the average 28.5% more power than the baseline control. A trade-off exists between energy captured and high-frequency loading on the rotor to find an optimal operating condition. Note that this will require a more detailed wind turbine

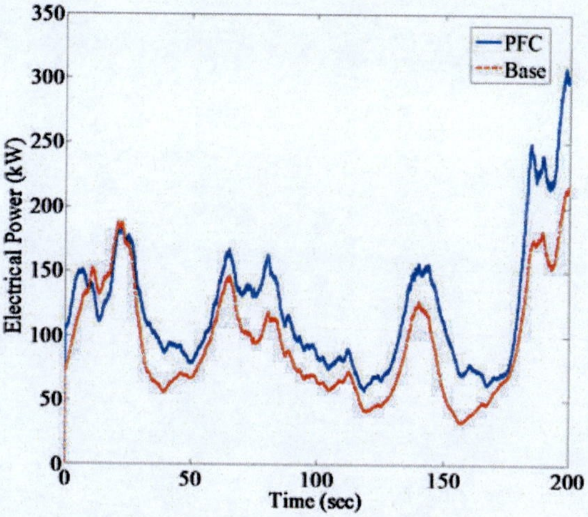

Fig. 13.5 Transient responses for generated electric power

aeroelastic simulator investigation [151] and verification of acceptable wind turbine loading conditions. This would also include the shaft torque load responses and other specific information associated with a detailed model. However, the initial results from this simple wind turbine model look promising.

13.5 Chapter Summary

In Chap. 13 an HSSPFC based nonlinear/adaptive power flow control design was investigated for a simplified CART-2 wind turbine system operating in Region II (below rated power). The nonlinear feedback algorithm included parameters that were allowed to accommodate robustness to variations in the dynamics. The new controller demonstrated an increase in energy capture in comparison with the conventional control scheme.

Chapter 14
Sustainability of Self-organizing Systems

14.1 Introduction

Exergy is the elixir of life. Exergy is that portion of energy available to do work. Elixir is defined as a substance held capable of prolonging life indefinitely, which implies sustainability of life. In terms of mathematics and engineering, exergy sustainability is defined as the continuous compensation of irreversible entropy production in an open system with an impedance and capacity-matched persistent exergy source. Irreversible and nonequilibrium thermodynamic concepts are combined with self-organizing systems theories as well as nonlinear control and stability analyses to explain this definition later in this chapter.

Exergy provides a missing link in the analysis of self-organizing systems: a tie between irreversible thermodynamics and Hamiltonian systems. As a result of this work, the concept of "on the edge of chaos" is formulated as a set of necessary and sufficient conditions for stability and performance of sustainable self-organizing systems. Also, the concepts of exergy sustainability and energy surety are combined to determine what is meant by optimality and scalability for self-organizing energy and power grids near the end of this chapter.

The Achilles heel or single point of failure of self-organizing systems is the requirement that exergy continuously flow into the system. The self-organizing system is continuously "shedding" irreversible entropy to the environment to keep itself organized and living as it consumes or dissipates the exergy flow.

Schrödinger (1945) [3] suggested that all organisms need to import "negative entropy" from their environment and export high entropy (for example, heat) into their environment in order to survive. This idea was developed into a general thermodynamic concept by Prigogine and his co-workers who coined the notion of "dissipative structures" (Prigogine, 1976 [9]; Prigogine and Stengers, 1984 [10]), structures of increasing complexity developed by open systems on the basis of energy exchanges with the environment. In the self-organization of dissipative structures, the environment serves both as a source of low-entropic energy and as a sink for the high-entropic energy which is necessarily produced [11].

Basically, self-organizing systems are attempting to balance and perform dialectic synthesis on evolving disordering and ordering pressures [152]. Said another

R.D. Robinett III, D.G. Wilson, *Nonlinear Power Flow Control Design*, Understanding Complex Systems,
DOI 10.1007/978-0-85729-823-2_14,

way, life is exergy dissipation (increasing entropy; disorder) and order production in an open system simultaneously. This process, which is the evolution of a complex adaptive system, is irreversible: the future is fundamentally different from the past, and it is impossible to reconstruct the past from the present [11].

Dissipation is the disordering power flow which is better known as consumption in economics and irreversible entropy production in thermodynamics. Exergy flow into a system is the ordering power flow that is better known as production in economics and exergy rate into an open system in thermodynamics. Balance between these competing power flows is key because these terms are relative to a goal and path through time which means that they can "flip over" or reverse roles. For example, the exergy flow into a system by a nuclear weapon is not "matched." It deposits exergy at a rate that destroys the system, which means that it is a disordering power flow increasing entropy. So, a mechanism must be inserted to "match" the input to the system if the goal is sustainability instead of destruction. Nuclear power is an attempt to match the exergy source to the exergy sink to move toward exergy sustainability.

The balance between these opposing power flows creates a sort of "equilibrium condition" for a self-organizing system. Ilya Prigogine described this as "far from thermodynamic equilibrium on the basis of energy dissipation" and, in cybernetics, it is often called an attractor [11, 153–156]. Most nonlinear self-organizing systems have several attractors, and the system moves between these attractors (reordering) due to variations (perturbations; noise; disorder) in the exergy flow and the system parameters. These system parameters are often called "control parameters" because their values determine the stability characteristics of the system. For example, the potential force field [157, 158] for a nonlinear spring system can be written in kinematic form (no dynamics) as

$$\dot{q} = kq + k_{\mathrm{NL}} q^3,$$

where k is the linear stiffness coefficient, and k_{NL} is the nonlinear stiffness coefficient. The potential function is defined as

$$\mathcal{V}(q) = \frac{1}{2} k q^2 + \frac{1}{4} k_{\mathrm{NL}} q^4. \tag{14.1}$$

This system changes its fundamental stability structure by changing $k > 0$ to $k < 0$ and $k_{\mathrm{NL}} > 0$. Figure 14.1 shows how the stable equilibrium state at $q = 0$ bifurcates into two symmetrical stable equilibrium states and becomes an unstable state.

These attractors are defined relative to a "fitness index." Some attractors are more likely to survive, more fit, than others. In the previous example, one attractor turned into two attractors which appear to be equally fit if the potential function is interpreted as the fitness surface. In fact, it is possible for the system to jump back and forth between these two attractors by varying the exergy flow and the control parameter through perturbations and noise (see Chap. 4).

As described earlier, nonlinear systems have several attractors and variations or "fluctuations" that reside between attractors in a system that will push the system to

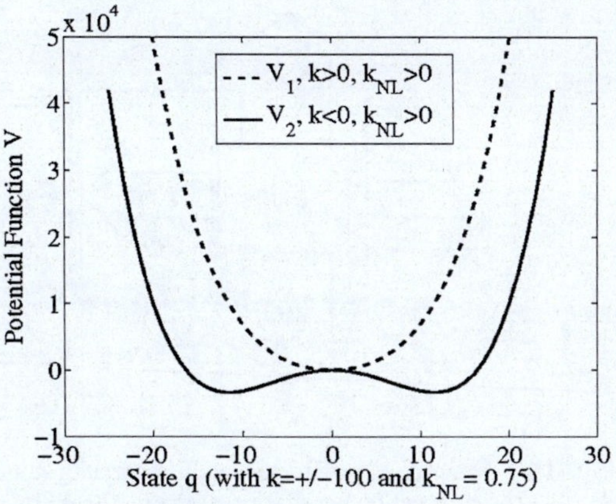

Fig. 14.1 Nonlinear spring potential function characteristics. Robinett III, R.D., Wilson, D.G., and Reed, A.W. [12], reprinted by permission of the publisher (New England Complex Systems Institute)

one or the other of the attractors. Positive feedback is necessary for random fluctuations to be amplified (generative) [159, 160]. Maintenance of the structured state in the presence of further fluctuations implies that some negative feedback is also present that dampens (dissipates) these effects [159, 160]. In Chaps. 2, 3, and 4, these are called a power generator and a power dissipator. Self-organization results from the interplay of positive and negative feedback [160]. In Chaps. 3 and 4, this is defined as the stability boundary and/or limit cycle. In more complex self-organizing systems, there will be several interlocking positive and negative feedback loops, so that changes in some directions are amplified while changes in other directions are suppressed [11, 159]. At the transition between order and disorder, a large number of bifurcations may be in existence which are analogous to the bifurcations of the previous potential function. Bifurcations may be arranged in a "cascade" where each branch of the fork itself bifurcates further and further, characteristic of the onset of the chaotic regime [4, 11, 159]. The system's behavior on this edge is typically governed by a "power law" where large adjustments are possible but are much less probable than small adjustments [11]. These concepts enable us to better understand nonlinear systems, also known as *complex adaptive systems*, that are on the "edge of chaos" or those systems that are in a domain between frozen constancy (equilibrium) and turbulent, chaotic activity [11]. The mechanism by which complex systems tend to maintain on this critical edge has also been described as *self-organized criticality* [11, 161].

This concept of "on the critical edge" can be described as "Yin and Yang control": the dialectic synthesis of opposing power flows. Yin and Yang theory [162] is a logic that is described as synthetic or dialectical: a part of a system can be understood only in its relation to the whole. There are five principles of Yin and Yang [162]:

1. All things have two aspects: a Yin aspect (decrease) and a Yang aspect (increase).
2. Any Yin and Yang aspect can be further divided into Yin and Yang.
3. Yin and Yang mutually create each other.
4. Yin and Yang control each other.

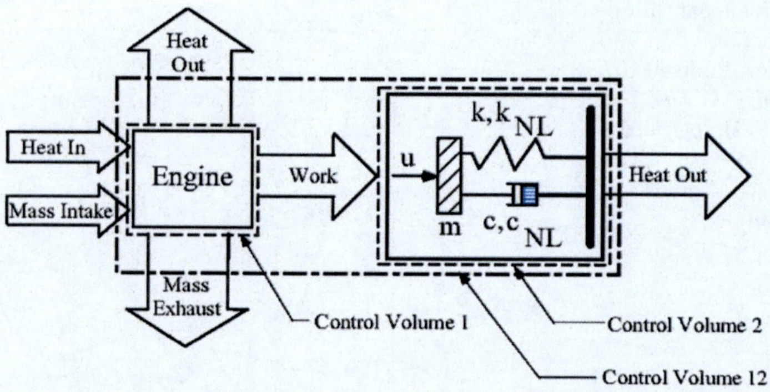

Fig. 14.2 Simplified nonlinear satellite model. Robinett III, R.D., Wilson, D.G., and Reed, A.W. [12], reprinted by permission of the publisher (New England Complex Systems Institute)

5. Yin and Yang transform into each other.

To specifically address these five characteristics with respect to the present concepts, the two opposing aspects are exergy generation and dissipation. The further division is the control volume analysis at any scale. The mutual creation is that the definition of generation is relative to dissipation. The control of each other is integral to the stability analysis. The transformation of one into the other was described in the nuclear weapon example.

Chapter 14 is divided into five sections. Section 14.1 is the introduction which follows the basic format from Heylighen [11] with support from Haken [157] and Buenstorf [159]. Section 14.2 utilizes self-organizing systems concepts to analyze the sustainability of a simplified nonlinear system model which represents a satellite in space, for example, the Earth. Section 14.3 provides the definition of a lifestyle with respect to a self-organizing system. Section 14.4 defines exergy sustainability within the context of energy surety. Section 14.5 gives the summary and concluding remarks of the chapter.

14.2 Simple Nonlinear Satellite System

With this brief background, it is time to analyze a simplified nonlinear satellite system (see Fig. 14.2) to develop the definition of exergy sustainability. For purposes of clarity, each control volume is subdivided into two subregions that contain the physical components. The component mass is constant. A single constant temperature characterizes each component. The component subregion is surrounded by an outer zone that characterizes the interaction between the component (at temperature T) and the environment (reservoir) characterized by temperature T_0.

14.2.1 Conservation Equations for the Engine Component (Control Volume 1)

By performing a control volume analysis [43] the following energy, entropy, and exergy equations result in:

$$\begin{aligned}
\dot{\mathcal{E}}1_{\text{component}} &= [\dot{\mathcal{Q}}_{\text{in}_1} - \dot{\mathcal{Q}}_{\text{out}_1}] + \dot{m}_1[h_{\text{int}}(T_{\text{int}}, P_{\text{int}}) - h_{\text{exh}}(T_{\text{exh}}, P_{\text{exh}})] - \dot{W},\\
\dot{\mathcal{S}}1_{\text{component}} &= \left[\frac{\dot{\mathcal{Q}}_{\text{in}_1} - \dot{\mathcal{Q}}_{\text{out}_1}}{T_1}\right] + \dot{m}_1[s_{\text{int}}(T_{\text{int}}, P_{\text{int}}) - s_{\text{exh}}(T_{\text{exh}}, P_{\text{exh}})]\\
&\quad + \dot{\mathcal{S}}_{\text{irr}1_{\text{component}}},\\
\dot{\Xi}1_{\text{component}} &= \left[\dot{\Xi}_{\text{in}_1} - \left(1 - \frac{T_0}{T_1}\right)\dot{\mathcal{Q}}_{\text{out}_1}\right] + \dot{m}_1[\zeta_{\text{int}}(T_{\text{int}}, P_{\text{int}})\\
&\quad - \zeta_{\text{exh}}(T_{\text{exh}}, P_{\text{exh}})] - \dot{W} - T_0\dot{\mathcal{S}}_{\text{irr}1_{\text{component}}}.
\end{aligned} \tag{14.2}$$

Note that the subscript "exh" (meaning exhaust) implies that the exiting quantities are associated with mass leaving the control volume. Similarly, the subscript "int" (for intake) implies the entering quantities are associated with the mass entering the control volume. The mass of the control volume does not change with time. Therefore, the mass flow rate exiting the control volume is equal to the mass flow rate entering the volume.

For the surface heat interaction(s), the energy, entropy, and exergy equations are:

$$\begin{aligned}
\dot{\mathcal{E}}1_{\text{interact_}\mathcal{Q}} &= 0 = [\dot{\mathcal{Q}}_{\text{out}_1} - \dot{\mathcal{Q}}_{\text{out}_1}],\\
\dot{\mathcal{S}}1_{\text{interact_}\mathcal{Q}} &= 0 = \dot{\mathcal{Q}}_{\text{out}_1}\left[\frac{1}{T_1} - \frac{1}{T_0}\right] + \dot{\mathcal{S}}_{\text{irr}1_{\text{interact_}\mathcal{Q}}},\\
\dot{\Xi}1_{\text{interact_}\mathcal{Q}} &= 0 = \left[\left(1 - \frac{T_0}{T_1}\right)\dot{\mathcal{Q}}_{\text{out}_1} - \left(1 - \frac{T_0}{T_0}\right)\dot{\mathcal{Q}}_{\text{out}_1}\right] - T_0\dot{\mathcal{S}}_{\text{irr}1_{\text{interact_}\mathcal{Q}}}\\
&= \left(1 - \frac{T_0}{T_1}\right)\dot{\mathcal{Q}}_{\text{out}_1} - T_0\dot{\mathcal{S}}_{\text{irr}1_{\text{interact_}\mathcal{Q}}}.
\end{aligned} \tag{14.3}$$

For the exhaust stream expansion and interaction(s), the thermodynamics of component mixing will be ignored, and the focus will be solely upon the final temperature of the exhaust gases. The thermodynamics of the mixing of the exhaust gases with the environment will be ignored once the gases cool to the ambient temperature. This gives:

$$\begin{aligned}
\dot{\mathcal{E}}1_{\text{interac_}m} &= 0 = \dot{m}_1[h_{\text{exh}}(T_{\text{exh}}, P_{\text{exh}}) - h_{\text{exh}}(T_0, P_0)] - \dot{\mathcal{Q}}1_{\text{interact_}m},\\
\dot{\mathcal{S}}1_{\text{interact_}m} &= 0 = \dot{m}_1[s_{\text{exh}}(T_{\text{exh}}, P_{\text{exh}}) - s_{\text{exh}}(T_0, P_0)] + \dot{\mathcal{S}}_{\text{irr}1_{\text{interact_}m}} - \frac{\dot{\mathcal{Q}}1_{\text{interact_}m}}{T_0},\\
\dot{\Xi}1_{\text{interact_}m} &= 0 = \dot{m}_1[\zeta_{\text{exh}}(T_{\text{exh}}, P_{\text{exh}}) - \zeta_{\text{exh}}(T_0, P_0)] - T_0\dot{\mathcal{S}}_{\text{irr}1_{\text{interact_}m}}.
\end{aligned} \tag{14.4}$$

Adding the interaction fluxes (14.3) and (14.4) to the component fluxes (14.2) produces the complete equations for Control Volume 1 as

$$\dot{\mathcal{E}}1_{\text{total}} = [\dot{\mathcal{Q}}_{\text{in}_1} - \dot{\mathcal{Q}}_{\text{out}_1}] + \dot{m}_1[h_{\text{int}}(T_{\text{int}}, P_{\text{int}}) - h_{\text{ext}}(T_{\text{exh}}, P_{\text{exh}})] - \dot{W} + \dot{m}_1[e_{\text{exh}}(T_{\text{exh}}, P_{\text{exh}}) - e_{\text{exh}}(T_0, P_0)] - \dot{\mathcal{Q}}1_{\text{interact\_}m},$$

$$\dot{\mathcal{S}}1_{\text{total}} = \left[\frac{\dot{\mathcal{Q}}_{\text{in}_1} - \dot{\mathcal{Q}}_{\text{out}_1}}{T_1}\right] + \dot{m}_1[s_{\text{int}}(T_{\text{int}}, P_{\text{int}}) - s_{\text{exh}}(T_{\text{exh}}, P_{\text{exh}})] + \dot{\mathcal{S}}_{\text{irr}1_{\text{component}}} + \dot{\mathcal{Q}}_{\text{out}_1}\left[\frac{1}{T_1} - \frac{1}{T_0}\right] + \dot{\mathcal{S}}_{\text{irr}1_{\text{interact\_}\mathcal{Q}}} + \dot{m}_1[s_{\text{exh}}(T_{\text{ext}}, P_{\text{exh}}) - s_{\text{exh}}(T_0, P_0)] + \dot{\mathcal{S}}_{\text{irr}1_{\text{interact\_}m}},$$

$$\dot{\Xi}1_{\text{total}} = \left[\dot{\Xi}_{\text{in}_1} - \left(1 - \frac{T_0}{T_1}\right)\dot{\mathcal{Q}}_{\text{out}_1}\right] + \dot{m}_1[\zeta_{\text{int}}(T_{\text{int}}, P_{\text{int}}) - \zeta_{\text{exh}}(T_{\text{exh}}, P_{\text{exh}})] - \dot{W} - T_0\dot{\mathcal{S}}_{\text{irr}1_{\text{component}}} + \left(1 - \frac{T_0}{T_1}\right)\dot{\mathcal{Q}}_{\text{out}_1} - T_0\dot{\mathcal{S}}_{\text{irr}1_{\text{interact\_}\mathcal{Q}}} + \dot{m}_1[\zeta_{\text{exh}}(T_{\text{exh}}, P_{\text{exh}}) - \zeta_{\text{exh}}(T_0, P_0)] - T_0\dot{\mathcal{S}}_{\text{irr}1_{\text{interact\_}m}}.$$

Next, simplifying yields:

$$\dot{\mathcal{E}}1_{\text{total}} = [\dot{\mathcal{Q}}_{\text{in}_1} - \dot{\mathcal{Q}}_{\text{out}_1} - \dot{Q}1_{\text{interact\_}m}] + \dot{m}_1[h_{\text{int}}(T_{\text{int}}, P_{\text{int}}) - h_{\text{exh}}(T_0, P_0)] - \dot{W}, \tag{14.5}$$

$$\dot{\mathcal{S}}1_{\text{total}} = \left[\frac{\dot{\mathcal{Q}}_{\text{in}_1}}{T_1} - \frac{\dot{\mathcal{Q}}_{\text{out}_1}}{T_0}\right] + \dot{m}_1[s_{\text{int}}(T_{\text{int}}, P_{\text{int}}) - s_{\text{exh}}(T_0, P_0)] + [\dot{\mathcal{S}}_{\text{irr}1_{\text{component}}} + \dot{\mathcal{S}}_{\text{irr}1_{\text{interact\_}\mathcal{Q}}} + \dot{\mathcal{S}}_{\text{irr}1_{\text{interact\_}m}}], \tag{14.6}$$

$$\dot{\Xi}1_{\text{total}} = \dot{\Xi}_{\text{in}_1} + \dot{m}_1[\zeta_{\text{int}}(T_{\text{int}}, P_{\text{int}}) - \zeta_{\text{exh}}(T_0, P_0)] - \dot{W} - T_0[\dot{\mathcal{S}}_{\text{irr}1_{\text{component}}} + \dot{\mathcal{S}}_{\text{irr}1_{\text{interact\_}\mathcal{Q}}} + \dot{\mathcal{S}}_{\text{irr}1_{\text{interact\_}m}}], \tag{14.7}$$

where

$$\dot{\mathcal{Q}}1_{\text{interact\_}m} = \dot{m}_1[e_{\text{exh}}(T_{\text{exh}}, P_{\text{exh}}) - e_{\text{exh}}(T_0, P_0)],$$

$$\dot{\mathcal{S}}_{\text{irr}1_{\text{interact\_}\mathcal{Q}}} = \dot{\mathcal{Q}}_{\text{out}_1}\left[\frac{1}{T_0} - \frac{1}{T_1}\right],$$

$$\dot{\mathcal{S}}_{\text{irr}1_{\text{interact\_}m}} = \frac{\dot{\mathcal{Q}}1_{\text{interact\_}m}}{T_0} - \dot{m}_1[s_{\text{exh}}(T_{\text{exh}}, P_{\text{exh}}) - s_{\text{exh}}(T_0, P_0)].$$

In the final energy equation, (14.5) for Control Volume 1, the heat loss from the "engine" ($\mathcal{Q}_{\text{out}_1}$) is differentiated from the heat loss from the exhaust stream ($\mathcal{Q}1_{\text{interact\_}m}$).

In the final entropy equation, (14.6), the total entropy generation is the sum of the entropy generation that occurs within the "engine" ($\mathcal{S}_{\mathrm{irr}_{1_{\mathrm{component}}}}$), that occurs due to the heat interactions between the engine and the environment ($\mathcal{S}_{\mathrm{irr}_{1_{\mathrm{interact}\_m}}}$), and that occurs due to the exhaust stream coming into equilibrium with the environment ($\mathcal{S}_{\mathrm{irr}_{1_{\mathrm{interact}\_Q}}}$).

Now look at the final exergy equation, (14.7). The first term is the incoming exergy stream via a heat interaction ($\dot{\Xi}_{\mathrm{in}_1}$). An example of this is the exergy flux of incident solar radiation on a solar collector. The second term is the product of the mass flux and the difference between the specific exergies of the exhaust and intake streams. This might be the intake and exhaust streams of a gas turbine. The third term is the work produced by the engine. The fourth term is the exergy waste due to the inefficiencies of the engine and its interaction with the environment.

The maximum amount of work [43] obtainable from an exergy supply (first two terms) occurs when the irreversible entropy generation is reduced to zero. This limit is an idealization that can never be realized in the "real" world. However, improvements in efficiency are attainable by reducing the irreversibilities. The itemization of the sources of irreversibility shows three paths for this reduction.

1. The thermodynamic efficiency of the "engine" can be improved. For example, increasing the maximum temperature that turbine blades can tolerate will result in a more efficient turbine cycle.
2. The heat interaction with the environment can be reduced. This is why high-temperature refrigerators are insulated.
3. The exergy of the exhaust stream can be reduced. A bottoming cycle can be added to gas turbines which produces work from the waste stream.

There are many technologies already in existence that can be used to increase the production of work from existing "engine" technologies. However, many of those technologies are not (currently) economically competitive.

14.2.2 *Conservation Equations for the Machine (Control Volume 2)*

Control Volume 2 consists of a nonlinear mass/spring/damper system (with Duffing oscillator/Coulomb friction contributing to the nonlinear effects). Work is supplied to the system which results in the acceleration of the mass. The work is dissipated by the damper. This increases the temperature of the components of the system, which is characterized by a single temperature, T_2. The thermal energy is then transferred to the environment which has a temperature of T_0. The transfer is realized solely by a heat interaction. No mass enters or leaves Control Volume 2. Performing the energy, entropy, and exergy analyses [43] yields:

$$\dot{\mathcal{E}}2_{\text{component}} = \dot{W} - \dot{\mathcal{Q}}2_{\text{out}},$$

$$\dot{\mathcal{S}}2_{\text{component}} = \dot{\mathcal{S}}_{\text{irr}_2} - \frac{\dot{\mathcal{Q}}_{\text{out}2}}{T_2},$$

$$\dot{\Xi}2_{\text{component}} = \dot{W} - T_0\dot{\mathcal{S}}_{\text{irr}_2} - \left(1 - \frac{T_0}{T_2}\right)\dot{\mathcal{Q}}_{\text{out}_2}.$$

For the surface heat interaction(s) the energy, entropy, and exergy equations are:

$$\dot{\mathcal{E}}2_{\text{interact\_}\mathcal{Q}} = 0 = \left[\dot{\mathcal{Q}}_{\text{out}_2} - \dot{\mathcal{Q}}_{\text{out}_2}\right],$$

$$\dot{\mathcal{S}}2_{\text{interact\_}\mathcal{Q}} = 0 = \dot{\mathcal{Q}}_{\text{out}_2}\left[\frac{1}{T_2} - \frac{1}{T_0}\right] + \dot{\mathcal{S}}_{\text{irr}2_{\text{interact\_}\mathcal{Q}}},$$

$$\begin{aligned}\dot{\Xi}2_{\text{interact\_}\mathcal{Q}} = 0 &= \left[\left(1 - \frac{T_0}{T_2}\right)\dot{\mathcal{Q}}_{\text{out}_2} - \left(1 - \frac{T_0}{T_0}\right)\dot{\mathcal{Q}}_{\text{out}_1}\right] - T_0\dot{\mathcal{S}}_{\text{irr}2_{\text{interact\_}\mathcal{Q}}} \\ &= \left(1 - \frac{T_0}{T_2}\right)\dot{\mathcal{Q}}_{\text{out}_2} - T_0\dot{\mathcal{S}}_{\text{irr}2_{\text{interact\_}\mathcal{Q}}}.\end{aligned}$$

Adding the interaction fluxes to the component fluxes produces the complete equations for Control Volume 2 as:

$$\dot{\mathcal{E}}2_{\text{total}} = \dot{W} - \dot{\mathcal{Q}}_{\text{out}_2},$$

$$\dot{\mathcal{S}}2_{\text{total}} = \dot{\mathcal{S}}_{\text{irr}_2} - \frac{\dot{\mathcal{Q}}_{\text{out}_2}}{T_2} + \dot{\mathcal{Q}}_{\text{out}_2}\left[\frac{1}{T_2} - \frac{1}{T_0}\right] + \dot{\mathcal{S}}_{\text{irr}2_{\text{interact\_}\mathcal{Q}}},$$

$$\dot{\Xi}2_{\text{total}} = \dot{W} - T_0\dot{\mathcal{S}}_{\text{irr}_2} - \left(1 - \frac{T_0}{T_2}\right)\dot{\mathcal{Q}}_{\text{out}_2} + \left(1 - \frac{T_0}{T_2}\right)\dot{\mathcal{Q}}_{\text{out}_2} - T_0\dot{\mathcal{S}}_{\text{irr}2_{\text{interact\_}\mathcal{Q}}},$$

and simplifying yields:

$$\dot{\mathcal{E}}2_{\text{total}} = \dot{W} - \dot{\mathcal{Q}}_{\text{out}_2},$$

$$\dot{\mathcal{S}}2_{\text{total}} = -\frac{\dot{\mathcal{Q}}_{\text{out}_2}}{T_0} + \left[\dot{\mathcal{S}}_{\text{irr}_2} + \dot{\mathcal{S}}_{\text{irr}2_{\text{interact\_}\mathcal{Q}}}\right],$$

$$\dot{\Xi}2_{\text{total}} = \dot{W} - T_0\left[\dot{\mathcal{S}}_{\text{irr}_2} + \dot{\mathcal{S}}_{\text{irr}2_{\text{interact\_}\mathcal{Q}}}\right],$$

where

$$\dot{\mathcal{S}}_{\text{irr}2_{\text{interact\_}\mathcal{Q}}} = \dot{\mathcal{Q}}_{\text{out}_2}\left[\frac{1}{T_0} - \frac{1}{T_2}\right].$$

14.2.3 Conservation Equations for the Total System (Control Volume 12)

Now the equations for Control Volumes 1 and 2 are added to give:

$$\begin{aligned}\dot{\mathcal{E}}_{\text{total}} &= \big[\dot{\mathcal{Q}}_{\text{in}_1} - \dot{\mathcal{Q}}_{\text{out}_1} - \dot{\mathcal{Q}}_{1_{\text{interact\_}m}}\big] + \dot{m}_1[h_{\text{int}}(T_{\text{int}}, P_{\text{int}}) - h_{\text{exh}}(T_0, P_0)] - \dot{W}\\ &\quad + \dot{W} - \dot{\mathcal{Q}}_{\text{out}_2},\\ \dot{\mathcal{S}}_{\text{total}} &= \left[\frac{\dot{\mathcal{Q}}_{\text{in}_1}}{T_1} - \frac{\dot{\mathcal{Q}}_{\text{out}_1}}{T_0}\right] + \dot{m}_1[s_{\text{int}}(T_{\text{int}}, P_{\text{int}}) - s_{\text{exh}}(T_0, P_0)] - \frac{\dot{\mathcal{Q}}_{\text{out}_2}}{T_0}\\ &\quad + \big[\dot{\mathcal{S}}_{\text{irr}_{1_{\text{component}}}} + \dot{\mathcal{S}}_{\text{irr}_{1_{\text{interact\_}\mathcal{Q}}}} + \dot{\mathcal{S}}_{\text{irr}_{1_{\text{interact\_}m}}}\big] + \big[\dot{\mathcal{S}}_{\text{irr}_2} + \dot{\mathcal{S}}_{\text{irr}_{2_{\text{interact\_}\mathcal{Q}}}}\big],\\ \dot{\Xi}_{\text{total}} &= \dot{\Xi}_{\text{in}_1} + \dot{m}_1[\zeta_{\text{int}}(T_{\text{int}}, P_{\text{int}}) - \zeta_{\text{exh}}(T_0, P_0)] - \dot{W}\\ &\quad - T_0\big[\dot{\mathcal{S}}_{\text{irr}_{1_{\text{component}}}} + \dot{\mathcal{S}}_{\text{irr}_{1_{\text{interact\_}\mathcal{Q}}}} + \dot{\mathcal{S}}_{\text{irr}_{1_{\text{interact\_}m}}}\big] + \dot{W}\\ &\quad - T_0\big[\dot{\mathcal{S}}_{\text{irr}_2} + \dot{\mathcal{S}}_{\text{irr}_{2_{\text{interact\_}\mathcal{Q}}}}\big],\end{aligned}$$

and simplifying yields:

$$\begin{aligned}\dot{\mathcal{E}}_{\text{total}} &= \big[\dot{\mathcal{Q}}_{\text{in}_1} - \dot{\mathcal{Q}}_{\text{out}_1} - \dot{\mathcal{Q}}_{1_{\text{interact\_}m}} - \dot{\mathcal{Q}}_{\text{out}_2}\big] + \dot{m}_1[h_{\text{int}}(T_{\text{int}}, P_{\text{int}}) - h_{\text{exh}}(T_0, P_0)],\\ \dot{\mathcal{S}}_{\text{total}} &= \left[\frac{\dot{\mathcal{Q}}_{\text{in}_1}}{T_1} - \frac{\dot{\mathcal{Q}}_{\text{out}_1} + \dot{\mathcal{Q}}_{\text{out}_2}}{T_0}\right] + \dot{m}_1[s_{\text{int}}(T_{\text{int}}, P_{\text{int}}) - s_{\text{exh}}(T_0, P_0)]\\ &\quad + \big[\dot{\mathcal{S}}_{\text{irr}_{1_{\text{component}}}} + \dot{\mathcal{S}}_{\text{irr}_{1_{\text{interact\_}\mathcal{Q}}}} + \dot{\mathcal{S}}_{\text{irr}_{1_{\text{interact\_}m}}} + \dot{\mathcal{S}}_{\text{irr}_2} + \dot{\mathcal{S}}_{\text{irr}_{2_{\text{interact\_}\mathcal{Q}}}}\big],\\ \dot{\Xi}_{\text{total}} &= \dot{\Xi}_{\text{in}_1} + \dot{m}_1[\zeta_{\text{int}}(T_{\text{int}}, P_{\text{int}}) - \zeta_{\text{exh}}(T_0, P_0)]\\ &\quad - T_0\big[\dot{\mathcal{S}}_{\text{irr}_{1_{\text{component}}}} + \dot{\mathcal{S}}_{\text{irr}_{1_{\text{interact\_}\mathcal{Q}}}} + \dot{\mathcal{S}}_{\text{irr}_{1_{\text{interact\_}m}}} + \dot{\mathcal{S}}_{\text{irr}_2} + \dot{\mathcal{S}}_{\text{irr}_{2_{\text{interact\_}\mathcal{Q}}}}\big],\end{aligned}$$

where

$$\begin{aligned}\dot{\mathcal{Q}}_{1_{\text{interact\_}m}} &= \dot{m}_1[e_{\text{exh}}(T_{\text{exh}}, P_{\text{exh}}) - e_{\text{exh}}(T_0, P_0)],\\ \dot{\mathcal{S}}_{\text{irr}_{1_{\text{interact\_}\mathcal{Q}}}} &= \dot{\mathcal{Q}}_{\text{out}_1}\left[\frac{1}{T_0} - \frac{1}{T_1}\right],\\ \dot{\mathcal{S}}_{\text{irr}_{1_{\text{interact\_}m}}} &= \frac{\dot{\mathcal{Q}}_{1_{\text{interact\_}m}}}{T_0} - \dot{m}_1[s_{\text{exh}}(T_{\text{exh}}, P_{\text{exh}}) - s_{\text{exh}}(T_0, P_0)],\\ \dot{\mathcal{S}}_{\text{irr}_{2_{\text{interact\_}\mathcal{Q}}}} &= \dot{\mathcal{Q}}_{\text{out}_2}\left[\frac{1}{T_0} - \frac{1}{T_2}\right].\end{aligned}$$

By following the derivations of the previous sections, this simplified nonlinear model reduces within Control Volume 2 to

$$\mathcal{H} = \frac{1}{2}m\dot{x}^2 + \frac{1}{2}kx^2 + \frac{1}{4}k_{\text{NL}}x^4,$$

$$
\begin{aligned}
\dot{\mathcal{H}} &= \left[m\ddot{x} + kx + k_{\mathrm{NL}}x^3\right]\dot{x} \\
&= \dot{W} - T_0\dot{\mathcal{S}}_i \\
&= u\dot{x} - C\dot{x}^2 - C_{\mathrm{NL}}\,\mathrm{sgn}(\dot{x})\dot{x} \\
&= \dot{x}\left[-K_P x - K_I\int_0^t x\,d\tau - K_D\dot{x} - C\dot{x} - C_{\mathrm{NL}}\,\mathrm{sgn}(\dot{x})\right],
\end{aligned}
$$

where

$$
\begin{aligned}
u &= \text{PID feedback controller (Implemented force input)} \\
&= -K_P x - K_I\int_0^t x\,d\tau - K_D\dot{x}.
\end{aligned}
$$

Then the following exergy terms are identified as

$$
\begin{aligned}
\dot{W} &= \left[-K_I\int_0^t x\,d\tau\right]\dot{x}, \\
T_0\dot{\mathcal{S}}_i &= \left[-K_D\dot{x} - C\dot{x} - C_{\mathrm{NL}}\,\mathrm{sgn}(\dot{x})\right]\dot{x}, \\
T_0\dot{\mathcal{S}}_{\mathrm{rev}} &= \left[m\ddot{x} + kx + K_P x + k_{\mathrm{NL}}x^3\right]\dot{x}.
\end{aligned}
$$

The generalized stability boundary is given as a balance between "positive and negative feedback" (exergy generation and exergy dissipation)

$$
\begin{aligned}
&[\dot{W}]_{\mathrm{ave}} = [T_0\dot{\mathcal{S}}_i]_{\mathrm{ave}}, \\
&\left[-K_I\int_0^t x\,d\tau\cdot\dot{x}\right]_{\mathrm{ave}} = \left[(K_D + C)\dot{x}^2 + C_{\mathrm{NL}}\,\mathrm{sgn}(\dot{x})\cdot\dot{x}\right]_{\mathrm{ave}}.
\end{aligned}
\tag{14.8}
$$

The "shape" of the resulting limit cycle is constrained to the Hamiltonian surface which determines the accessible bifurcated structure for $k < 0$ or the paraboloid for $k > 0$ as a function of exergy level (see Figs. 14.1 and 14.3).

Notice that the trajectories are constrained to move along the Hamiltonian surface. By analogy, the Hamiltonian surface enables a "lifestyle" defined by "population" (mass), "investment/infrastructure" (stiffness), "production" (exergy input), and "consumption" (irreversible entropy production). The production and consumption are analogous to supply and demand that are enabled by an infrastructure which supports a population. These concepts will be used in the next section to discuss the sustainability of a lifestyle.

14.3 Lifestyle Definition

Next, this model is interpreted in terms of a "lifestyle" of the mass–spring–damper system within the satellite. First, the lifestyle is defined by a cyclic path, attractor, or

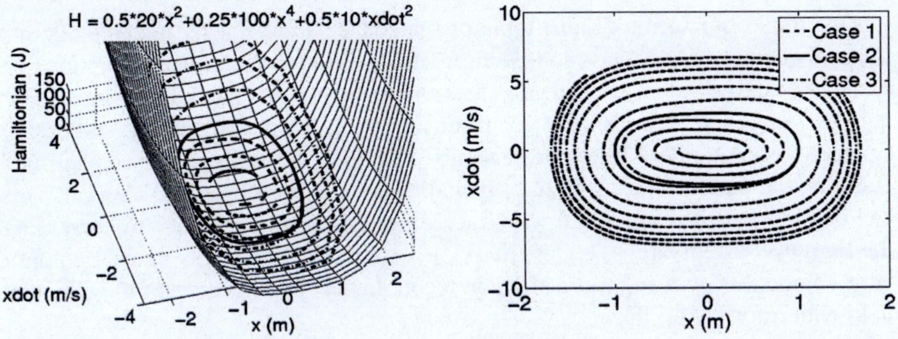

Fig. 14.3 All cases: mass–spring–damper with Duffing oscillator/Coulomb friction model numerical results: Hamiltonian 3D surface (*left*) and phase plane 2D projection (*right*). Robinett III, R.D., Wilson, D.G., and Reed, A.W. [12], reprinted by permission of the publisher (New England Complex Systems Institute)

limit cycle in the phase plane that is constrained to the Hamiltonian surface $\mathcal{H} = V$ (left), projected onto the phase plane (right) in Fig. 14.3. This path occurs as a result of satisfying (14.8). This interpretation directly provides the definition of exergy sustainability: the continuous compensation of irreversible entropy production in an open system with an impedance and capacity-matched persistent exergy source. In other words, the cyclic lifestyle will persist indefinitely as long as (14.8) is satisfied and m, k, K_P, and k_{NL} are constants.

Second, this lifestyle will change if the "population" (mass), "investment/infrastructure" (stiffness), "production" (exergy input), and/or "consumption" (irreversible entropy production) are changed independently because all of these parameters are interconnected through the system. Since the lifestyle path over time is constrained to the Hamiltonian surface, if the population is changed, then the stiffness can be changed to hold the lifestyle constant while holding production and consumption constant. The infrastructure is expanded to accommodate the increasing population. On the other hand, if the population increases, then the consumption increases, which means that the production must increase as well. In addition, if the lifestyle increases by creating more services (more exergy consumption), then the production must increase to offset consumption which is enabled through an expanded infrastructure. Bottom-line: lifestyle is directly related to exergy consumption and how exergy sources are matched to exergy sinks through the infrastructure.

Third, this simplified lifestyle presents the mass–spring–damper system as an exergy parasite on the satellite with respect to the sun. By digging a little deeper into the model, the parasites become humans on the earth with respect to the sun. The goal of exergy sustainability now includes striving to become a symbiotic versus a destructive parasite, such as the GAIA approach [163], because cleaning up the disordering effects of a destructive parasite will only consume more exergy through an impedance mismatch. In fact, the human race has done an impressive job of overcoming the impedance mismatch between population growth and the carrying capacity of the biosphere with fossil fuels; we effectively eat fossil fuel.

One final observation on this topic of impedance and capacity matching is that the goal of war and economic competition is to create a production/consumption rate which is sustainable for you and generates an impedance mismatch that is unsustainable for your enemy/competitor. The ultimate goal is to cut-off, destroy, and/or dissipate your competitor's exergy reserves by changing/deforming your competitor's Hamiltonian surface (infrastructure, including population of the work force). This can be accomplished in several ways including: (1) pick-up the pace by increasing the limit cycle frequency (i.e., less mass), (2) accelerate the exergy consumption of your competitor by using more efficient technologies, and (3) deform the potential field with information flow.

14.3.1 Deformation of Potential Field with Information Flow

The deformation of the potential field with information flow is the most seductive because it potentially requires the least amount of additional physical infrastructure. The INTERNET is the most obvious example. A direct application of this idea is by utilizing the techniques of reference [35] in the present context.

The team of robots in Chap. 7 and [35] created a "virtual potential field" by flowing information through a distributed decentralized sensor and feedback control network. The Hamiltonian (Lyapunov function) of the ith robot is deformed by

$$V_i = \mathcal{H}_i = \mathcal{T}_i + \mathcal{V}_{c_i} > 0 \quad \forall \mathbf{x}_i \neq \mathbf{x}_i^*, \quad \dot{\mathbf{x}}_i \neq \mathbf{0}, \quad \mathcal{V}_{c_i}(\mathbf{x}_i^*) = 0, \quad \text{and} \quad \mathcal{T}_i(0) = 0,$$

where

$$\mathcal{T}_i = \frac{1}{2} m_i \dot{\mathbf{x}}_i^T \dot{\mathbf{x}}_i,$$

$$\mathcal{V}_{c_i} = \hat{c}_i(\mathbf{x}_i) - \hat{c}_i(\mathbf{x}_i^*) = \frac{1}{2} \hat{c}_{1_i}^T \hat{c}_{2_i}^{-1} \hat{c}_{1_i} + \hat{c}_{1_i}^T \mathbf{x}_i + \frac{1}{2} \mathbf{x}_i^T \hat{c}_{2_i} \mathbf{x}_i,$$

and the equation of motion (from Lagrange's method) is

$$m_i \ddot{\mathbf{x}}_i + \frac{\partial \mathcal{V}_{c_i}}{\partial \mathbf{x}_i} = \mathbf{u}_i.$$

Therefore, for N robots, the time derivative of the Hamiltonian becomes

$$\dot{\mathcal{H}} = \sum_{i=1}^{N} \dot{\mathcal{H}}_i = \sum_{i=1}^{N} \left[m_i \ddot{\mathbf{x}}_i + \frac{\partial \mathcal{V}_{c_i}}{\partial \mathbf{x}_i} \right]^T \dot{\mathbf{x}}_i = \sum_{i=1}^{N} [\mathbf{u}_i]^T \dot{\mathbf{x}}_i.$$

The estimator/guidance algorithm for finding the source/target is

$$\hat{c}_i(\mathbf{x}_i) = \hat{c}_{0_i} + \mathbf{x}_i^T \hat{c}_{1_i} + \frac{1}{2} \mathbf{x}_i^T \hat{c}_{2_i} \mathbf{x}_i.$$

The feedback controller is

$$\mathbf{u}_i = -\mathbf{K}_{I_i} \int_0^t \mathbf{x}_i \, d\tau - \mathbf{K}_{D_i} \dot{\mathbf{x}}_i,$$

and the stability boundary becomes

$$\sum_{i=1}^{N} [[\mathbf{K}_{D_i} \dot{\mathbf{x}}_i]^T \dot{\mathbf{x}}_i]_{\text{ave}} = \sum_{i=1}^{N} \left[\left[-\mathbf{K}_{I_i} \int_0^t \mathbf{x}_i \, d\tau \right]^T \dot{\mathbf{x}}_i \right]_{\text{ave}},$$

which determines the limit cycle behavior constrained to the deformed Hamiltonian surface. The collective performance is analyzed with the vector Lyapunov technique as

$$V = \sum_{i}^{N} \rho_i V_i$$

with V positive definite for static stability and

$$\dot{V} = \sum_{i}^{N} \rho_i \dot{V}_i = \sum_{i=1}^{N} \rho_i \left[m_i \ddot{\mathbf{x}}_i + \frac{\partial \mathcal{V}_{c_i}}{\partial \mathbf{x}_i} \right]^T \dot{\mathbf{x}}_i = \sum_{i=1}^{N} \rho_i [\mathbf{u}_i]^T \dot{\mathbf{x}}_i$$

with $\dot{V}$ negative definite for dynamic stability. The collective passivity controller design criteria becomes

$$\sum_{i=1}^{N} \rho_i \big[[\mathbf{K}_{D_i} \dot{\mathbf{x}}_i]^T \dot{\mathbf{x}}_i \big] > \sum_{i=1}^{N} \rho_i \left[\left[-\mathbf{K}_{I_i} \int_0^t \mathbf{x}_i \, d\tau \right]^T \dot{\mathbf{x}}_i \right].$$

The collective Hamiltonian can be deformed in order to enhance your exergy usage or manipulate your competitor's exergy usage.

A simple example is to manipulate (14.1) with proportional feedback to reverse the bifurcation of $k < 0$. For

$$V = \mathcal{H} = \mathcal{T} + \mathcal{V} + \mathcal{V}_c,$$

we have

$$\begin{aligned} \mathcal{V} + \mathcal{V}_c &= -\frac{1}{2} k x^2 + \frac{1}{4} k_{\text{NL}} x^4 + \frac{1}{2} K_P x^2 \\ &= \frac{1}{2} [K_P - k] x^2 + \frac{1}{4} k_{\text{NL}} x^4 \end{aligned}$$

and $K_P \geq k$. To reemphasize the effect that the proportional controller gain K_P has on the system, Hamiltonian phase plane plots are generated. By investigating a system with negative stiffness and by adding enough K_P to result in an overall positive net stiffness, the shape of the Hamiltonian surface changes from a saddle point surface (see Fig. 14.4) to a positive bowl surface (see Fig. 14.5). A two-dimensional

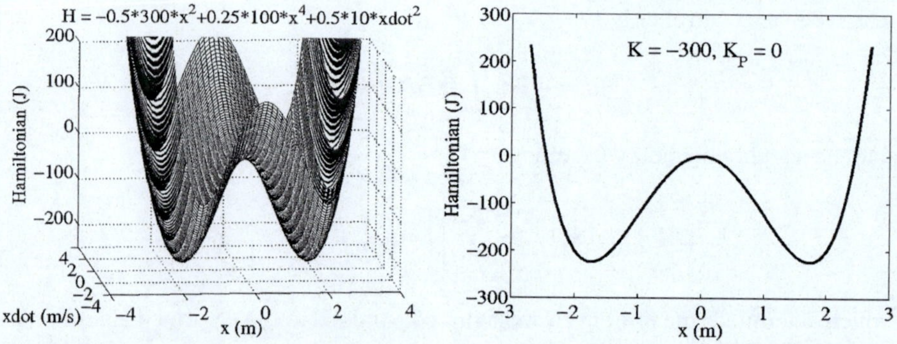

Fig. 14.4 Three-dimensional (*left*) Hamiltonian phase plane plot negative stiffness produces a saddle surface. The two-dimensional cross-section plot (*right*) is at $\dot{x} = 0$. Robinett III, R.D. and Wilson, D.G. [60], reprinted by permission of the publisher (©2006 IEEE)

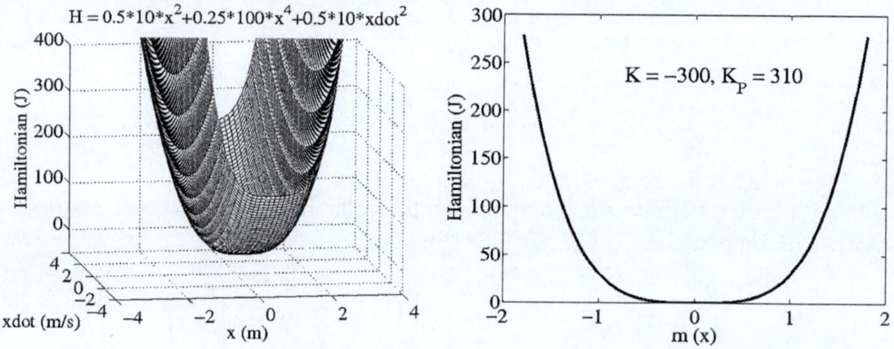

Fig. 14.5 Three-dimensional (*left*) Hamiltonian phase plane plot where the net positive stiffness produces a positive bowl surface. The two-dimensional cross-section plot (*right*) is at $\dot{x} = 0$. Robinett III, R.D. and Wilson, D.G. [60], reprinted by permission of the publisher (©2006 IEEE)

cross-section of the Hamiltonian versus the position shows the characteristics of the overall storage or potential functions. The operating point at $(\mathcal{H}, \dot{x}, x) = (0, 0, 0)$ changes from being unstable to stable, for small values of $|x| > 0$, when enough additional K_P is added, a net positive stiffness for the system results.

14.4 Exergy Sustainability: An Energy Surety Approach

Reiterating the basic problem, the most common problems and challenges the US and other countries are facing is the integration of green renewable power resources into existing aging Electric Power Grid (EPG) infrastructures. Faced with fast approaching deadlines, from renewable portfolio standards (RPS) many are trying to retrofit and patch in their renewables the best that they are able. Many of the proposed "future" smart grids are overlaying/marrying information networks with existing EPG infrastructures. What is needed is a paradigm shift in our current thoughts and practices in power engineering. At the heart of EPG is coordination and control of the generation to meet customer loads. To formally address the green grid of the future with distributed variable generation, buying/selling power (bi-directional flow), and decentralization of the EPG, a new problem will be required to be solved.

To compare the value of different energy sources to be integrated into new green EPGs, such as coal-burning power plants, wind turbines, solar photovoltaics, and storage a new metric needs to be defined. Typical metrics of costs/profits or entropy (measure of disorder in energy system) have been unsuccessful as a unifying metric for all the diverse sources being evaluated. It turns out that a unifying metric defined as exergy [12, 45] coupled with a novel controls/optimization technique (nonlinear power flow) [13] to be applied throughout the EPG from the wind turbines (photovoltaics) through the generators (power inverters) through the distribution and transmission lines to the load will successfully focus on the goal of power engineering: reliable, cost effective, secure, safe, and sustainable power in the form of a linear limit cycle [46].

From a thermodynamic point of view, exergy sustainability [12] can be described as continuously being able to match exergy sources with exergy needs. Exergy is energy available to do useful work, considering the energy available from a given source within its particular environmental surroundings [12, 45]. To assess all possible avenues of exergy utilization, one must choose from the best mix of sources and apply conservation principles to all steps, starting with energy production and ending with final use, even considering what would normally be characterized as waste heat and mass. A high-exergy source (e.g., nuclear) is optimal for conversion to electricity, while a low-exergy source (e.g., waste heat) might be used for space heating, for example. The optimal exergy solution may not always be the most satisfying economic solution [164].

The objective of this section is to address the following: *what is optimal with respect to energy surety?* In the following sections, definitions and descriptions such as optimality and scalability will be presented with respect to the electric power grid portion of a much larger energy nucleus. Traditionally, this has consisted of large portions of generation that is called upon in an "open-loop" fashion to be dispatched to service random load needs. The grid of the future will require much improved automation with an efficiently integrated "closed-loop" configuration.

14.4.1 Optimality

The basic question to ask is *what is optimal with respect to energy surety?* Energy surety is an approach to an "ideal" energy system that, when satisfied, enables the system to function properly while allowing it to resist stresses that could result in unacceptable losses. The attributes of the energy surety model include safety, security, reliability, sustainability, and cost effectiveness [164] as shown in Table 14.1.

Table 14.1 Energy surety requirements

Elements	Definition
Safe	Safely supplies energy to end user
Secure	Resists natural and manmade disruptions
Reliable	Maintains delivery when & where needed
Sustainable	Matches resources with needs
Cost effective	Energy at lowest predictable cost

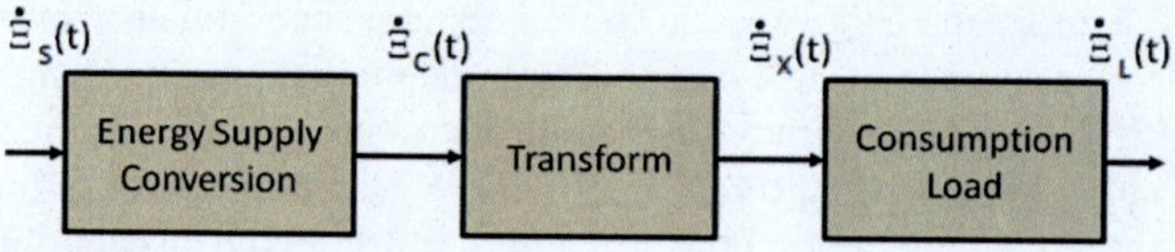

Fig. 14.6 Energy system input–output model

An energy system is optimal with respect to energy surety if it provides the required exergy with perfect safety and security, 100% reliability, an indefinite supply that has no environmental impact, and zero cost. The reality is that one cannot obtain or afford such a system. Consequently, the actual goal will be to minimize the residual risk of the system.

On the other hand, it is instructive to provide these optimal boundaries to better understand the design space and find the dominant control parameters. The first step is to formulate this problem as an input–output model as depicted in Fig. 14.6.

A constraint on the system becomes immediately apparent. For one's lifestyle to persist,

$$\dot{\Xi}_S(t) \geq \dot{\Xi}_L(t) \quad \forall t \geq 0, \tag{14.9}$$

where Ξ is exergy, the derivative is the rate of change, and the subscripts S and L represent *Supply* and *Load*, respectively. This constraint can be interpreted as meaning that the exergy supply rate must be greater than or equal to the exergy consumption rate. This constraint leads to three classes of control parameters: (i) energy supplies, (ii) transformation mechanisms, and (iii) consumption loads.

The electric grid provides an instructive example (see Table 14.2 for optimal grid metrics and descriptions). The predominant energy supplies are coal and natural gas because of their highly effective hydrocarbon storage medium which provide the utilities with *base-load and dispatchable power*. A more detailed version of Fig. 14.6 is presented in Fig. 14.7 for the fossil fuel process with waste mass flows and waste exergy flows. However, the increased attention on CO_2 issues and renewable portfolio standards have led to the increasing penetration of renewable and nuclear power supplies. This, in turn, is causing problems with meeting equation (14.9) since these new energy supplies are transient in nature and/or do not provide the base-load and dispatchable power required by the utilities and present grid. This situation leads to the modification of the transformation mechanisms for the power grid, by adding different energy storage mechanisms and/or a smart grid. In

Table 14.2 Optimums and metrics/energy surety with respect to EPG

Metric	Equation	Description
Transformation:	$\dot{\Xi}_S = \dot{\Xi}_L$	Min. transformation steps
Transmission:	$\dot{\Xi}_S = \dot{\Xi}_L$	Max. exergy/power density
Matching:	$\dot{\Xi}_S = \dot{\Xi}_L$	Compensate for impedance/ capacity mismatch
Persistent source:	$\dot{\Xi}_S = \dot{\Xi}_L$ $\forall t \geq 0, t \rightarrow \infty$	Maximize efficiency/sustainability
Minimize:	$\dot{\Xi}_L$	Maximize efficiency
Economic:		Minimize cost: \$/kW h
		Maximize profit
		Lifestyle/What we value? (kW h/person)
Political:		Control/influence:
		Population, religion, etc.
Security:		Network vulnerabilities
		Unpredictable source
		Corrupted information flow
		Reliable infrastructure

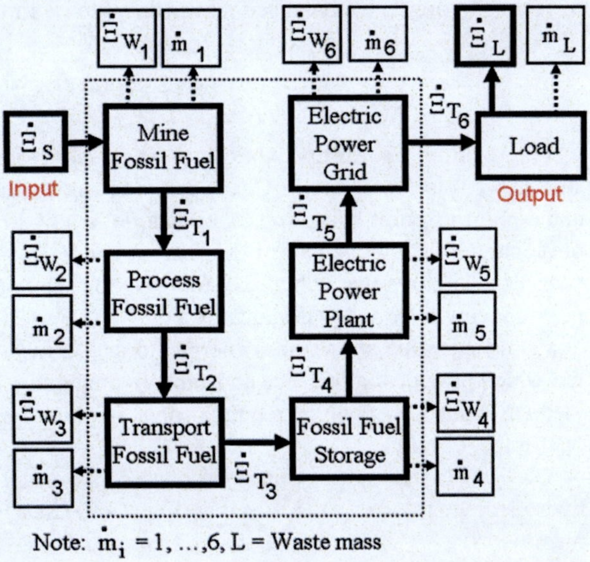

Fig. 14.7 Fossil fuel exergy process

addition, the consumption load can be modified by using real-time pricing through smart metering and dispatchable loads (follow the supply) with intrinsic storage such as municiple water systems.

The second step in the problem formulation is to investigate the equality of (14.9). The goal is to match the energy supply to the consumption load with maximal efficiency or

$$\dot{\Xi}_S(t) = \dot{\Xi}_L(t) \quad \forall t \geq 0 \tag{14.10}$$

and reversibly such that

$$\dot{S}_i(t) = 0 \quad \forall t \geq 0 \tag{14.11}$$

(where $\dot{S}_i$ is the irreversible entropy production rate), which implies that all of the exergy supply is made available for consumption: a unity transformation gain in Table 14.2 (see Chap. 2 for thermodynamic developments). This implies that all processes are 100% efficient and reversible at any time scale. Once again, this goal is not possible since reversible processes are quasi-static and no system is 100% efficient.

On the other hand, (14.10) and (14.11) provide some valuable insights. First, minimizing $\dot{S}_i(t)$ produces more efficient energy systems, which reduces the required $\dot{\Xi}_S(t)$. Second, $\dot{S}_i(t)$ will always be created, so shedding it to persistent energy supplies (i.e., renewables and nuclear with reprocessing) leads to a more sustainable energy system. Third, exergy is consumed, and it is a scarce resource that should be used in economic models instead of energy which is conserved by the First Law of Thermodynamics.

The third step in the problem formulation is to rewrite (14.10) as a line integral over time to produce a balanced power flow over a limit cycle or

$$\oint_\tau \dot{\Xi}_S(t)\, dt = \oint_\tau \dot{\Xi}_L(t)\, dt \quad \forall t \geq 0, \tag{14.12}$$

which requires an exergy storage mechanism. Note that (14.12) does not ensure (14.10) is met since in Chap. 3 it was shown that linear systems only guarantee point-to-point balancing. An example is a hydropower plant that charges up or recharges during one part of the year and generates enough power the rest of the year. Likewise, a pumped hydrosystem provides an opportunity for load-leveling or peak-shaving over a day as shown in Fig. 14.8. The energy supply generates power at a constant value and utilizes energy storage to match the consumption load profile which provides impedance and capacity matching. In reality, (14.12) must be an inequality like (14.9) which requires some amount of excess capacity in generation, or storage, or both.

One way to deal with a transient power supply such as wind is to utilize pumped hydro storage to satisfy (14.12) as in Fig. 14.9. Clearly, the technique of choice to meet

$$\oint_\tau \dot{\Xi}_S(t)\, dt \geq \oint_\tau \dot{\Xi}_L(t)\, dt \quad \forall t \geq 0 \tag{14.13}$$

depends on being safe, secure, reliable, sustainable, and cost effective with an acceptable risk (energy surety [164]).

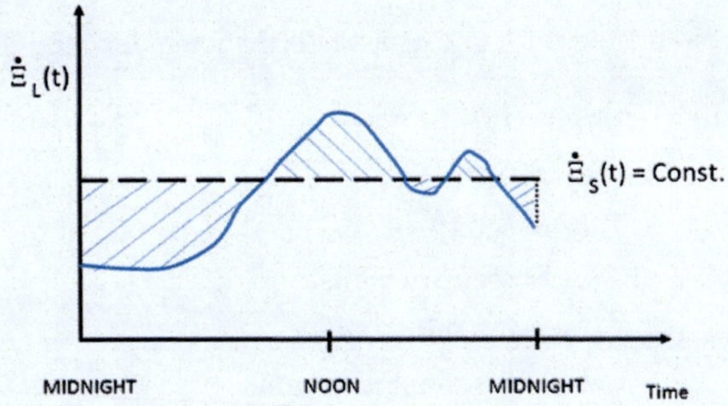

Fig. 14.8 Load-leveling over limit cycles

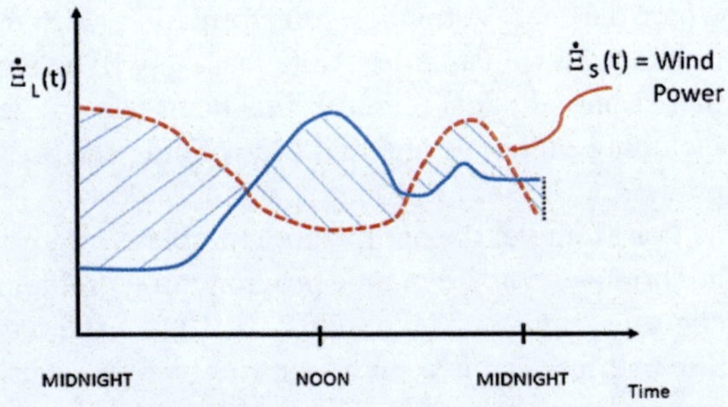

Fig. 14.9 Load-leveling for wind power

Notice that (14.13) is very different from (14.9). If (14.9) is satisfied, then (14.13) is satisfied; however the converse is not true. All of the loads will be met by (14.13), but not necessarily at all load profiles. Basically, the exergy consumption is met, but the power profiles are modified.

So far the discussion has addressed optimality from the supply side including the transformation process of the grid with exergy storage. The other side, consumption loads, can be controlled as well, but appears to have a high-level of complexity. The consumption loads are driven by lifestyle, population, economic growth, religion, and politics. One can define for a grid system a *nominal* consumption load profile (multiscale limit cycle) by defining the desired economic growth rate, population growth rate, and lifestyle changes. Basically, these three metrics are driven by politics and religion even though the carrying capacity of biosphere will ultimately dictate the economic, population, and lifestyle changes unless alternative energy supplies can be integrated into the infrastructure as effectively as fossil fuels.

The last step in the problem formulation is to solve the optimization problem. The optimization problem can be formulated as a dynamic programming problem. The five surety metrics (safety, security, reliability, cost, sustainability) can be implemented with one as a performance index, and the other four as constraints or

$$\begin{aligned}
&\text{Minimize } J = cost\\
&\text{subject to:}\\
&\quad c_1 = \text{safety value},\\
&\quad c_2 = \text{security value},\\
&\quad c_3 = \text{reliability value},\\
&\quad c_4 = \text{sustainability value},\\
&\quad c_{5_n} = \oint_\tau \Big[\dot{\Xi}_{S_n} + \sum_i \dot{\Xi}_{S_{n,i}} - \sum_j \dot{\Xi}_{S_{j,n}} \Big] dt - \oint_\tau \dot{\Xi}_{L_n}\, dt \geq 0,
\end{aligned} \tag{14.14}$$

where the energy supplies, transformation mechanisms, and load profiles are the optimized control variables. Note that c_{5_n} will be derived in the next subsection. Once these optimal paths (control time histories) are defined, the closed-loop controllers including economic and real-time pricing can be designed and refined in an iterative process.

For example, the optimization problem for grid-integration of a wind turbine can be formulated as a dynamic programming problem. A typical wind turbine feedback control system is shown in Fig. 14.10 as a reference for defining relevant optimization metrics. The five surety metrics (safety, security, reliability, cost, sustainability) for this problem are (see Chaps. 10 and 13):

Minimize $J = cost$

subject to:

$c_1 =$ wind turbine blade nonlinear stall flutter suppression margins determined from nonlinear aeroservoelastic analysis,

$c_2 =$ operational ability of the system to serve load demand in the presence of disturbances, which is dynamic stability or $\int_0^{\tau_c} \dot{\mathcal{H}}\, dt = \int_0^{\tau_c} [G - L]\, dt < 0$,

$c_3 =$ planning ability of the system to serve load demand in presence of disturbances, which is static stability or $\mathcal{H} > 0\ \forall x \neq x^*,\ \dot{x} \neq \dot{x}^*,$ where $V(x^*) + V_c(x^*) = 0$,

$c_4 =$ wind turbine controller designed to maximize power output of wind turbine while UPFC controller designed to smooth power output of wind turbine to create nominal 60-Hz AC signal into microgrid.

Notice that the c_{5_n} constraint is eliminated due to a single wind turbine. Also, this is one of many optimal formulations that can be evaluated to meet the desired cost/benefit trade-offs.

Returning to the load side, a step toward controlling consumption loads can be made by adding real-time pricing through smart metering. This approach has the potential to modify the consumption load profile to better match the transient power

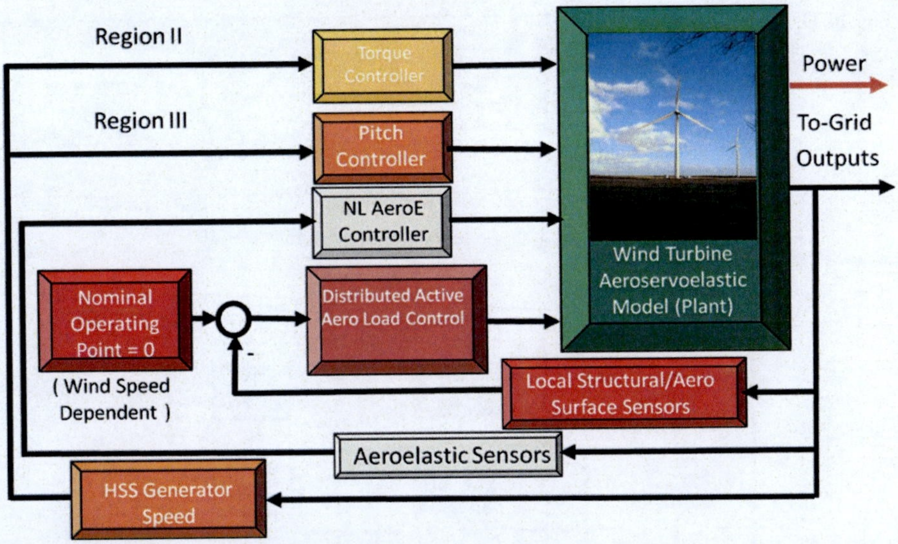

Fig. 14.10 Typical wind turbine control system block diagram

supplies. In particular, a smart grid can enable the load profile to follow the supply profile. This is the first step toward an integrated grid.

The control of this system by closing the loop on load to supply can lead to some interesting stability issues since the grid is nominally open-loop. The update rate of the existing *closed-loop* market selection process is getting closer to the *natural frequencies* of the grid. Also, the highly interconnected communication network is creating unintentional feedback loops. Any approach to satisfying (14.13) will require changes to the existing electric power grid. The grid will become *smarter*, more integrated, and include more transient power supplies. Given this situation, it is time to introduce a scalable construct, the *Energy Surety Microgrid* (ESM), to analyze and design the power grid of the future.

14.4.2 Scalability

To complete this chapter, it is time to derive the $c5_n$ optimization constraint of (14.14). In general, optimization can be applied to: (i) subgrids (sectors), (ii) power flow, and/or (iii) distributed decentralized power flow control that are the basic building blocks of an ESM. Therefore, the electric power grid can be investigated on many different scales and levels. The ESM is defined mathematically in terms of *control volumes* analogous to thermodynamics (see Chap. 2). Figure 14.11 presents a typical microgrid layout. The constraint on control volume 1 is

$$\oint_\tau \left[\dot{\Xi}_{S_1}(t) + \dot{\Xi}_{S_{12}} - \dot{\Xi}_{S_{21}}(t)\right] dt \geq \oint_\tau \dot{\Xi}_{L_1}(t)\, dt \quad \forall t \geq 0, \tag{14.15}$$

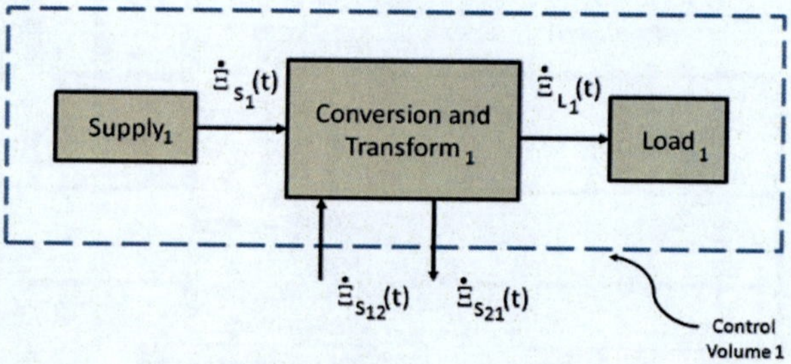

Fig. 14.11 Control volume 1

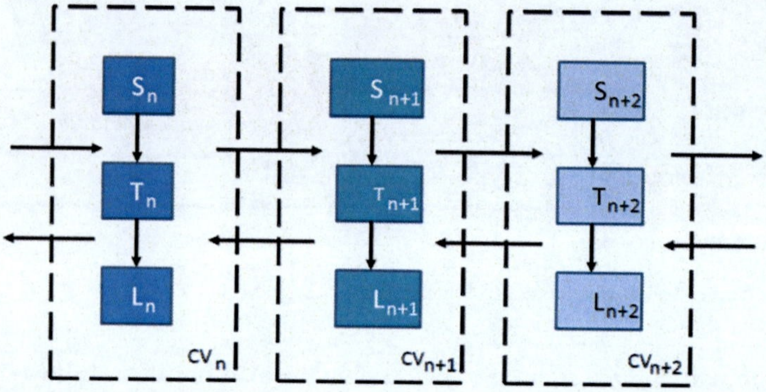

Fig. 14.12 Multiple control volumes

where $\dot{\Xi}_{S_{12}}$ is the exergy supply from control volume 2 to control volume 1 and vise-versa for $\dot{\Xi}_{S_{21}}$. Multiple control volumes are shown in Fig. 14.12.

A generalized, decentralized constraint for control volume n is

$$\oint_\tau \left[\dot{\Xi}_{S_n}(t) + \dot{\Xi}_{S_{n,n-1}}(t) + \dot{\Xi}_{S_{n,n+1}}(t) - \dot{\Xi}_{S_{n-1,n}}(t) + \dot{\Xi}_{S_{n+1,n}}(t)\right] dt \geq \oint_\tau \dot{\Xi}_{L_n}(t)\, dt. \tag{14.16}$$

This constraint provides the basic interaction rules for a decentralized, distributed power grid based on a collection of scalable micrograms and completes the formulation of (14.14) to investigate the sustainability of the future EPG with respect to self-organizing systems.

14.5 Chapter Summary

This chapter has developed a new definition of exergy sustainability: the continuous compensation of irreversible entropy production in an open system with an impedance and capacity-matched persistent exergy source. The development of this definition has led to a missing link in the analysis of self-organizing systems: a tie between irreversible thermodynamics and Hamiltonian systems. This tie was exploited through nonlinear control theory to define necessary and sufficient conditions for stability of nonlinear systems that were employed to formulate the concept of “on the edge of chaos.” Also, an equivalence was developed between physical stored exergy and information-based exergy which can be exploited to change a lifestyle and enhance one’s economic competitiveness or performance on the battlefield. Finally, exergy sustainability and energy surety were applied to the EPG to formulate an example of sustainability of self-organizing systems.

References

1. Murray, R.M. (ed.): Control in an Information Rich World: Report of the Panel on Future Directions in Control, Dynamics, and Systems. SIAM, Philadelphia (2003)
2. Greven, A., Keller, G., Warnecke, G.: Entropy. Princeton University Press, Princeton (2003)
3. Schrödinger, E.: What Is Life? Cambridge University Press, Cambridge (1992). Reprint edn.
4. Kondepudi, D., Prigogine, I.: Modern Thermodynamics: From Heat Engines to Dissipative Structures. Wiley, New York (1999)
5. Nicolis, G., Prigogine, I.: Exploring Complexity: An Introduction. Freeman, New York (1989)
6. Meirovitch, L.: Methods of Analytical Dynamics. McGraw-Hill, New York (1970)
7. Robinett III, R.D., Wilson, D.G.: Collective systems: physical and information exergies. Sandia National Laboratories, SAND2007 Report (March 2007)
8. Robinett III, R.D.: Collective Groups and Teams. Draft book (January 2007)
9. Nicolis, G., Prigogine, I.: Self-organization in Non-equilibrium Systems. Wiley, New York (1977)
10. Prigogine, I., Stengers, I.: Order out of Chaos. Bantam Books, New York (1984)
11. Heylighen, F.: The science of self-organization and adaptivity. Principia Cybernetic Web. http://perspmcl.vub.ac.be/
12. Robinett III, R.D., Wilson, D.G., Reed, A.W.: Exergy sustainability for complex systems. InterJournal Complex Systems, 1616, New England Complex Systems Institute (2006)
13. Robinett III, R.D., Wilson, D.G.: Exergy and irreversible entropy production thermodynamic concepts for nonlinear control design. Int. J. Exergy **6**(3), 357–387 (2009)
14. Chetayev, N.G.: The Stability of Motion. Pergamon, Elmsford (1961)
15. Junkins, J.L., Kim, Y.: Introduction to Dynamics and Control of Flexible Structures. AIAA, Washington (1993)
16. Gopinath, A.K., Beran, P.S., Jameson, A.: Comparative analysis of computational methods for limit-cycle oscillations. In: 47th AIAA Structures, Structural Dynamics and Materials Conference, Newport, Rhode Island, May 2006
17. Hall, K.C., Thomas, J.P., Clark, W.S.: Computation of unsteady nonlinear flows in cascades using a harmonic balance technique. AIAA J. **40**(5), 879–886 (2002)
18. Price, S.J., Lee, B.H.K., Alighanbari, H.: Post-instability behavior of a two-dimensional airfoil with a structural nonlinearity. J. Aircr. **31**(6), 1395–1401 (1994)
19. Ko, J., Kurdila, A.J., Strganac, T.W.: Nonlinear control of a prototypical wing section with torsional nonlinearity. J. Guid. Control Dyn. **20**(6), 1181–1189 (1997)
20. Lee, B.H.K., Jiang, L.Y., Wong, Y.S.: Flutter of an airfoil with a cubic nonlinear restoring force. AIAA paper AIAA-98-1725
21. Clark, R.L., Dowell, E.H., Frampton, K.D.: Control of a three-degree-of-freedom airfoil with limit-cycle behavior. J. Aircr. **37**(3), 533–536 (2000)

R.D. Robinett III, D.G. Wilson, *Nonlinear Power Flow Control Design*, Understanding Complex Systems,
DOI 10.1007/978-0-85729-823-2,

22. Yingsong, G., Zhichun, Y.: Aeroelastic analysis of an airfoil with a hysteresis non-linearity. In: 47th AIAA/ASME/ASCE/AHS/ASC Structures, Structural Dynamics, and Materials Conference, Newport, Rhode Island, May 2006
23. Sabatini, M.: Limit cycle's uniqueness for a class of plane systems. Technical Report UTM 662, Mathematica, University of Trento, Italy (February 2004)
24. Carletti, T., Villari, G.: A note on existence and uniqueness of limit cycles for Liénard systems. Preprint submitted to Journal of Mathematical Analysis and Applications, August 2007
25. Boyce, W.E., DiPrima, R.C.: Elementary Differential Equations and Boundary Value Problems, 8th edn. Wiley, New York (2005)
26. Weinstock, R.: Calculus of Variations with Applications to Physics and Engineering. Dover, New York (1974)
27. Dowell, E.H., Crawley, E.F., Curtiss, H.C. Jr., Peters, D.A., Scanlan, R.H., Sisto, F.: A Modern Course in Aeroelasticity. Sijthoff & Noordhoff, Rockville (1978)
28. Flanigan, F.J.: Complex Variables—Harmonic and Analytic Functions. Dover, New York (1983)
29. Boas, M.L.: Mathematical Methods in the Physical Sciences. Wiley, New York (1966)
30. Robinett III, R.D., Wilson, D.G.: Exergy and entropy thermodynamic concepts for control system design: slewing single axis. In: AIAA Guidance, Navigation, and Control Conference and Exhibit, Keystone, CO, August 2006
31. Abramson, H.N.: An Introduction to the Dynamics of Airplanes. Ronald Press, New York (1958)
32. Ogata, K.: Modern Control Engineering. Prentice-Hall, Englewood Cliffs (1970)
33. Robinett III, R.D., Wilson, D.G.: Collective plume tracing: a minimal information approach to collective control. In: Proceedings of the 2007 American Control Conference, New York, July 2007
34. Robinett III, R.D., Wilson, D.G.: Decentralized exergy/entropy thermodynamic control for collective robotic systems. In: Proceedings of the ASME 2007 International Mechanical Engineering Congress and Exposition, Seattle, WA, November 2007
35. Robinett III, R.D., Hurtado, J.E.: Stability and control of collective systems. J. Intell. Robot. Syst. **39**, 43–55 (2004)
36. Hurtado, J.E., Robinett III, R.D., Dohrmann, C.R., Goldsmith, S.Y.: Decentralized control for a swarm of vehicles performing source localization. J. Intell. Robot. Syst. **41**, 1–18 (2004)
37. Robinett III, R.D., Wilson, D.G.: Hamiltonian surface shaping with information theory and exergy/entropy control for collective plume tracing. Int. J. Systems, Control and Communications **2**(1/2/3) (2010)
38. Berg, T.: Fisher information: its flow, fusion, and coordination. Sandia National Laboratories, SAND2002-1969 Report (June 2002)
39. Schwartz, M.: Information Transmission, Modulation, and Noise: A Unified Approach to Communication Systems, 3rd edn. McGraw-Hill, New York (1980)
40. Shannon, C.E.: A mathematical theory of communication. Bell Syst. Tech. J. **27**, 379–423 (1948), 623–656
41. Frieden, B.R.: Physics from Fisher Information: A Unification. Cambridge University Press, Cambridge (1999)
42. Cooper, J.A., Robinett III, R.D.: Structured communication and collective cohesion measured by entropy. J. Syst. Safety **41**(4) (2005) (electronic)
43. Gyftopoulos, E.P., Beretta, G.P.: Thermodynamics, Foundations and Applications. MacMillan & Co., New York (1991)
44. Scott, D.S.: Links and lies. Int. J. Hydrog. Energy **28**, 473–476 (2003)
45. Scott, D.S.: Exergy. Int. J. Hydrog. Energy **28**, 369–375 (2003)
46. Robinett III, R.D., Wilson, D.G.: What is a limit cycle? Int. J. Control **81**(12), 1886–1900 (2008)
47. Robinett III, R.D., Parker, G.G., Schaub, H., Junkins, J.L.: Lyapunov optimal saturated control for nonlinear systems. J. Guid. Control Dyn. **20**(6), 1083–1088 (1997)

48. Anthony, K.-H.: Hamilton's action principle and thermodynamics of irreversible processes—a unifying procedure for reversible and irreversible processes. J. Non-Newton. Fluid Mech. **96**, 291–339 (2001)
49. Smith, R.J.: Circuits, Devices, and Systems: A First Course in Electrical Engineering, 3rd edn. Wiley, New York (1976)
50. Robinett III, R.D., Wilson, D.G.: Collective plume tracing: a minimal information approach to collective control. Int. J. Robust Nonlinear Control (2009). doi:10.1002/rnc.1420
51. Lyapunov, A.M.: Stability of Motion. Academic Press, New York (1966)
52. LaSalle, J., Lefschetz, S.: Stability by Liapunov's Direct Method with Applications. Academic Press, San Diego (1961)
53. Scanlan, R.H., Rosenbaum, R.: Aircraft Vibration and Flutter. Dover, New York (1968)
54. Anderson, J.D. Jr.: Introduction to Flight. McGraw-Hill, New York (1978)
55. Robinett III, R.D.: A unified approach to vehicle design, control, and flight path optimization. PhD Dissertation, Texas A&M University (1987)
56. Robinett III, R.D., Wilson, D.G.: Transient stability and control of renewable generators based on Hamiltonian surface shaping and power flow control: Part I—Theory. In: IEEE MSC Conference, Yokohama, Japan, September 2010
57. Robinett III, R.D., Peterson, B.J., Fahrenholtz, J.C.: Lag-stabilized force feedback damping. J. Intell. Robot. Syst. **21**, 277–285 (1998)
58. Yechout, T.R., Morris, S.L., Bossert, D.E., Hallgren, W.F.: Introduction to Aircraft Flight Mechanics: Performance, Static Stability, Dynamic Stability, and Classical Feedback Control. AIAA, Washington (2003)
59. Saaty, T.L., Bram, J.: Nonlinear Mathematics. McGraw-Hill, New York (1964)
60. Robinett III, R.D., Wilson, D.G.: Exergy and irreversible entropy production thermodynamic concepts for control design: nonlinear systems. In: 14th Mediterranean Conference on Control and Automation, Ancona, Italy, June 28–30, 2006
61. Robinett III, R.D., Dohrmann, C., Eisler, G.R., Feddema, J.T., Parker, G.G., Wilson, D.G., Stokes, D.: Flexible Robot Dynamics and Controls. IFSR International Series on Systems Science and Engineering, vol. 19. Kluwer Academic/Plenum, New York (2002)
62. Robinett III, R.D., Wilson, D.G., Eisler, G.R., Hurtado, J.E.: Applied Dynamic Programming for Optimization of Dynamical Systems. Advances in Design and Control. SIAM, Philadelphia (2005)
63. Gelfand, I.M., Fomin, S.V.: Calculus of Variations. Prentice-Hall, New York (1963)
64. Oldham, K.B., Spanier, J.: The Fractional Calculus—Theory and Application of Differentiation and Integration to Arbitrary Order. Academic Press, San Diego (1974)
65. Podlubny, I.: Fractional Differential Equations—An Introduction to Fractional Derivatives, Fractional Differential Equations, to Methods of Their Solution and Some of Their Applications. Academic Press, San Diego (1999)
66. Robinett III, R.D., Wilson, D.G.: Exergy and entropy thermodynamic concepts for nonlinear control design. In: ASME 2006 International Mechanical Engineering Congress & Exposition, Chicago, IL, November 5–10, 2006
67. Robinett III, R.D., Wilson, D.G.: Nonlinear slewing spacecraft control based on exergy, power flow, and static and dynamic stability. Journal of the Astronautical Sciences **57**(4) (2009)
68. Craig, R.R. Jr.: Structural Dynamics—An Introduction to Computer Methods. Wiley, New York (1981)
69. van der Pol, B.: Radio Rev. **1**, 704–754 (1920)
70. van der Pol, B.: Philos. Mag. **3**, 65 (1927)
71. Tatikonda, S.C.: Control under communication constraints. PhD Dissertation, EECS, Massachusetts Institute of Technology, Boston, MA (September 2000)
72. Ilaya, O.: Multi-objective decentralized model predictive control for cooperative multi-UAV systems. In: AIAA Guidance, Navigation and Control Conference and Exhibit, Hilton Head, SC, August 2007
73. Izzo, D., Pettazzi, L.: Autonomous and distributed motion planning for satellite swarm. J. Guid. Control Dyn. **30**(2), 449–459 (2007)

74. Feddema, J.T., Robinett III, R.D., Byrne, R.H.: An optimization approach to distributed controls of multiple robot vehicles. In: IEEE/RSJ International Conference on Intelligent Robots and Systems, Workshop on Control and Cooperation of Intelligent Miniature Robots, Las Vegas, NV, October 2003
75. Feddema, J.T., Lewis, C., Schoenwald, D.A.: Decentralized control of robotic vehicles: theory and application. IEEE Trans. Robot. Autom. **18**, 852–864 (2002)
76. Lewis, C., Feddema, J.T., Klarer, P.: Robotic perimeter detection system. Proc. SPIE **3577**, 14–21 (1998)
77. Feddema, J.T., Schoenwald, D.A.: Stability analysis of decentralized cooperative controls. In: Shultz, A.C., Parker, L.E. (eds.) Multi-Robot Systems: From Swarms to Intelligent Automata. Kluwer Academic, Dordrecht (2002)
78. Byrne, R.H., Eskridge, S.E., Hurtado, J.E., Savage, E.L.: Algorithms and analysis for underwater vehicle plume tracing. Sandia National Laboratories, SAND2003-2643 Report (July 2003)
79. Byrne, R.H., Adkins, D.R., Eskridge, S.E., Harrington, J.J., Heller, E.J., Hurtado, J.E.: Miniature mobile robots for plume tracking and source localization research. J. Micromechatronics **1**(3), 253–261 (2002)
80. Hurtado, J.E., Robinett III, R.D.: Convergence of Newton's method via Lyapunov analysis. Journal of Guidance, Control, and Dynamics **28**(2) (2005)
81. Vincenti, W.G., Kruger, C.H. Jr.: Introduction to Physical Gas Dynamics. Krieger, Melbourne (1982)
82. Reif, F.: Fundamentals of Statistical and Thermal Physics. McGraw-Hill, New York (1965)
83. Goldsmith, S.Y., Robinett III, R.D.: Collective search by mobile robots using alpha-beta coordination. In: Collective Robotics Workshop, Agent World 98, May 1998 (1998)
84. Gasiorowicz, S.: Quantum Physics. Wiley, New York (1974)
85. Robinett III, R.D., Wilson, D.G.: Nonlinear aeroelastic power flow control for wind turbines. In: 51st AIAA/ASME/ASCE/AHS/ASC Structures, Structural Dynamics, and Materials Conference, Orlando, FL, April 12–15, 2010
86. Robinett III, R.D., Wilson, D.G.: Maximizing the performance of wind turbines with nonlinear aeroservoelastic power flow control. In: 2010 IEEE MSC Conference, Yokohama, Japan, September 2010
87. Abourida, S., Dufour, C., Belanger, J., Lapointe, V.: Real-time, PC-based simulator of electric systems and drives. In: International Conference on Power Systems Transients 2003, New Orleans, USA (2003)
88. The Mathworks: SimPowerSystems: For Use with Simulink, Hydro-Quebec TransEnergie Technologies, User's Guide, Version 4. The Mathworks, Natick (1998–2007)
89. Kilani, K.B., Schleuter, R.A.: Trends in model development for stability studies in power systems. Electr. Power Syst. Res. **53**, 207–215 (2000)
90. Pavella, M., Ernst, D., Ruiz-Vega, D.: Transient Stability of Power Systems: A Unified Approach to Assessment and Control. Kluwer Academic, Dordrecht (2000)
91. Grisby, L.L. (ed.): Power System Stability and Control, Electric Power Engineering Handbook, 2nd edn. CRC Press, Boca Raton (2007)
92. Wang, Y., Cheng, D., Li, C., Ge, Y.: Dissipative Hamiltonian realization and energy-based L_2-disturbance attenuation control of multi-machine power systems. IEEE Trans. Automatic Control **48**(8) (2003)
93. Hao, J., Wang, J., Chen, C., Shi, L.: Nonlinear excitation control of multi-machine power systems with structure preserving models based on Hamiltonian system theory. Electr. Power Syst. Res. **74**, 401–408 (2005)
94. Robinett III, R.D., Wilson, D.G.: Nonlinear power flow control applied to power engineering. In: SPEEDAM 2008, International Symposium on Power Electronics, Electrical Drives, Automation and Motion, pp. 1420–1427, Ischia, Italy, June 11–13, 2008
95. Slotine, J.-J.E., Li, W.: Applied Nonlinear Control. Prentice Hall, Englewood Cliffs (1991)
96. http://www.sandia.gov/SAI/files/SEGIS%20Concept%20Paper-071025.pdf
97. http://www.eere.energy.gov/windandhydro/wind_2030.html

98. Vittal, V., Michel, A.N., Fouad, A.A.: Power system transient stability analysis: formulation as nearly Hamiltonian systems. Journal of Circuits, Systems, and Signal Processing **3**(1) (1984)
99. Alberto, L.F.C., Bretas, N.G.: Application of Melnikov's method for computing heteroclinic orbits in a classical SMIB power system model. IEEE Trans. Circuits Syst. I, Fundam. Theory Appl. **47**(7), 1085–1089 (2000)
100. Ghandhari, M.: Control Lyapunov functions: a control strategy for damping of power oscillations in large power systems. Doctoral Dissertation, TRITA-EES-0004, ISSN 1100-1607, Royal Institute of Technology, Stockholm, Sweden (2000)
101. Gross, C.A.: Power System Analysis. Wiley, New York (1986)
102. Wilson, D.G., Robinett III, R.D.: Nonlinear power flow control applications to conventional generator swing equations subject to variable generation. In: International Symposium on Power Electronics, Electrical Drives, Automation and Motion, SPEEDAM 2010, Pisa, Italy, June 14–16, 2010
103. Pai, M.A.: Power System Stability—Analysis by the Direct Method of Lyapunov. North-Holland, Amsterdam (1981)
104. Sauer, P.W., Pai, M.A.: Power System Dynamics and Stability. Stipes Publishing Co. (1997)
105. Wilson, D.G., Robinett III, R.D.: Transient stability and control of wind turbine generation based on Hamiltonian surface shaping and power flow control. In: 9th International Workshop on Large-Scale Integration of Wind Power into Power Systems as Well as on Transmission Networks for Offshore Wind Power Plants, Quebec City, Quebec, Canada, October 2010
106. Winchester, J.: Grumman X-29, X-Planes and Prototypes. Amber Books, London (2005)
107. Taliaferro, S.D.: An inversion of the Lagrange–Dirichlet stability theorem. Arch. Ration. Mech. Anal. **73**, 183–190 (1980)
108. Hagedorn, P., Mawhin, J.: A simple variational approach to a converse of the Lagrange-Dirichlet theorem. Arch. Ration. Mech. Anal. **120**, 327–335 (1992), Springer-Verlag
109. Willems, J.C.: Dissipative dynamical systems part I: General theory; part II: Linear systems with quadratic supply rates. Arch. Ration. Mech. Anal. **45**, 321–393 (1972)
110. Kokotovic, P., Arcak, M.: Constructive nonlinear control: a historical perspective. Preprint submitted to Elsevier, August 2000
111. Bloch, A.N., Chang, D.E., Leonard, N.E., Marsden, J.E.: Controlled Lagrangians and the stabilization of mechanical systems: potential shaping. IEEE Trans. Autom. Control **46**(10), 1556–1571 (2001)
112. Ortega, R., Garcia, E.: Energy-shaping stabilization of dynamical systems. Laboratoire des Signaux et Systemes, SUPELEC, Gif-sur-Yvette, France (2004)
113. Takegaki, M., Arimoto, S.: A new feedback method for dynamic control of manipulators. Journal of Dynamic Systems, Measurement, and Control **102**(119) (1981)
114. Robinett III, R.D., Wilson, D.G.: Exergy and irreversible entropy production thermodynamic concepts for control system design: regulators. In: 2006 IEEE International Conference on Control Applications (CCA), Munich, Germany, October 2006, pp. 2249–2256 (2006)
115. Robinett III, R.D., Wilson, D.G.: Exergy and irreversible entropy production thermodynamic concepts for control system design: robotic servo applications. In: 2006 IEEE International Conference on Robotics and Automation, Orlando, FL, May 15–19, 2006, pp. 3685–3692 (2006)
116. Robinett III, R.D., Wilson, D.G.: Exergy and irreversible entropy production thermodynamic concepts for control system design: nonlinear regulator systems. In: The 8th IASTED International Conference on Control and Applications, Montreal, Quebec, Canada, May 24–26, 2006
117. Qu, Z., Dawson, D.M.: Robust Tracking Control of Robot Manipulators. IEEE Press, New York (1996)
118. Fung, Y.C.: An Introduction to the Theory of Aeroelasticity. Dover, New York (1969)
119. Guckenheimer, J., Holmes, P.: Nonlinear Oscillations, Dynamical Systems, and Bifurcations of Vector Fields. Springer, New York (1983)

120. Hughes, P.C.: Spacecraft Attitude Dynamics. Dover, New York (2004)
121. Srivener, S.L., Thompson, R.C.: Survey of time-optimal attitude maneuvers. J. Guid. Control Dyn. **17**(2), 225–233 (1994)
122. Tsiotras, P.: Stabilization and optimality results for attitude control problem. J. Guid. Control Dyn. **9**(4), 772–779 (1996)
123. Subbarao, K.: Nonlinear PID-like controllers for rigid-body attitude stabilization. J. Astronaut. Sci. **52**(1 and 2), 61–74 (2004)
124. Tsiotras, P.: Further passivity results for the attitude control problems. IEEE Trans. Autom. Control **43**(11), 1597–1600 (1998)
125. Cavallo, A., DeMaria, G.: Attitude control of large angle maneuvers. In: Proceedings of the IEEE Workshop Variable Structure Control, pp. 232–236 (1996)
126. Vadali, S.R.: Variable-structure control of spacecraft large-angle maneuvers. AIAA Journal of Guidance, Control, and Dynamics 235–239 (1986)
127. Robinett III, R.D., Parker, G.G.: Spacecraft Euler parameter tracking of large angle maneuvers via sliding mode control. J. Guid. Control Dyn. **19**, 702–703 (1996)
128. Robinett III, R.D., Parker, G.G.: Least squares sliding mode control tracking of spacecraft large angle maneuvers. J. Astronaut. Sci. **45**(4), 433–450 (1997)
129. Singh, S.N.: Nonlinear attitude control of spacecraft. IEEE Trans. Aerosp. Electron. Syst. **23**, 371–379 (1987)
130. Tandale, M.D., Valasek, J.: Adaptive dynamic inversion control with actuator saturation constraints applied to tracking spacecraft maneuvers. J. Astronaut. Sci. **52**(4), 517–530 (2004)
131. Tewari, A.: Optimal nonlinear spacecraft attitude control through Hamiltonian–Jacobi formulation. J. Astronaut. Sci. **50**(1), 99–112 (2002)
132. Joshi, S.M., Kelkar, A.G., Wen, J.T.Y.: Robust attitude stabilization of spacecraft using nonlinear quaternion feedback. IEEE Trans. Autom. Control **40**, 1800–1803 (1995)
133. Wie, B., Barba, P.M.: Quaternion feedback for spacecraft large angle maneuvers. J. Guid. Control Dyn. **3**(3), 360–365 (1985)
134. Dalsmo, M., Egeland, O.: State feedback H_∞-suboptimal control of a rigid spacecraft. IEEE Trans. Autom. Control **42**, 1186–1189 (1997)
135. Kang, W.: Nonlinear control and its applications to rigid spacecraft. IEEE Trans. Autom. Control **40**, 1281–1285 (1995)
136. Kristic, M., Tsiotras, P.: Inverse optimal stabilization of rigid spacecraft. IEEE Trans. Autom. Control **44**, 1042–1049 (1999)
137. Yang, C.D., Kung, C.C.: Nonlinear H_∞ flight control of general six degree-of-freedom motions. J. Guid. Control Dyn. **23**(2), 278–288 (2000)
138. Show, L.-L., Juang, J.-C., Jan, Y.-W.: An LMI-based nonlinear attitude control approach. IEEE Trans. Control Syst. Technol. **11**(1), 73–83 (2003)
139. Pacheco, R.P., Steffen, V. Jr.: Orthogonal function techniques for the identification of nonlinear mechanical systems. Mater. Sci. Forum **440–441**, 59–68 (2003). Trans Tech Pubs, Switzerland
140. Ekelund, T.: Modeling and linear quadratic optimal control of wind turbines. Ph.D. Thesis, Chalmers University of Technology, Sweden (1997)
141. Connor, B., Leithead, W.E., Grimble, M.J.: LQG control of a constant speed horizontal axis wind turbine. In: Proceedings of the Third IEEE Conference on Control Applications, Glasgow, Scotland, August 24–26, 1994, vol. 1, pp. 251–252 (1994)
142. Abdin, E.S., Xu, W.: Control design and dynamic performance analysis of a wind turbine-induction generator unit. IEEE Trans. Energy Convers. **15**(1), 91–96 (2000)
143. Wright, A.D., Fingersh, L.J.: Advanced control design for wind turbines, Part I: Control design, implementation, and initial tests. NREL Technical Report NREL/TP-500-42437 (March 2008)
144. Bongers, P.: Modeling and identification of flexible wind turbines and a factorization approach to robust control. Ph.D. Thesis, Delft University of Technology, Netherlands (1994)
145. Connor, D., Iyer, S.N., Leithead, W.E., Grimble, M.J.: Control of a horizontal axis wind turbine using H_∞ control. In: Proceedings of the First IEEE Conference on Control Applications, Dayton, OH, September 13–16, 1992

146. Battista, H.D., Mantz, R.J., Christiansen, C.F.: Dynamical sliding mode power control of wind driven induction generators. IEEE Trans. Energy Convers. **15**(14), 451–457 (2000)
147. Song, Y.D., Dhinakaran, B., Bao, X.Y.: Variable speed control of wind turbines using nonlinear and adaptive algorithms. J. Wind Eng. Ind. Aerodyn. **85**, 293–308 (2000)
148. Boukhezzar, B., Siguerdidjane, H., Hand, M.: Nonlinear control of variable speed wind turbines for load reduction and power optimization. In: 44th AIAA Aerospace Science Meeting and Exhibit, Reno, NV, January 2006
149. Leithead, W.E., Connor, D.: Control of variable speed wind turbines: design task. Int. J. Control **73**, 1173–1188 (2000)
150. Jonkman, B.J., Buhl, M.L. Jr.: TurbSim user's guide for version 1.40. NREL Technical Report, NREL/TP-xxx (September 2008)
151. Jonkman, J.M., Buhl, M.L. Jr.: FAST user's guide. NREL Technical Report, NREL/TP-500-35816 (July 2004)
152. Cooper, J.A., Robinett III, R.D.: Deriving sustainable ordered surety by overcoming persistent disorder pressures. Journal of Systems Safety **42**(4), 27–34 (2006)
153. Principia Cybernetica Web, http://pespmc1.vub.ac.be. Principia Cybernetica tries to tackle age-old philosophical questions with the help of the most recent cybernetic theories and technologies
154. Alonso, A.A., Fernandez, C.V., Banga, J.R.: Dissipative systems: from physics to robust nonlinear control. Int. J. Robust Nonlinear Control **14**, 157–179 (2004)
155. Hill, D., Moylan, P.J.: The stability of nonlinear dissipative systems. IEEE Transactions on Automatic Control **21**(5), 708–711 (1976)
156. Alonso, A.A., Ydstie, B.E.: Stabilization of distributed systems using irreversible thermodynamics. Automatica **37**, 1739–1755 (2001)
157. Haken, H.: Advanced Synergetics: Instability Hierarchies of Self-Organizing Systems and Devices. Springer, New York (1983)
158. Haken, H.: Synergetics: An Introduction, 3rd edn. Springer, New York (1983)
159. Buenstorf, G.: Self-organization and sustainability: energetics of evolution and implications for ecological economics. Ecol. Econ. **33**, 119–134 (2000)
160. Allen, P.M.: Cities and Regions a Self-Organizing Systems, Models of Complexity. Gordon and Breach, Amsterdam (1997)
161. Bak, P.: How Nature Works: The Science of Self-Organized Criticality. Springer, Berlin (1996)
162. Kaptchuk, T.J.: The Web That Has No Weaver: Understanding Chinese Medicine. Congdon and Weed, New York (1983)
163. Lovelock, J., Margulis, L.: The GAIA hypothesis. www.mountainman.com.au/gaia.html
164. Tatro, M.L., Covan, J.M., Kuswa, G.W., Jones, S.A., Robinett III, R.D., Menicucci, D.: Toward an energy surety future. Sandia Report, SAND2005-6281 (October 2005)

Index

R.D. Robinett III, D.G. Wilson, *Nonlinear Power Flow Control Design*,
Understanding Complex Systems,
DOI 10.1007/978-0-85729-823-2, © Springer-Verlag London Limited 2011